THE SOCIAL SCIENCES EMPOWERED

PROCEEDINGS OF THE 7TH INTERNATIONAL CONGRESS ON INTERDISCIPLINARY BEHAVIOR AND SOCIAL SCIENCES 2018 (ICIBSOS 2018), 21-22 JULY, BANGKOK, THAILAND, 22-23 SEPTEMBER, BALI, INDONESIA, 6-7 OCTOBER, BALI, INDONESIA, 24-25 NOVEMBER 2018, YOGYAKARTA, INDONESIA

# The Social Sciences Empowered

*Editors*

Ford Lumban Gaol

*Bina Nusantara University, Jakarta, Indonesia*

Fonny Hutagalung & Chew Fong Peng

*University of Malaya, Kuala Lumpur, Malaysia*

CRC Press
Taylor & Francis Group
Boca Raton London New York

CRC Press is an imprint of the
Taylor & Francis Group, an **informa** business
A BALKEMA BOOK

Published by:
CRC Press/Balkema
Schipholweg 107C, 2316 XC Leiden, The Netherlands



ISBN 13: 978-0-367-00090-5 (hbk)
ISBN 13: 978-0-429-44456-2 (ebk)
DOI: https://doi.org/10.1201/9780429444562

Typeset by Integra Software Services Pvt. Ltd., Pondicherry, India

*The Social Sciences Empowered – Lumban Gaol, Hutagalung & Fong Peng (eds)*
*© 2020 Taylor & Francis Group, London, ISBN 978-0-367-00090-5*

# Table of contents

*The Social Sciences Empowered – Lumban Gaol, Hutagalung & Fong Peng (eds)*
*© 2020 Taylor & Francis Group, London, ISBN 978-0-367-00090-5*

# Preface

The 7th International Congress on Interdisciplinary Behavior & Social Science 2018 has the theme "Empowered The Social Sciences: Making Knowledge, Innovation and Progress Accessible to All".

As cross-cutting issues of broad relevance, Social Sciences and Humanities are fully integrated to enhance industrial leadership and to tackle every societal challenge. ICIBSoS 2018 provided the economic and social analysis necessary for reforming Humanities issues such as Education, Sociology, Anthropology, Politics, History, Philosophy and Psychology, as well as food security.

Contributions to ICIBSoS 2018 provide a range of insights into the cultural and human dimensions in such diverse subject areas as transport, climate change, energy or agriculture. ICIBSoS 2018 papers also analyse the cultural, behavioural, psychological, social and institutional changes that transform people's behaviour and the global environment. ICIBSoS 2018 authors propose new ideas, strategies and governance structures for overcoming the crisis from a global perspective, for innovating the public sector and business models, for promoting social innovation and fostering creativity in development of services and product design.

One example that was discussed in ICIBSoS 2018 was the evolution of the concept of Ecosystem to a concept that embraces the structures and systems of Society 5.0, that will profoundly contribute to Society and Environment. Issues such as Sharing Economics and the Economics of Disruptive Technology were also discussed during ICIBSoS 2018.

Hence, in ICIBSoS 2018 a discussion was held about shaping new emerging directions, including the knowledge that relates to conception and philosophy issues that in turn concern understanding Social Sciences issues and the relations between people. Constructive interdisciplinary dialogues in support of the development of innovative frameworks and terminologies have become the goal and contribution of ICIBSoS 2018.

Finally, we hope that this book will make a valuable contribution to the fields of social sciences and humanities.

Best regards,

Ford Lumban Gaol
*Bina Nusantara University, Indonesia*

Fonny Hutagalung
*University of Malaya, Malaysia*

Chew Fong Peng
*University of Malaya, Malaysia*

*The Social Sciences Empowered – Lumban Gaol, Hutagalung & Fong Peng (eds)*
*© 2020 Taylor & Francis Group, London, ISBN 978-0-367-00090-5*

# Editorial board

*The Social Sciences Empowered – Lumban Gaol, Hutagalung & Fong Peng (eds)*


# The techniques employed in teaching technical terms: A study on educators at a tertiary educational institution

Noraziah Mohd Amin
*Academy of Language Studies, Universiti Teknologi MARA (UiTM), Pulau Pinang, Malaysia*

Noor Azam Abdul Rahman
*Language & Social Science Unit, College of Computer Science & Information Technology, Universiti Tenaga Nasional, Putrajaya, Malaysia*

Norhasni Mohd Amin
*Sekolah Kebangsaan Seri Wangsa, Bukit Selambau, Kedah, Malaysia*

Wan Noorli Razali
*Academy of Language Studies, Universiti Teknologi MARA (UiTM), Pulau Pinang, Malaysia*

ABSTRACT: It is imperative to examine the strategies or techniques that instructors use in teaching technical vocabulary in order to discover the perception of the teachers towards the teaching methods executed. To address this issue, the present study then aimed at investigating the strategies employed by 29 educators elected as facilitators at *Institut Kemahiran Tinggi* PERDA-TECH, Nibong Tebal in teaching technical terms to the students assigned by their institute for them. Based on the questionnaire survey conducted, the findings indicated that most respondents agreed that they involved the use of Malay translations in teaching technical vocabulary such as the use of Malay synonyms (item 4), which was discovered to be the most preferred method as 69% (M = 3.97, SD = .731) of the respondents revealed practicing this technique. Nevertheless, 41.1% of the respondents (M = 3.03, SD = 1.180) agreed with item 5 about practicing Malay translations sometimes still could not help the respondents' students understand the technical words, particularly difficult words. It is obvious that the techniques preferred by the respondents depend on the effectiveness of them- either they are frequently used or not.

*Keywords*: Technical words, teaching techniques, English language

## 1 INTRODUCTION

Technical words are argued by Nation (2001) should be treated as high-frequency words, and in doing so, they should be taught in various methods. With respect to the attempt to discover the preferences for the techniques employed in teaching technical terms among educators at a higher learning institution, the present study was conducted on several lecturers from Institut Kemahiran Tinggi PERDA-TECH. This study intends to investigate if these educators possess any tendency towards practicing certain techniques and if the methods of imparting knowledge about technical terms to their respective students are effective. The results of the study can determine the actions to be taken in improving the pedagogical approach of the teachers in dealing with technical vocabulary.

### 1.1 *Problem statement*

It is argued by Chung and Nation (2003) that teachers basically face two main difficulties in their attempts to help learners deal with technical terms namely; 1) teachers occasionally do not possess specialist intelligence of their learners' technical fields, 2) technical terms should be worked on in dealing with the specialized areas. As for *Institut Kemahiran Tinggi* PERDA-TECH (PERDA-TECH Higher Skill Institute), Nibong Tebal, Penang, Malaysia, until the last July 2017 session, the problem most frequently had been faced by the lecturers was the failure of their students to complete their log book tasks due to several factors such as there were no additional efforts from their teachers to help them with learning the technical words in the log book as well as there were no stern actions imposed on the students for failing to do the tasks in the book. The lack of vocabulary and self-confidence has been identified as some of the factors contributing to *Institut Kemahiran Tinggi* PERDA-TECH students' weak English. One of the initiatives implemented to improve the students' English proficiency is the introduction of English Log Book Program in the curriculum for the purpose of assisting the students in learning English in terms of expanding their technical vocabulary, as well as improving their understanding of the meanings of some selected words, and their ability to use the words. The program is run with the help of the lecturers appointed as facilitators specifically for the implementation of it. Since teaching and learning technical words are part of the program, it is thus imperative to discover the techniques used by the instructors at the institute in teaching that category of words. With the data obtained, probably the institute could provide the materials or devices needed for the execution of the techniques.

### 1.2 *Research objectives*

The objective of a particular study serves as the direction of the study. As for the present study, it attempted to achieve the following objective:

1. To investigate the techniques practiced by the facilitators at *Institut Kemahiran Tinggi* PERDA-TECH in teaching technical terms.

### 1.3 *Research questions*

A research is incomplete without any issue to unravel or question to answer where the research revolves around. Thus, this study intends to seek the answer to the following research question:

1. What are the techniques practiced by the facilitators at *Institut Kemahiran Tinggi* PERDA-TECH in teaching technical terms?

### 1.4 *Significance of the study*

This study expects to clarify the issues with technical terms particularly with respect to the techniques to use in teaching technical words. Besides, it is hoped that the data obtained from this research can be another source of knowledge and enlightenment with regard to the issues highlighted.

## 2 LITERATURE REVIEW

### 2.1 *Definition of "technical terms" and the size of technical vocabulary*

Nation (2001) classifies vocabulary into four levels namely; high frequency words, academic vocabulary, technical vocabulary, and low frequency words. In describing the types of words used in oral and written texts, Nation (2001) explains that technical words are the words that are very closely related to the subject domains of the texts. "Technical words" are described as words identified with a few features such as being in the category of low frequency words and being limited to a particular discipline where they are used frequently (Nation, 2001; Chung & Nation, 2004). Besides, technical words are also a portion of the taxonomy of information in a particular field (Chung & Nation, 2004). According to Tongpoon-Patanasorn (2018), "Technical words are words whose meanings are related to one specific and specialized subject area, such as Business English (BE)" (p. 45). Technical terms are defined as words that appear in specialized texts and their definitions are specific to certain domains in the academic area (Nation, 2001). Nation (2001) believes that technical words can refer to words that rarely appears in other subject areas (e.g: "cabotage", "amortization") or words that are in the category of high frequency words but they own specialized definitions such as "demand", "supply" and "cost" (used in Economics). Nation (2001) provides the examples of "indigenous", "regeneration", "podocarp", "beech", "rimu" (a New Zealand tree) and "timber" for technical words which are common in this subject field. It is claimed by Nation (2001) with respect to the amount of technical words, it can be assumed that roughly technical dictionaries such as those of Geography, Biology and Applied Linguistics each consists of less than 1000 words.

### 2.2 *The importance of learning technical words*

It is proposed that semantic nuances and pragmatic uses of technical words are vital for learners to be well-versed in in order for them to be competent in a certain discipline (Tongpoon-Patanasorn, 2018). Being able to use technical words appropriately can be the benchmark for the degree or level of one's subject knowledge and this is also argued can be the foundation in evaluating a learner's subject competence (Gablasova, 2014). In addition, Gablasova (2014) claims that being knowledgeable in specialized or technical vocabulary has the tendency to contribute to greater impact on the learners' development academically.

### 2.3 *The teaching and learning of technical terms*

Incidental vocabulary learning can occur with respect to learning technical words pertaining to academic study and this frequently includes an activity which is reading to learn (Grabe, 2009). With respect to learning technical words from text books, different lexical familiarization techniques in texts are used by textbook writers in order to familiarize such words to students, thus making the words more understandable and learnable (Gablasova, 2014). With regard to learning technical terms in textbooks and lectures especially, providing definitions are common methods of lexical familiarization (Lessard–Clouston, 2009; Nation, 2001). It is claimed by most textbook writers that learning the definitions of new technical terms is a fundamental part of learning new subject knowledge (Bravo & Cervetti, 2009; Woodward–Kron, 2008).

Chung and Nation (2003) believe that learning common collocations and grammatical elements of the technical words can assist learners in their efforts in learning the words productively. Some techniques that can be employed by teachers in helping their learners deal with technical terms are providing the means for the learners to obtain the skills of recognizing technical vocabulary, interpreting meanings, linking senses to central meanings, and learning word fragments (Chung & Nation, 2003). According to McLaughlin and Parkinson (2018), in their research on some trainees at a polytechnic institution in New Zealand, with regard to their learning of specialized language of carpentry, it is inferred that this type of learning occurs as part of learners' daily communication on the building site instead of formally learning the language in class.

## 3 METHODOLOGY

### 3.1 *Research design*

The study employed a descriptive and quantitative research design through the questionnaire survey on the techniques used by the lecturers at *Institut Kemahiran Tinggi* PERDA-TECH, Nibong Tebal in teaching technical words to their students under the English Log Book Program.

### 3.2 *Participants*

The respondents involved the present study were 29 Malay instructors teaching at *Institut Kemahiran Tinggi* PERDA-TECH and they specialize in several disciplines. The demographic details of the respondents are presented in Table 1.

The respondents were also the facilitators chosen for the English Log Book Program. The content of the English Log Book is based on the concept of vocabulary enrichment which requires the students to complete five levels throughout their duration of study at *Institut Kemahiran Tinggi* PERDA-TECH. The English Log Book involves 5 levels with respect to the levels of difficulty of several selected words, while Level 6 contains some reading materials for the students. The division of the log book is shown in the table below:

Table 2. Content of english log book.

| Level | Content/Types of Words | Number of Words |
|---|---|---|
| Level 1 & 2 | Common words included in the general studies | 100 (50 words for every level) |
| Level 3 & 4 | Words related to instructors' respective fields (technical words) | 100 (50 words for every level) |
| Level 5 & 6 | Common phrases and conversations | Depending on necessity of use |

### 3.3 *Instrument*

The instrument used in the present research was a 16-item questionnaire involving self-reporting statements pertaining to the techniques practiced by the respondents in teaching technical words. Questions regarding the respondents' demographic profiles were also included in the designing of the questionnaire. The questionnaire developed for this research offered item measurement mainly in the form of percentages in terms of to what extent the respondents' level of agreement or disagreement with the statements

Table 1. Demographic profiles of the respondents.

| Respondents (29 PERDA-TECH Lecturers) | | | | | | |
|---|---|---|---|---|---|---|
| | | Gender | | Teaching experience | | |
| Specialization (discipline) | | Male | Female | Year(s) of teaching | Number of lecturers | Age |
| Multimedia | 2 | 11 | 18 | 2 | 1 | 21-47 |
| Information Technology | 3 | | | 3 | 3 | |
| Manufacturing Engineering | 4 | | | 4 | 4 | |
| Mathematics | 2 | | | 5 | 7 | |
| Physics | 1 | | | 6 | 2 | |
| English | 1 | | | 7 | 3 | |
| Agriculture Engineering | 3 | | | 8 | 5 | |
| Automotive | 3 | | | 10 | 1 | |
| Mechanical | 1 | | | 12 | 2 | |
| Welding technology | 1 | | | 16 | 1 | |
| Computer Technology | 2 | | | | | |
| General Studies | 3 | | | | | |
| Fashion Design | 3 | | | | | |
| **TOTAL** | **29** | **29** | | | **29** | |

(whichever that applied to the respondents) could be revealed by them. The levels of their agreeing and disagreeing with the 16 statements in the questionnaire were represented by the 5-point Likert scale as follows: Strongly disagree = 1; Disagree = 2; Undecided = 3; Agree = 4; and, Strongly agree = 5.

### 3.4 *Data collection method*

The respondents completed the questionnaire administered to them within the duration of one hour with the presence of the researchers should the respondents needed any explanation or clarification regarding the items in the questionnaire. The researchers explained to the respondents the purpose of the study and data collection prior to the administration of the questionnaire to make sure the respondents were clear about the nature of the research and that they were prepared for answering the questionnaire.

### 3.5 *Data analysis*

The data obtained was computed into the Statistical Package for Social Sciences (SPSS) version 20 for numerical data generation. Data such as frequencies and percentages, mean scores and standard deviation scores were extracted from the software for a more comprehensive statistical analysis of the findings of the present study.

## 4 FINDINGS AND ANALYSES

### 4.1 *Internal consistency of the pilot test*

Table 3. Cronbach's alpha reliability coefficient for the construct investigated and reliability statistics: Items about the respondents' techniques in teaching technical terms.

| N = 10 | | |
|---|---|---|
| Cronbach's Alpha | Cronbach's Alpha Based on Standardized Items | N of Items |
| .753 | .782 | 16 |

A pilot test was done for the objective of assessing the reliability of the items in the questionnaire involving 10 respondents of 5 male and 5 female facilitators. The Cronbach's alpha value, 0.753 produced by SPSS represented an acceptable value that indicated the questionnaire was reliable to be utilized for the study. Besides, this value also implied that the questionnaire was generally true about the items it intended to measure and that the items were comprehensible for the respondents. The alpha coefficient for the 16 items was .753, suggesting that the items possessed relatively high internal consistency. A reliability coefficient of .70 or higher is considered "acceptable" in most Social Science studies and this value is also a standard benchmark. Nunnaly (1978) proposed 0.7 to be an acceptable reliability coefficient but lower thresholds are also occasionally used in the literature.

### 4.2 *Frequency (f) distribution (%) for the items of the respondents' techniques in teaching technical terms*

Table 4 presents the Skewness values range from -1.177 to .951 and the Kurtosis values range -1.400 to 5.284. These values of Skewness and Kurtosis obtained were within the recommended values of -2 to 2 to (Burns & Burns, 2008). These values generated implied that the respondents had given acceptable feedback in terms of the relationship between the items in the questionnaire and the variable investigated to form a bell curve.

Table 5 shows the findings that represent the items investigated to answer the research question. A large number of the respondents, 19 (65.5%, M = 2.24, SD = .786) disagreed with item 1 ("Teaching technical terms to my students is something that is difficult for me"). The majority of the participants, 15 or 51.7% again assigned "disagree" to item 2 ("I often have problems explaining technical terms involving abstract concepts"). This item also had the most respondents (11 or 37.9%) that assigned "undecided" to it of all the items in the questionnaire.

As for the techniques preferred to be used by the respondents, it was discovered that 15 (51.7%) and 9 (31%) of them respectively revealed they agreed and disagreed with item 3 that constituted using examples of sentences in teaching technical words. Most respondents agreed that they involved the use of the students' mother tongue, Malay in teaching technical vocabulary. For example, the Malay word equivalent usage (item 4) was the most preferred technique as 69% (M = 3.97, SD = .731) of the respondents revealed practicing this method. However, 41.1% (M = 3.03, SD = 1.180) agreed with the statement, "Malay translations sometimes still cannot help my students understand the technical terms, especially the terms that are hard to understand" (item 5). 37.9% (M = 3.17, SD = 1.197) of the respondents assigned "agree" to item 14 ("I find the majority of my students find it easier to remember the English technical words taught in Malay compared to English").

With respect to using teaching aids or materials, the majority of the respondents, 13 (44.8%) agreed that they frequently used dictionary in the classroom when teaching technical terms. Besides, more than half of the participants, 15 people or 51.7% agreed they usually used graphic aids with their students. In contrast, 15 respondents (M = 1.66, SD = .614) disagreed with item 8 which indicated they did not solely depend on the materials supplied by their

Table 4. Independent samples test.

N = 29

| Items | Mean | Std. Dev | Kurtosis Stat | Kurtosis Std. Error | Skewness Stat | Skewness Std. Error |
|---|---|---|---|---|---|---|
| 1. Teaching technical terms to my students is something that is difficult for me. | 2.24 | .786 | .959 | .845 | .951 | .434 |
| 2. I often have problems explaining technical terms involving abstract concepts. | 2.48 | .688 | .021 | .845 | .420 | .434 |
| 3. I always use examples of sentences for the technical terms that I teach my students to make them understand the words better. | 4.07 | .842 | .813 | .845 | -.907 | .434 |
| 4. I always use the same Malay synonyms for English technical terms I teach (e.g: 1) English: "carburettor", Malay: "karburator"; 2) English: "mechanic", Malay: "mekanik"). | 3.97 | .731 | 2.431 | .845 | -1.124 | .434 |
| 5. Malay translations sometimes still cannot help my students understand the technical terms, especially the terms that are hard to understand. | 3.03 | 1.180 | -1.359 | .845 | -.071 | .434 |
| 6. I always ask my students to refer to the dictionary when I teach technical terms in the classroom. | 3.28 | 1.099 | -.016 | .845 | -.769 | .434 |
| 7. I usually use graphic aids (e.g: pictures, diagrams etc) when teaching technical terms to my students. | 4.14 | .833 | 1.298 | .845 | -1.068 | .434 |
| 8. I only rely on materials that have been provided by my institution without any additional materials when teaching technical terms to my students. | 1.66 | .614 | -.556 | .845 | .349 | .434 |
| 9. I ask my students to memorize each of the technical words that they have learned along with the meanings of the words. | 4.03 | .731 | 2.944 | .845 | -1.232 | .434 |
| 10. I use the method of assigning tasks to small groups with my students to improve their understanding of the technical words taught. | 3.90 | .817 | 1.370 | .845 | -1.067 | .434 |
| 11. I use teaching aids like PowerPoint presentations in teaching technical words to my students. | 4.28 | .649 | 4.055 | .845 | -1.174 | .434 |
| 12. Sometimes I bring students to real locations such as workshops or laboratories to show the real technical equipment when teaching certain technical words. | 4.17 | .889 | .604 | .845 | -1.015 | .434 |
| 13. I like to use teaching tools in their forms of word games to enhance my students' understanding of the technical words taught. | 3.52 | 1.090 | -.477 | .845 | -.314 | .434 |
| 14. I find the majority of my students find it easier to remember the English technical words taught in Malay compared to English. | 3.17 | 1.197 | -.835 | .845 | -.356 | .434 |
| 15. I ask my students to write the words taught in their respective notebooks. | 3.83 | .759 | .944 | .845 | -.746 | .434 |
| 16. I give technical word exercises to my students. | 4.00 | .802 | 1.148 | .845 | -.893 | .434 |

institution without making any initiative for additional materials. As for using PowerPoint presentations as teaching tools (item 11), 62% (M = 4.28, SD = .649) revealed they preferred such electronic materials. Vocabulary games were also preferred teaching tools by most of the respondents as 15 (9+6) of them agreed and strongly agreed with item 13.

The other techniques mostly employed by the respondents were lexical memorization as 20 or 69% of them agreed with item 9, small group tasks (item 10) since a large number of the participants, 19 (M = 3.90, SD = .817) agreed they practiced this pedagogical method, and trips to real premises where the same number of respondents, 12 or 41.4% respectively assigned "agree" and "strongly agree" to item 12. Other teaching techniques employed by the respondents were writing technical terms in notebooks and word exercises where 18 (62.1%) and 17 (58.6%) respondents agreed with item 15 and 16 respectively.

Table 5. Independent samples test.

N = 29

| Items | Response options | | | | | | | | | |
|---|---|---|---|---|---|---|---|---|---|---|
| | SD | | D | | U | | A | | SA | |
| | *f* | % | *f* | % | *f* | % | *f* | % | *f* | % |
| 1. Teaching technical terms to my students is something that is difficult for me. | 3 | 10.3 | 19 | 65.5 | 4 | 13.8 | 3 | 10.3 | 0 | 0 |
| 2. I often have problems explaining technical terms involving abstract concepts. | 1 | 3.4 | 15 | 51.7 | 11 | 37.9 | 2 | 6.9 | 0 | 0 |
| 3. I always use examples of sentences for the technical terms that I teach my students to make them understand the words better. | 0 | 0 | 2 | 6.9 | 3 | 10.3 | 15 | 51.7 | 9 | 31 |
| 4. I always use the same Malay synonyms for English technical terms I teach (e.g: 1) English: "carburettor", Malay: "karburator"; 2) English: "mechanic", Malay: "mekanik"). | 0 | 0 | 2 | 6.9 | 2 | 6.9 | 20 | 69.0 | 5 | 17.2 |
| 5. Malay translations sometimes still cannot help my students understand the technical terms, especially the terms that are hard to understand. | 2 | 6.9 | 11 | 37.9 | 2 | 6.9 | 12 | 41.1 | 2 | 6.9 |
| 6. I always ask my students to refer to the dictionary when I teach technical terms in the classroom. | 3 | 10.3 | 3 | 10.3 | 8 | 27.6 | 13 | 44.8 | 2 | 6.9 |
| 7. I usually use graphic aids (e.g: pictures, diagrams etc) when teaching technical terms to my students. | 0 | 0 | 2 | 6.9 | 2 | 6.9 | 15 | 51.7 | 10 | 34.5 |
| 8. I only rely on materials that have been provided by my institution without any additional materials when teaching technical terms to my students. | 12 | 41.4 | 15 | 51.7 | 2 | 6.9 | 0 | 0 | 0 | 0 |
| 9. I ask my students to memorize each of the technical words that they have learned along with the meanings of the words. | 0 | 0 | 2 | 6.9 | 1 | 3.4 | 20 | 69.0 | 6 | 20.7 |
| 10. I use the method of assigning tasks to small groups with my students to improve their understanding of the technical words taught. | 0 | 0 | 3 | 10.3 | 2 | 6.9 | 19 | 65.5 | 5 | 17.2 |
| 11. I use teaching aids like PowerPoint presentations in teaching technical words to my students. | 0 | 0 | 1 | 3.4 | 0 | 0 | 18 | 62.1 | 10 | 34.5 |
| 12. Sometimes I bring students to real locations such as workshops or laboratories to show the real technical equipment when teaching certain technical words. | 0 | 0 | 2 | 6.9 | 3 | 10.3 | 12 | 41.4 | 12 | 41.4 |
| 13. I like to use teaching tools in their forms of word games to enhance my students' understanding of the technical words taught. | 1 | 3.4 | 4 | 13.8 | 9 | 31.0 | 9 | 31.0 | 6 | 20.7 |
| 14. I find the majority of my students find it easier to remember the English technical words taught in Malay compared to English. | 3 | 10.3 | 6 | 20.7 | 6 | 20.7 | 11 | 37.9 | 3 | 10.3 |
| 15. I ask my students to write the words taught in their respective notebooks. | 0 | 0 | 2 | 6.9 | 5 | 17.2 | 18 | 62.1 | 4 | 13.8 |
| 16. I give technical word exercises to my students. | 0 | 0 | 2 | 6.9 | 3 | 10.3 | 17 | 58.6 | 7 | 24.1 |

## 5 DISCUSSION

It can be deduced from the findings that the respondents in majority are more likely to favour teacher-centred teaching techniques as they admitted implementing a variety of teaching methods on their part as teachers. Besides, the teachers in majority also admitted the use of their first language, Malay which is also the official language of Malaysia. With respect to using Malay equivalents for the English technical terms taught, apparently using translation in both the teachers' and students' mother tongue is a popular technique. According to Nation

(2005), one of the methods that can contribute to providing attention to words is to rapidly supply the definitions of the words in L1 translations. In addition, it seems that the lecturers of *Institut Kemahiran Tinggi* PERDA-TECH really make use of various teaching aids available in teaching technical terms to their students either electronic or non-electronic ones.

## 6 CONCLUSION

Nation (2001) believes that technical terms require specialist knowledge of certain domains and teachers need to employ the strategies that can help their students understand and remember the words more effectively. Based on the results of the present study, some techniques such as the use of Malay language in teaching technical words and the use of sample sentences for the technical terms taught are perceived to be effective in making the students remember and understand this category of words better. This is probably due to the fact that Malay is the lecturers' and students' mother tongue and thus code-switching to the language they are familiar with can help in the students' learning of technical words. As for the examples of how the technical terms are used in sentences, these sentences perhaps involve objects or ideas that the students can relate to that can make them understand the terms better.

As it is vital for educators to be well-informed of which strategies of teaching technical words are effective for them, according to Nation (2001), instructors should practice the strategies that are able to make their students understand and remember the words better. Therefore, further research could perhaps be performed in order to discover which technical term teaching strategies could benefit learners successfully in making learning technical vocabulary more effective. In conclusion, effective teaching strategies should always be studied and tried with learners for the objective of finding the ones that can suit the learners the best.

## REFERENCES


Bravo, M. A., & Cervetti, A. (2009). Teaching vocabulary through text and experience in content areas. In M. F. Graves (Ed.), Essential readings on vocabulary instruction (pp. 141–152). Newark, NJ: International Reading Association.

Burns, R. B., & Burns, R. A. (2008). Business Research Methods and Statistics Using SPSS. Thousand Oaks, California: Sage Publications Inc.

Chung, T, M., & Nation, P. (2003). Technical vocabulary in specialised texts. Reading in a Foreign Language. 15(2), pp. 103–116.

Gablasova, D. (2014). The Modern Language Journal, 98, 4, Wiley Periodicals, Inc. on behalf of the National Federation of Modern Language Teachers Associations, Inc.

Grabe, W. (2009). Reading in a second language: Moving from theory to practice. New York: Cambridge University Press.

Lessard–Clouston, M. (2006). Breadth and depth specialized vocabulary learning in theology among native and non-native English speakers. The Canadian Modern Language Review/La Revue Canadienne des Langues Vivantes, 63, 175–198.

McLaughlin, E., & Parkinson, J. (2018). 'We learn as we go': How acquisition of a technical vocabulary is supported during vocational training. English for Specific Purposes, 50, 14–27.

Nation, I. S. P. (2001). Learning vocabulary in another language. Cambridge, England: Cambridge University Press.

Nation, I. S. P. (2005). Teaching Vocabulary. Asian EFL Journal. Available at: http://www.asian-efl-journal.com/sept_05_pn.pdf

Nunnaly, J. (1978). Psychometric theory. New York: McGraw-Hill.

Tongpoon-Patanasorn, A. (2018). Developing a frequent technical words list for finance: A hybrid approach. English for Specific Purposes, Volume 51. Elsevier Ltd. Pages 45–54.

Woodward-Kron, R. (2008). More than just jargon: The nature and role of specialist language in learning disciplinary knowledge. Journal of English for Academic Purposes, 7(4), 234–249. Available at: http://dx.doi.org/10.1016/j.jeap.2008.10.004

*The Social Sciences Empowered – Lumban Gaol, Hutagalung & Fong Peng (eds)*
*© 2020 Taylor & Francis Group, London, ISBN 978-0-367-00090-5*

# Development of a digital economy ecosystem in Russia

Irina B. Teslenko
*Federal state-funded educational institution of higher education "Vladimir State University named after Alexander and Nikolay Stoletovs," Vladimir, Russia*

Olga B. Digilina
*RUDN University, Moscow, Russia*

Nizami V. Abdullaev & Nadezhda V. Muravyeva
*Federal state-funded educational institution of higher education "Vladimir State University named after Alexander and Nikolay Stoletovs," Vladimir, Russia*

ABSTRACT: The article contains analysis of the main directions of digital economy ecosystem development in the Russian Federation. The authors define the ecosystem of digital business development as a number of companies that use intellectual, production, information, and other resources on a regular basis. Because the digital economy ecosystem is the main condition of implementation of information and communication technologies in all spheres of life for people, state regulation of cybersecurity, stimulation and support of promotion of information technology abroad, development of infrastructure of access and data storage, and cashless payments, all types of mass digital communication and services are necessary.

## 1 INTRODUCTION

The key component of the institutional environment of the digital economy is its ecosystem. In 1993, J. Moore began to use "ecosystem" with regard to business, transferring the term from biology to economics. The Business Trends report published by Deloitte Consulting states that ecosystems are dynamic and jointly developing communities consisting of a variety of entities that create and receive new content during interaction and competition [7].

On the one hand, ecosystems are stable and durable, but, on the other hand, they are characterized by transformations, the constant internal dynamic processes. While some systems (for example, software) are developed towards increasingly subtle individualization, other systems (for example, technological infrastructure) on the contrary are combined, absorb smaller players, and create large firms that offer resources, information, and platforms to others.

Such a dynamic structure consisting of affiliates (small enterprises, corporations, universities, scientific organizations, non-profit organizations, etc.), where cooperative and competitive processes are carried out, became known as the ecosystem. R. U. Ayres noted that the purpose of the ecosystem is to improve the company's interaction with its partners and increase competitive advantages in terms of creation of innovative products, which will make it a leader in its industry. According to him, companies need to implement innovations and build new relationships for the "healthy" ecosystem [17]. The "ecosystem" term has entered the Russian business practice more recently [12].

According to the "Strategy for the Development of the Information Society in the Russian Federation for 2017–2030" (approved by RF Presidential Decree No. 203 dated May 9, 2017), the digital economy ecosystem is a partnership of organizations that *provides* constant interaction of their technological platforms, applied Internet services, analytical systems, information systems of state authorities of the Russian Federation, organizations, and citizens. Such a definition is given in a dictionary of financial and legal terms.

By "ecosystem," Sberbank means a network of organizations that are created around a single technological platform and use its services to make offers to customers and gain access to them, and plans to create a platform that will be a Russian analogue of Alibaba, Amazon, Facebook, and Google ecosystems.

The scientific literature provides the following definition of the business ecosystem. An ecosystem is a stable open system of various natural elements of the environment, between which there are regular constant processes of resource interchange, thus providing for the stable functioning at the corporate, regional, national, and integration levels [3].

In general, the significance of use of the term ecosystem in scientific literature is to focus on the establishment of close cooperation ties and interaction between organizations in order to actively create and implement innovations. In present-day conditions, foreign experts give preference to interaction and partnership.

Indeed, the digital economy ecosystem is a set of companies that share intellectual, production, information and other resources, and access to them. A shared-use resource provides all participants with the access to the benefits of the ecosystem scope [1]. Along with cooperation and interaction, there is objective competition between organizations in the ecosystem due to the characteristics of the digital economy.

According EDMA (the Electronics Developers and Manufacturers Association) experts, the digital economy is characterized by global standardization of technologies, structures, and processes resulting in increased price competition and increased requirements on the range of activities. Companies that make easily scalable and replicable successful decisions do not share the market with medium- and poorly effective companies, they occupy it entirely. In each segment, a leader that controls 40% of the market takes more than 80% of the profit. A huge, ever-increasing gap emerges in the investment opportunities of the leader and other participants. There is market monopolization, emergence of oligopolistic structures, and the specialization of countries is fixed in the world market.

The ecosystem has a complex structure. It can be represented as a set of subsystems, such as technological (small innovative enterprises, start-ups, technology parks, business incubators, accelerators, clusters, associations, etc.), financial (development institutions, VC investors, VC funds, business angel investors, crowdfunding platforms, banks, etc.), science and education (universities, research institutes, research laboratories, research departments of large companies), information (information centers and portals, conferences, trade fairs, advisory agencies, etc.), retraining and innovation transfer (technology transfer centers, engineering centers, experimental machine design bureaus, patent, inspection and certification organizations, etc.), etc. Such an approach seems sound to the authors and really reflects the essence of any infrastructure.

## 2 STATEMENT

In this article, the digital economy ecosystem of Russia refers to the market segments where value added is created using digital (information) technologies. We can distinguish seven key components (hubs) in such ecosystem: the state and society, marketing and advertising, finance and trade, infrastructure and communications, media and entertainment, cyber security, and education and human resources [14]. So-called slicers are distinguished inside the components. There are 10 of them: development and design, analytics and data, AI and big data, hardware, business models, Internet of Things, mobile, platforms, regulation, and start-ups and investments.

The purpose of our study is to determine the degree of development of the digital economy ecosystem in Russia at present, as well as the role of the state in the process of its creation.

If we take this approach to structuring the digital economy ecosystem, the following data show the state and development of its structural elements.

Currently, there are 3.773 billion active Internet users, 3.448 billion active mobile Internet users, and 2.907 billion active social media users in the world. In general, 50% of the world population uses the Internet and 46% uses the mobile Internet. In 2016, 2.3 million employees were employed in this field of information technology.

The index of digital literacy of the Russian population was 5.42 in 2016. As much as 65% of the Russian population uses the mobile Internet. Some 87.6 million people (71% of the population) use the Internet at least once a month, and 84.4 million people (69%) use the Internet at least once a week.

According to Mediascope, 12–24-year-olds most often go on the Internet using a smartphone (89.8% of the population), while representatives of the 25–44 age group prefer to go on the Internet using a computer or laptop (75.4% of the population). However, 74.7% in this age group goes on the Internet using a smartphone.

Representatives of the 45+ age group choose a computer or laptop to get online (45.8% of the population). Residents of cities with a population of more than 100,000 often go on the Internet using a smartphone; residents of cities of less than 100,000 also choose smartphones [14]. Russian statistics show the superiority of the mobile over the desktop: 65% versus 53%.

In 2016, marketing and advertising were estimated at 171 billion rubles, the Internet advertising market at 136 billion rubles, digital content at 63 billion rubles, and e-commerce at 1,238 billion rubles. In 2016, 36% of users made purchases in online stores, retail online exports amounted to $2 billion (including digital goods and services); infrastructure and software were estimated at 2,000 billion rubles (.RU is the 5th among national domains in the world (ccTLD), .RU is the 9th among all domains in the world (ccTLD, gTLD, new gTLD), and .RF is the world's largest IDN domain).

In 2016, online retail was estimated at 706 billion rubles (an increase of 18% compared to 2015); online travel at 363 billion rubles (an increase of 15%); Internet services at 169 billion rubles (an increase of 15%); and online payments at 686 billion rubles (an increase of 17%) [14].

The Top 100 online stores accounted for 68% of sales; at year-end 2016, two of every five purchases were made via mobile devices (including tickets and food). Curiously enough, there are a lot of mobile orders in those categories of goods and services that are not considered to be Information Technology: for example, DIY (Do It Yourself [14] is a kind of activity that includes self-production, repair, improvement of equipment, furniture, clothing, equipment and other consumer goods), clothing, etc. Some 66% of smartphone owners use applications for payments and purchases. The most rapidly increasing category is food delivery (with small absolute volumes).

The Russian market volume M2M/IoT (Internet of Things) reached 300 billion rubles as of the first half of 2016, having increased from 225 billion rubles in the first half of 2015 [2]. The market of DPC (Data-Processing Centre) grows by 10–15% per year [14].

According to Gartner, there will be about 21 billion IoT-enabled devices in the world by 2020. The overall market potential of smart metering devices in Russia in the private utility power and water consumption segment is more than 206 million smart meters and exceeds 400 billion rubles.

In the near future, the mobile economy will grow by 10.7% due to reducing costs, increasing efficiency, and providing new services through mobile devices and access [2]. According to the forecasts, by 2021, the contribution of the mobile economy to the economic growth will be 7.5% and the contribution to the country's GDP will be 4.7%.

Today, the mobile economy creates more than a million jobs, and by 2021 it will create one and a half million, more or less solving the problem of unemployment in the labor market. Owing to state investments in infrastructure and the availability of free Internet in Russia, the cost of data transmission is quite low.

In addition, the average cost of a smartphone in Russia is $168, and in other countries – $241. Russia is distinguished by a high level of competence of IT specialists; many Russian developers are in the Top 10 by revenue [4] at AppStore and Google Play app stores in other countries. Russia is on fifth place in the world in terms of application downloads [15].

As for industry platforms, including in Russia, their activities are very important. They provide the following: worldwide access to resources without intermediaries, rent of resources (human resources, business models, technologies, IP blocks, and finance), use of a volunteer model (Open Source Model is an Open Source Software), as well as on-demand sales through ecosystems [8].

The media and entertainment field is formed by digital content, games (about 72 million gamers are in Russia, which is about 65% of Internet users), social media, and books (the Russian e-book market has grown tenfold for the last four years) [5].

The number of vacancies for remote work and outsourcing has increased significantly. At the same time, the share of freelancers from the total number of employees in the Russian labor market is not more than 2% (up to 1.5 million people) [5].

As for start-ups and investments, there is a significant increase in indicators related to the purchase of start-ups in this field. There is also growth in the mergers and acquisitions market. According to researchers, the number of start-ups in the world will grow, despite increasing competition, due to the growing technological revolution. According to analysts of Dow Jones, their number has increased by 3.6 over the past two years.

Start-ups and small innovative enterprises have increasingly come from the ideas of individuals. In addition, no special production areas or significant numbers of workers are required today. On crowdfunding platforms you can raise money, start a business, and become a world leader in the future to support your idea, as has happened to other companies that hold the top lines of the world ranking: Apple ($800 billion), Alphabet ($550 billion), and Microsoft ($420 billion) [5].

Corporate innovations are increasing and corporate investment is developing actively due to such structures as accelerators (for example, IKEA, SAP, InspiRUSSIA, etc.) and corporate funds (for example, Sistema VC, Sistema Asia Fund, etc.). In addition, the number of business angel investors and their investments [15], as well as the p2p network and the number of crowdfunding platforms are increasing. Among the most active are Venture Club, StartTrack, the Investment Club of the Skolkovo School of Management, the Investment Club of the Internet Initiatives Development Fun, SmartHub, and others.

In 2017, the Central Bank of Russia monitored the crowdfunding market for the first time. The Bank's assessment of the market capacity is the first step towards its regulation.

In total, the regulator has analyzed the activities of the ten largest crowdfunding platforms, among which were Planeta.ru, Boomstarter.ru, Kroogi.com, Thankyou.ru, Rusini.org, and Smipon.ru, and came to the conclusion that crowdfunding had great growth potential.

According to the experts' forecasts, the volume of the crowdfunding market would have reached 4 billion rubles by the end of 2016, and then it would have to grow by 15–30% annually (for comparison: bank loans to individuals amounted to 7.2 trillion rubles in 2016) [13].

The development of the digital economy and its ecosystem makes it possible to predict that three quarters of Russians, 86.7 million people, will become Internet users by 2020 [15]. Until 2020, the labor market will maintain a stable demand for IT specialists: in general, the demand for them is more than 350 thousand people for the Russian Federation.

The goals and objectives of digital economy development until 2024 are defined in the Digital Economy of the Russian Federation program within the frameworks of the following five basic directions:

- statutory regulation;
- personnel and education;
- formation of research competences and technical scopes;
- information infrastructure; and
- information security [9].

As for the digital economy ecosystem, the challenge is to ensure successful functioning of the following:

- at least ten leading companies (ecosystem operators) competitive on global markets;
- at least ten industry (industrial) digital platforms for the key subject fields of the economy (including digital health, digital education and "smart city");
- at least 500 small and medium-sized enterprises in the field of digital technologies and platforms, as well as digital services [9].

The following growing tendencies will contribute to the development of the ecosystem.

1. Growth of investments.
   IDC, the international research consultancy, conducted a survey of IT executives about future investments. Some 24% of respondents noted that their organizations planned to increase expenditures for IT infrastructure by more than 10% and 21% planned to increase by 5–10% in 2018.
   The reasons for such a response were the following: growth of the data volume (investments in equipment for their storage are needed); security vulnerability (in 2016, 31% of companies in Russia faced DDOS attacks, or denial of service due to a hacker attack); insufficient network bandwidth (investments in network equipment are required); development of wireless communications.
   According to IDC, the return on sales of cloud infrastructure components increased by 25.8% year-on-year and reached $12.3 billion in the second quarter of 2017. Moreover, the share of the infrastructure return for public clouds accounted for 33.5% of the market [18].
2. Improvement of the data storage system.
   According to IDC, the data volume in the world will increase tenfold from 2013 to 2020 (up to 44 trillion gigabytes), so now the question arises as follows: where to store these data sets?
   Western Digital (WD), the hardware manufacturer, believes that by 2020, 70% of all data will be placed on HDD (hard disc drives) that the company is actively developing. WD is engaged in disk capacity extension without increasing media size. Relying on microwave-assisted magnetic recording (MAMR), the company is going to create HDDs of up to 40 TB. Samples of super-compatible hard disc drives will be available to corporate customers in 2019.
   IBM relies on flash memory. Its development is focused on reducing costs and accelerating the deployment of private clouds (with regard to the IBM FlashSystem 900).
3. Increase Internet speed.
   In the near future, we should expect an increase in the speed of data exchange between devices of the Internet of Things. (According to Gartner, their number will reach 20 billion by 2020.)
   Cisco, the company for the development and sale of network equipment, predicts that video will account for 80% of all Internet traffic by 2021, compared to 67% in 2016. Next-generation mobile networks capable of processing more data in a shorter time will be needed.
   Companies start to use 5G networks. Nokia teams up with Amazon Web Services for joint developments in the field of IoT and 5G. The creation of a development center for the 5G network in Russia is planned [18].
4. Works in the artificial intelligence (AI) industry.
   Gartner experts believe that the creation of systems that learn, adapt, and prepare to act autonomously will be the object of competition between technology suppliers until 2020.
   Artificial intelligence is beginning to be used to fight against cyber threats that, paradoxically, come from the same artificial intelligence. Harvard Business Review noted that the main hope for protection against attacks organized due to AI was AI.
5. Development of the hyper-converged infrastructure market.
   Experts of HPE (Hewlett Packard Enterprise) came to the conclusion that technologies that efficiently use and process data can solve the problems of information volume growth. A hyper-converged infrastructure is capable of implementing it.
   A converged infrastructure is a type of infrastructure that represents a ready-made solution from a manufacturer and is designed to accelerate infrastructure deployment. A hyper-converged infrastructure is an infrastructure where computing power, storage, servers, and networks are combined by software tools and are managed through a common administrative console. It is a software-defined technology, all components of which are integrated. For this reason, a single system administrator is sufficient to manage data stores and server hardware instead of a few IT specialists. This infrastructure is easily scaled: in order to increase capacity and performance, you need to add a new block. Instead of expanding capacity by increasing the number of disks, memory or processors, the performance is increased by adding new modules [6]. In 2017, sales of hyper-converged systems increased by 64.7% [18].
6. Increased popularity of cloud services (IaaS).
   According to IDC forecasts, 40% of IT spending for hardware, software, and services will be concentrated around the clouds by the end of 2018.
   In 2017, many companies migrated in the cloud. Start-ups also realized the advantages of the cloud, which become more flexible for their operation.
   The market of tools for enterprise transition in the clouds is expanding. It is expected that 2018 will clear the way to cloud technologies for a record number of organizations.

## 3 CONCLUSIONS

Summarizing the above, it can be noted that the ecosystem is developing and will be developed as the digital economy establishes. The ecosystem has a rather complex structure, in which elements are in constant interaction and compete with each other.

The digital economy ecosystem has its own unique features: it easily scales resources, new goods, services, and competencies; provides rapid promotion of basic technologies and entrepreneurial initiatives; significantly expands the market coverage; and is not limited to the state boundaries [1].

At the same time, digital economy development and ecosystem creation require state regulation, primarily concerning issues of cyber security; stimulation and support for the promotion of information technology abroad; the development of access and data storage infrastructure; non-cash payments and all types of mass digital communications and services [2]; and creation of comfortable conditions for IT companies to do business in Russia. The development of the appropriate tax legislation for the IT industry, regulatory documents related to big data, artificial intelligence, robotics, independent platforms, etc., is required [16].

State activities should be aimed at creating the institutional environment and the appropriate ecosystem that would ensure alignment of interests of all stakeholders of the digital economy and create conditions for ensuring the institutional equilibrium of the economic and social system.

## REFERENCES


[1] The Electronics Developers and Manufacturers Association (the EDMA) talked about the value of ecosystems for the digital economy development in Russia. [electronic source] http://arpe.ru/news/ARPE_rasskazala_o_tsennosti_ekosistem_dlya_razvitiya_tsifrovoy_ekonomiki_v_Rossii/

[2] Business in the runet amounted to 1.500 billion rubles for 2016. [electronic source]. http://www.tssonline.ru/newstext.php?news_id=116156

[3] Golokhvastovd v. ecosystem as a tool of balanced integration of interests of economic agents in the digital economy. [electronic source]. http://izron.ru/articles/sovremennyy-vzglyad-na-problemy-ekonomiki-i-menedzhmenta-sbornik-nauchnykh-trudov-po-itogam-mezhduna/sektsiya-2-ekonomika-i-upravlenie-narodnym-khozyaystvom-spetsialnost-08-00-05/ekosistema-kak-instrument-sbalansirovannoy-integratsii-interesov-ekonomicheskikh-agentov-v-usloviyakh/

[4] Investments in start-ups. [electronic source]. http://smfanton.ru/nuzhno-znat/startup.html

[5] Contribution of the mobile economy to the GDP of Russia. [electronic source]. http://vestnik-sviazy.ru dated august 13, 2017, http://www.vestnik-sviazy.ru/news/mobilnaya-ekonomika/

[6] Something about converged (and hyper-converged) IT infrastructure. [electronic source]. https://habrahabr.ru/company/it-grad/blog/281813/

[7] Review of new business trends or the labor market ecosystem. [electronic source]. https://habrahabr.ru/company/mbaconsult/blog/295502/

[8] Platform of industrial digital economy ecosystems. [electronic source]. http://www.eurasiancommission.org/ru/act/dmi/workgroup/Documents/3.BC.pdf

[9] The Digital Economy of the Russian Federation program. Decree of the Government of the Russian Federation No. 1632-r dated July 28, 2017. [electronic source]. http://www.sbras.ru/files/news/docs/programma_tsifrovaya_ekonomika.pdf, http://www.sbras.ru/ru/news/docs/government

[10] *Dictionary of Financial and Legal Terms*. [electronic source]. https://www.consultant.ru/cons/cgi/online.cgi?req=jt;div=LAW, http://www.garant.ru/files/5/4/1110145/1110145.zip, http://www.garant.ru/products/ipo/prime/doc/71570570/

[11] http://www.consultant.ru/law/ref/ju_dict/word/jekosistema_cifrovoj_jekonomiki/

[12] Formation of the University's business ecosystem: New challenges. [electronic source]. http://kopnov.livejournal.com/6597.html

[13] The Central Bank monitored the crowdfunding market. [electronic source]. http://web-payment.ru/newsitem/1103/cbr-monitoring-crowdfunding/

[14] Digital economy of Russia in 2016: Statistics and trends. [electronic source]. http://www.iksmedia.ru/news/5401378-czifrovaya-ekonomika-rossii-v-2016.html.

[15] Digital economy of Russia in 2017: Analytics, figures and facts. [electronic source]. https://www.shopolog.ru/metodichka/analytics/cifrovaya-ekonomika-rossii-2017-analitika-cifry-fakty/https://www.shopolog.ru/metodichka/analytics/cifrovaya-ekonomika-rossii-2017-analitika-cifry-fakty/

[16] Digital econy of Russia: The development program. [electronic source]. http://www.garantexpress.ru/statji/zifrovaya-ekonomika-rossii-programma-razvitiya/. Digital economy was approved by the government [electronic source]. http://www.comnews.ru/node/108966, http://www.comnews.ru/node/108966#ixzz4yExXK1DN

[17] Yakovleva, A. Yu. Factors and patterns of formation and development of innovation ecosystems. Ph.D. (Economics) Thesis, Higher School of Economics National Research University. M., 2012. [electronic source]. https://search.rsl.ru/ru/record/01005472744. http://dlib.rsl.ru, http://libed.ru/knigi-nauka/98679-1-faktori-modeli-formirovaniya-razvitiya-innovacionnih-ekosistem.php;

[18] 6 trends in IT infrastructure: Forecast for 2018. [electronic source]. https://habrahabr.ru/company/it-grad/blog/341374/

*The Social Sciences Empowered – Lumban Gaol, Hutagalung & Fong Peng (eds)*
*© 2020 Taylor & Francis Group, London, ISBN 978-0-367-00090-5*

# State-socicety integration with the use of information technologies in Russia

N.A. Mamedova & A.I. Urintsov
*Plekhanov Russian University of Economics, Russia*

K.S. Aleksandrova
*M.V. Lomonosov Moscow State University, Russia*

ABSTRACT: The article defines ways of interaction between the state and civil society institutions in accordance with the basic principles of interaction between the state and society. Examples of conditionally active interaction of the state with the institutes of civil society in Russia are presented and characterized. These examples are common throughout the state; their role in the process of informing society about the activities of the state is significant. However, the organization of feedback should be worked out in a qualitative way, so that the acts of interaction have a real, rather than a formal character.

## 1 INTRODUCTION

Basic principles of state–society interaction include fundamental premises that characterize the essence and activity of the state, and the place and role of society. Since these principles must consider the features and opportunities of mutual state–society impact, they exist in the form of compensation and the addition of one principle to another. In accordance with principles of state–society interaction, we look at how such interactions are organized in Russia (Henderson 2011; Richter 2009). The purpose of the study is to identify the most common and effective tools for interaction between the state and civil society institutions. The research hypothesis consists of establishing the relationship between the prevalence of interaction tools and the quality of feedback from society to the state.

Methods of interaction can be divided into conditionally active and conditionally passive. The use of the term "conditionally" means that within one method of interaction, individual actions contrary to the way of interaction can also occur.

Conditionally active interaction is defined as the way where the subject of the interaction initiates an active impact with the reaction of another subject. At the same time, the initiative of interaction does not apply to the implementation of the responsibilities of the subject (legislatively established). On the other hand, conditionally passive interaction is the method by which a subject of an interaction resorts to the implementation of a legally established procedure for actions.

Thus, to classify the way of interaction, it is necessary to define its nature (Crotty 2009), to establish how an initiative's level and the fact of interaction depend on the subject's free determination of interaction and commitment to participate in the development of a civil society (Zherebin 2010; Gorenburg 2013). That is, if an initiative and interaction can be correlated with the verbal formula "I do, because I think it's right," then this way of interaction is conditionally active. And if the verbal formula is "I do, because I must," then the way is conditionally passive.

## 2 MATERIALS AND DISCUSSION

Below we have described mechanisms of the conditionally active ways of state–society interactions. Conditionally passive modes of interaction do not need to be described in detail: they are the realization of the responsibilities of state and municipal bodies within their competence.

*Organization and attraction of public representatives to the work of public councils, commissions, working groups under the executive management bodies.* Such organizations are created and abolished by normative legal acts of the executive bodies of the administration. They operate on an ongoing basis and they are consultative and advisory bodies that carry out their activities on a voluntary basis.

The main principles of public councils and other similar organizations activities are: legality, publicity, professionalism, and voluntariness. The executive body ensures the organization of the activity of the public

(The research has been funded by the Russian President's grant for state support of leading scientific schools No. НШ-5449.2018.6, "Study of digital transformation of the economy.")

council, coordinates the interaction with it, and determines those who are responsible for the interaction. The role of consultative and advisory bodies includes assistance to the executive body in elaboration and preparation of proposals for legislative and regulatory framework, development of executive authority and society interactions, and the improvement of law enforcement practice in executive body activities.

The composition of tasks varies depending on the results of work, as well as on the state socioeconomic situation. Consultative and advisory bodies' work is organized by a provision regulating the purposes and tasks of the body, the authorities, the procedure for the body's formation, and the rights and duties of the body members. The list of public councils operating under executive management bodies is posted on the official website of the Public Chamber of the Russian Federation (https://www.oprf.ru/en/762/836/) and on the website of the Open Government (http://open.gov.ru/os/). The full list of commissions and councils under the President of the Russian Federation is available on the official website of the President of the Russian Federation (http://kremlin.ru/structure/commissions, http://kremlin.ru/structure/Councils). Information about the activities of each individual public council, commission, or working group is posted on the website of that body of executive power that initiated its creation and ensures its functioning.

*The organization and holding of public discussions on political and socioeconomic development issues.* Public discussions are held in order to inform the public and local authorities about the drafts of state decisions; and to determine public reaction and identify public opinion on the subject of discussion, draft proposals and recommendations on the discussion subject. The most common practice of this method of interaction is public discussions of draft laws; such as police law (2010), education law (2010–11, 2015), contract system law (2012–13, 2015). It is possible to hold several discussions at various stages of drafting the bill, and at various venues: on the websites, on the basis of public associations, or expert councils and commissions, in the framework of scientific and practical events.

Regardless of form, comments and suggestions are the substance of the discussion. This information is subject to being publicly disclosed, commented on by experts, discussed by the organizers of the discussion, and included in a special analytical report. The authorized body's duty is to inform the public about the subject of the discussion and to provide information on the content of the document submitted for discussion, including details of time, place, and procedures for registration and participating in the event. Public discussion can be conducted by accumulating comments and suggestions on the authorized body official website or a separately created website (for example, http://zakonoproekt2012.ru/), but also by citizens submitting written appeals.

All substantive comments and proposals obtained through public discussion are fixed in the final document, subjected to open placement, and are considered mandatory by the authorized body. Decisions taken by the authorized body are also posted in open sources, for example in a report form. Such sources can be printed publications, Internet sites of the exact body or the sites of public associations involved in the organization (Bettina 2013; Volkov 2012).

A list of issues, when bills are subjected to a public discussion, does not occur in the law. It is obvious that this way of interaction with the society is used for the most controversial and large-scale draft laws. Presidential Decree No. 167 of 09.02.2011 "On public discussion of federal constitutional laws and federal laws drafts" provides the right of the President of the Russian Federation to submit drafts of federal constitutional laws and federal laws affecting the main direction of state policy in the field of social and economic development of the Russian Federation for public discussion.

Accordingly, the bodies that drafted the bill and participated in its development place the text of the draft law, an explanatory note, and economical-financial justification for it on the Internet. Information on the procedure for sending comments and proposals on the draft law to the relevant website, and an indication of the time during which its public discussion will be conducted is also made available. Citizens who participate in the public discussion have the opportunity to review the comments and proposals received on the draft law from the website. After 90 days from the day of the bill's public discussion, the President of the Russian Federation is given a report on the results of the discussion.

*Organization of thematic dialogue with the public.* This type of interaction is currently actively developing. There are traditional forms, for example, a hotline or helpline, as well as new ones, particularly the "Direct Line" with the President of Russia or the BRICS Civil Forum. The purpose of this interaction method is to receive feedback from the public on the most pressing issues in a fairly short period of time (if it is a temporary service) or to carry out the interaction in a constant mode.

The most frequently used forms (hotline, helpline) now are an effective form of collecting and processing information from subscribers (users). A telephone line (mostly 24/7 and free in Russia) or e-mails can be arranged. This could provide for mandatory provision of applicants' personal data, as well as mandatory (recommended) forms of treatment. Appeals can be moderated, and actions are taken to establish the facts, information about which was contained in the appeal. Examples of such services are: a hotline on citizens complaints on alleged corruption in the Russian Federation Public Chamber, hotline of the Russian Ministry of Internal Affairs, the hotline of the President of the Russian Federation, the FAS Russia hotline on raising prices issues, and a hotline on the State final attestation issues.

A different mechanism operates when organizing and conducting a "Direct Line" with the President of Russia (http://zadat-vopros-putinu.ru/). Since 2001, this annual event, where the Russian President answers questions and reports from citizens in real-time, is broadcast on multiple TV channels, radio stations, and online on the Internet. If the functioning of the hotline or helpline services involves an applicant-operator communication, then the dialogue continues for several hours, during which the President of Russia answers a number of questions and reports. The selection of applicants to be announced is carried out by a single focal point for treatment of declarations. The selection criterion is the most interesting and relevant reports and requests.

Any person present in the studio can make an appeal, or it can be sent by phone, transmitted as a video, like SMS or MMS; direct inclusions from Russian cities are also envisaged. Preliminary appeals are accumulated and registered for treatment within a week preceding the event. The past years show that the number of applications received both during the event and registered before it significantly exceeds the possibilities of the event schedule, which, for its part, shows the public's high demand on this form of interaction. Civil forums as a way of state–society interaction are also extensively practiced in Russia. The historical prototype of a civil forum is city assemblies. It is wrong to assume that the civil forum is only a platform for expressing one's own opinion and conducting discussions on a number of socially significant issues and problems. The civil forum should be considered as a way of developing and choosing a common position in an open public dialogue.

Carrying out a series of discussions leads to a dialogue far beyond the opinion exchange: rather,, a public opinion arises, related to significant values and reflecting a balanced and stable position of a group of people, a scientific school, or any different community. The civil forum is aimed to develop specific administrative decisions. For that purpose, business games technologies, brainstorming, evaluating alternatives, mental experiments, and measuring public mood and priorities (including through a multilevel survey) are applied, all within the framework of a guided discussion. Examples of the civil society are: the Civil Forum of the Trans-Baikal Territory (held annually since 2002), and the Civil Forum BRICS (first held in 2015). Despite the general approach of interventions, each of these has its own characteristics.

The Civil Forum of the Trans-Baikal Territory is organized by the Government of the Transbaikal Region, the Legislative Assembly of the Trans-Baikal Territory, and the Public Chamber of the Trans-Baikal Territory. The forum is held on thematic venues, which are organized in various places of the administrative center of the region in Chita. On the basis of the forum, permanent working groups are formed. The venues are different in their scale, target audience, and events forms, but each of them provides free participation and an opportunity for direct contact with the organizers. This forum is aimed not only at accumulating positions on socially significant issues, but also at forming business contacts to enhance the investment attractiveness of the region and the flow of investment into it.

The BRICS Civil Forum is a platform for developing solutions at the national and world levels. Public participation in the event is viewed as a means to adequately broadcast decisions and commitments undertaken by the leaders of the BRICS Summit and the results of the BRICS activities to the world community. In 2015, the forum was established with the support of the official Russian Presidency of the BRICS and is positioned as a permanent format (from 2015 onwards) for the club's activities. The purpose of this forum is to organize a new point of interaction between such political clubs as Business 20, Labor 20, Youth 20, Civil 20, and Think 20, and thus form an independent superstructure for such a political club as the G20.

*Participation in civil society institutions creation and implementation.* Apart from organizing thematic platforms that ensure the state–society dialogue, the state also participates in civil society institutions' formation. These are independent official structures, which can be granted the right to exercise certain public authority. Such institutions have an explicit social orientation and usually operate on a national scale, and have a branched structure that is built in an analogy with the administrative vertical, linking the parent organization and its regional centers. Examples of such institutions are the Public Chamber of the Russian Federation, the Institute of the Commissioner for Human Rights in the Russian Federation, the Institute of the Ombudsman for the Rights of the Child of the Russian Federation, the Institute of the Commissioner for the Rights of Entrepreneurs of the Russian Federation, plenipotentiary representatives of the President of the Russian Federation in federal districts, and the Russian tripartite commission for regulating social and labor relations.

The Russian Federation Public Chamber (https://www.oprf.ru/) carries out interaction of citizens with state authorities and local self-government in order to take into account the needs and interests of citizens, to protect their rights and freedoms in the formation and implementation of public policy, and to implement public control over the activities of government bodies. The Chamber is elected every three years, and formed from forty citizens of the Russian Federation approved by the President of the Russian Federation, eighty-five representatives of the public chambers of the Russian regions, and forty-three representatives of all-Russian public associations and other nonprofit organizations.

The activities of the Chamber are replete with large-scale projects involving many people and organizations throughout Russia, for example, the project "Perspective - Successful Civil Practices," the forum "Community - Active Citizens Forum", "Public Councils at the Federal Executive Authority," "Zero reading of bills," "Ukraine: Monitoring the Situation," and others. In its annual report on activities, the Chamber consolidates data on all areas of work, assesses various sociopolitical solutions and trends, and is a valuable source of statistical, analytical, and forecast data.

The authorized representative (for human rights, for the rights of the child, for the rights of entrepreneurs) is a public service institution, meaning that the post of commissioner is a public office established to provide guarantees of state protection of rights and freedoms (legitimate interests) of a person and citizen (organization) in accordance with the norms of the Russian legislation, as well as universally recognized principles and norms of international law and international treaties of Russia. The activity of the commissioner does not cancel the activity of authorized persons in this or that area of state bodies of power and administration, nor does it entail a review of their competence and subject matter. It should be pointed out that the institution of the authorized representative continues to function even during the imposition of an emergency or martial law regime.

The institution of the authorized representative is established both at the federal and regional levels by the relevant legislative acts. The activity of the commissioner supplements the existing forms and means of protecting pre-existing rights and freedoms. The commissioner must not only respond to appeals from citizens and organizations within his competence, but also must monitor the situation, identifying cases of violation of the rights and freedoms of citizens (organizations), and qualifying the actions of other authorized bodies for observing rights and freedoms (legitimate interests).

The authorized representative informs the state bodies and the public about their activities and about the status of affairs in the priority area of their activities in the country, and interacts with federal and regional government bodies and with other commissioners and commissions in constituent entities of the Russian Federation, nongovernmental human rights organizations, and the media. The interaction of the commissioner with state authorities is carried out on the basis of bilateral agreements on cooperation, participation in collective commissions, collective working groups, and so on. The most noticeable lever of influence of the authorized representative on public opinion is the annual report on activities, which is brought to the attention of the general public, officials of state, and municipal bodies. In addition to the annual report, authorized representatives prepare special reports, which reflect the vision of the situation on selected topical issues.

A striking example of the fact that the institution of an authorized representative is a key link in the system of state–society interaction is that by identifying and summarizing the practice of the most typical cases of violation of rights, freedoms (legitimate interests) of a person, citizen (organization) based on the results of working with appeals and monitoring the situation, the authorized representative discovers the need to amend the current legislation. Such work contributes to the discovery of corruption, abuses of power, of facts of inaction and negligence. Significant work is done by the authorized representative on legal education, providing legal advice, including to citizens wishing to exercise their constitutional right to apply to international bodies.

The Institute of the Plenipotentiary of the President of the Russian Federation (http://www.gov.ru/main/page3.html, date of the request 15.10.2015) can be attributed to nonstate structures, since the plenipotentiary is an employee of the Administration of the President of the Russian Federation, which does not belong to the authorities. The plenipotentiary representative, fulfilling the instructions of the President of the Russian Federation, helps him to fulfill the head of the state duties.

The Russian tripartite commission for the regulation of social and labor relations (RTC) is an organization of three parties: resentatives of all-Russian associations of trade unions, all-Russian employers' associations, and the Government of the Russian Federation. The commission functions in accordance with the Federal Law of 01.05.1999 No. 92-FZ. Each party confines its representatives to the commission to no more than 30 people. Activities of each party are organized by the party coordinator. In addition, the President of the Russian Federation appoints the Commission Coordinator, who is not a member of the commission, and who does not interfere in the activities or participate in the voting.

At the regional level, tripartite commissions can also be formed to regulate social and labor relations, whose activities are carried out in accordance with the laws of the subject of the Russian Federation. The model of their activity actually corresponds to the model of the RTK. At the sectoral level, bilateral commissions are usually formed for collective bargaining and preparation of draft industry (interindustry) agreements and their conclusions. But there could be commissions on a tripartite basis with the participation of relevant executive authorities. However, in the signing of sectoral agreements, these bodies do not participate directly as their functions include only coordinating the position of the commission with the position of the executive body and providing information and advisory support to the sectoral commission activities.

*Support for nonprofit organization activities.* Budgetary resources that are distributed through nonprofit organizations are also called public finances. Noncommercial organizations provide additional services to citizens that are not provided by state and municipal institutions They also have the opportunity to provide benefits to certain groups of citizens in provision of public goods and services (Peregudov 2009). Their outreach on separate target groups contributes to more effective monitoring of the socioeconomic situation of the country. In addition, nonprofit organizations provide the most significant contribution to the system of public control over the activities of state and municipal bodies. The model of distribution of budgetary funds among nonprofit organizations operating in Russia provides support for projects with proven effectiveness for the faster dissemination of best practices and new projects on a competitive basis. Nonprofit organizations are considered as an instrument for solving social and economic problems.

According to expert opinion, the potential of the nonprofit sector is currently not being used sufficiently. In addition, state support for the development infrastructure of the nonprofit sector is also inadequate. At the same time, it should be noted that the state mainly subsidizes socially oriented nonprofit organizations (SONPO). There are two main subjects of financing: the Ministry of Economic Development of Russia (contests for co-financing of regional support programs for SONPO) and the Presidential Affairs Office of the Russian Federation (grants). The procedure for contests for subsidizing regional programs is more open than the procedure for granting presidential grants. The Ministry of Economic Development not only publishes the criteria for evaluating bids, but applications are also published on the portal www.nko.economy.gov.ru, as well as reports on granting subsidies. Every year the Ministry of Economic Development of Russia prepares a report to the Government of the Russian Federation on the results of subsidizing SONPO and regional support programs for SONPO. Information on the criteria for selecting grantees, applications, reports, and generalized data on the financing of NPOs Office of the President of the Russian Federation is not available to the public.

A number of funds have been created on the initiative of the President of the Russian Federation; through these funds, socially significant projects are financed in the form of grants. The Fund "Russian World," the Foundation for Support of Public Diplomacy of A.M. Gorchakov. Thus, due to distributed budget funds, support is provided for public and nonprofit organizations, professional associations, and scientific and educational institutions. This approach ensures the popularization of certain areas of public policy. Other state bodies at the federal and regional levels are also involved in the process of allocating budgetary funds through NGOs. On a competitive or targeted basis, in the form of grants, state assignments, government agencies, and funds established for this purpose support NPOs. Every year, the Ministry of Economic Development of Russia, the Ministry of Health of Russia, the Ministry of Labor of Russia, the Ministry of Education and Science of Russia, the Ministry of Transport of Russia, the Ministry of Sport of Russia, and the Ministry of Culture of Russia provide support to NPOs at the expense of budgetary funds.

Specialized funds established at the governmental level (the Russian Humanitarian Scientific Foundation, the Russian Foundation for Basic Research, the Foundation for Advanced Studies, the Industrial Development Fund, the Foundation for Promoting the Development of Small Forms of Enterprises in the Scientific and Technical Sphere, etc.) also participate in this process. Information on the procedure for allocating budgetary funds has varying degrees of openness. In some cases, only the final decisions on granting funding (grants of the Presidential Administration of the Russian Federation) are openly available, and information about the experts involved and selection criteria accompanied by recommendations on training Applications (subsidies of the Ministry of Economic Development of Russia) is not.

In order to claim the subsidy from the Ministry of Economic Development of the Russian Federation for co-financing regional support programs for SONCO, the region must, first of all, have the means to support the SONCO in the regional budget, and second of all, provide information on the results of NPO support and interaction with NPOs, civil society, and experts. According to the experts of the All-Russian Civil Forum in 2015, 73 subjects of the Russian Federation were implementing support programs for SONPO. Traditional directions of rendering social services to the population through NPOs are: the development of sports and creative leisure, support for persons with disabilities, care for the elderly, assistance to the poor, work with problem families and hard-to-educate adolescents, and social adaptation of children and adults. Among the promising areas of work of NGOs with the population are: promoting environmental protection and energy conservation, preserving the historical and cultural heritage, disseminating the ideas of volunteerism and charity, preventing alcoholism and drug addiction, and preventing violations.

An independent direction for subsidizing the activities of SONPO is a competition for information, consulting, and methodological support of other SONPO. Thus, on a competitive basis, the accumulated experience and transfer of management technologies and best practices are disseminated. The main direction for this subsidy is to conduct educational and public awareness activities to manage the NPOs activities and individual projects, attract volunteers, and resources research for projects implementation and promotion. By distributing budgetary funds through NPOs, the state seeks to diversify the activities of authorized institutions rendering services to the population. The policy on denationalization of this sphere will contribute to the improvement of the quality and volume of the provision of social services, diversity, including through the partnership of NPOs and commercial organizations.

## 3 CONCLUSION

According to the number of subjects involved, the most representative are the following: the work of social groups in the executive bodies of administration and the organization of a thematic dialogue with the public. The most costly forms are recognized: participation in the activities of civil society institutions and support for the activities of nonprofit organizations. The most labor-intensive forms are recognized: the organization and conduct of public discussions and the organization and provision of public relations.

The logic of the continuation of the research is to collect and analyze statistical data on the operation of each of the forms of interaction between society and the state, and to identify forms that are not conditioned by the influence of the state but are purely an initiative of the society.

## REFERENCES


Bettina, R., & Sullivan, J. 2013. Electronic Resources and the Study of Political Elites in Russia. *Europe-Asia Studies*. https://doi: 10.1080/09668136.2013.848646

Crotty, Jo. 2009. Making a Difference? NGOs and Civil Society Development in Russia. *Europe-Asia Studies*. https://doi: 10.1080/09668130802532936

Gorenburg, D. 2013. Civil Society in Russian Politics: Editor's Introduction. *Russian Politics and Law*. https://doi: 10.2753/RUP1061-1940510200

Henderson, S. 2011. Civil Society in Russia: State-Society Relations in the Post-Yeltsin Era. *Problems of Post-Communism*. https://doi: 10.2753/PPC1075-8216580302

Peregudov, S. 2009. Business and State Bureaucracy in Russia Dynamics of Interaction. *Russian Politics and Law*. https://doi: 10.2753/RUP1061-1940470403

Richter, J., Ghodsee, K. 2009. The State in Putin's Russia: Introduction. *Problems of Post-Communism*. https://doi: 10.2753/PPC1075-8216560601

Volkov, D. 2012. The Internet and Political Involvement in Russia (Based on the Data of Mass Surveys). *Russian Education and Society*. https://doi: 10.2753/RES1060-9393540903

Zherebin, V., Ermakova, N., & Makhrova, O. 2010. Consumption in the Information Society. *Russian Education and Society*. https://doi: 10.2753/RES1060-9393520304

*The Social Sciences Empowered – Lumban Gaol, Hutagalung & Fong Peng (eds)*
 *ISBN 978-0-367-00090-5*

# Driving factors of low cost housing in Jakarta

B.P.K. Bintoro, T. Susanto, E.J. Mihardja, J.C.P. Putra, M. Ihsan & M.D. Novianti
*Universitas Bakrie, Jakarta, Indonesia*

ABSTRACT: Low cost housing is one of the basic needs for human livings. This study serves as an initial comprehension regarding low cost housing in Jakarta in terms of its driving factors. The interview approach was used to collect data from respondents. The results of interview were presented in several figures to characterize its results. The findings indicated that house pricing, buyers' profile, purchasing power, maintenance cost, first-time homebuyers, type of low cost housing, newlyweds, high land prices and office location are the factors that mostly influence the low cost housing in Jakarta.

## 1 INTRODUCTION

Housing is a basic needs of man. It is as important as clothing and food. Housing is one of the indexes that measure a person's standard of living. It is also an important health indicator. Nevertheless, access to quality housing remains a major challenge in Jakarta.

Several studies have been carried out by various researchers to curb this menace. Winarso and Firman (2002) conducted a research on the Residential Land Development in Jabotabek. The studies revealed that the excess land development in Jabotabek focuses only on the rich and this in turn has triggered economic instability.

Besides being the country's capital, Jakarta is one of the rapidly urbanizing cities in Asia and as such it is a source of attraction to many people. Consequently, the population growth in Jakarta is on a rapid increase. The Figure 1 below shows the increasing population over the past years and the future projections of the population. It is expected that the increase in population rate should lead to a subsequent increase in the population.

Subsequently, Hammer et al. (2015) developed an approach for classifying the age of a certain group. He stated that the age group can be classified into three categories. These categories are: child age from 0 to 14 years, productive age which is from 15 to 64 years while the elderly ranges from 65 and above. The projection of productive age is seen in Figure 2 below.

Based on Figure 2, it shows that the value of productive projection is in the range of 71.7% to 72.8%. This trend is believed to be one of the factors that contribute to the housing need in Indonesia. Therefore, providing the low cost housing could be a solution to the housing need in the capital. This also means that the buyer's profile is an important consideration in the sales of low cost housing.

Hoek-Smith (2005) in his study on House Financing Sector in Indonesia, considered the structure of housing finance in Indonesia. He found that the prevalent housing need is to accommodate new household formation. The result of this is that households might need a huge sum of money to buy a house. However, buyer's target, buyer's profile and home buyers purchasing are the main factors that influence the lending decisions of other parties.

Moreover, it is important to note that newlyweds or recent university graduates are the main targets for housing developers to market their products. Yet, the prices of the housing are very high leading to a reduction in the purchasing power of these groups. Hence, this paper aims to extract the point of view of developers which consist of government developers, private developers, and consultants in a number of factors that drive these parties to produce low cost housing.

In addition, Hoek-smith, (2005) studied housing finance with mortgage mode in order to overcome the issues of housing finance structure as follows:

### 1.1 *Demand constraints for long-term mortgage borrowing*

A novel mortgage model was proposed by Justiniano *et al.*, (2015). They described the fact that there is a constraint in borrowing when there is a credit demand trigger. Also they modeled the lending limitation as a credit supply shifter. Their model indicated that investment as inflow from foreign funds into mortgage output is considerably isomorphic in terms of reducing the restriction of domestic loans limitation. This refers to the availability of funds to borrower to buy a house.

Hoek-smith (2005) also considered several factors that influence the constraint for long-term mortgage

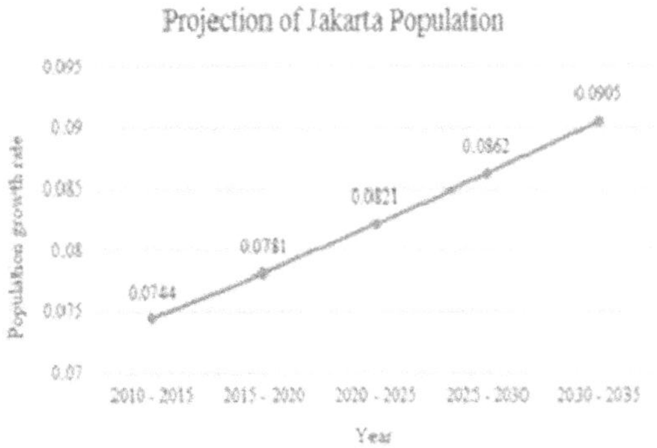


Figure 1. Projection of Jakarta Population 2010 - 2035 (Proyeksi Penduduk Indonesia, Badan Perencanaan Pembangunan Nasional, Badan Pusat Statistik, 2013).

Figure 2. Projection of productive age at Jakarta (Proyeksi Penduduk Indonesia, Badan Perencanaan Pembangunan Nasional, Badan Pusat Statistik, 2013).

and found out that the workers status, high interest rates and the number of buyers profile come from middle and lower-middle income.

Thus, these approaches are expected to be solutions for housing finance with long-term mortgage borrowing mode.

### 1.2 *Supply constraints in expanding mortgage lending*

The major factor to consider in this approach is the relatively high cost of housing. This is as a result of the lack of supply of service land and the monotonous permitting procedures that becomes unprofitable for the developers. Also the banks fail to give financial support to constructors with poor past performances. Therefore, most of the projects are left to run on the owner's funds.

Furthermore, mortgage finance is still faced with more rising challenges that threaten its expansion. One of such is the increase in nominal interest due to inflation. As a result of this, long term investors have moved part of the deposits they held in the banking system into short term deposits. Mortgage finance also has a relatively high risk especially when there is lack of information on it.

Successful action of the secondary mortgage corporation involves clarification of tax rules and adjustments of the limits on deadlines. However, finance companies could also play a role by mortgage lending in a medium term. Regardless, inadequate housing market information can lead to mortgage finance failures.

### 1.3 *Need to develop shorter term, non mortgage based housing finance option*

This subsidy scheme is developed by considering a number of financing problems such as new house, existing home, and home improvements. It is expected that this scheme can fuel the development of housing sector by providing guarantees as a part of the loan.

Conversely, Aswin et al, (2013) conducted a study to cross examine factors that influence cost of housing in Jakarta Metropolitan Region. This study was carried out using the interview method to collect data required. The findings gotten form the research was then used to develop a conceptual model. Prior to this, the preceding set of questions were designed as interview questions by studying previous researches regarding factors influencing house price in Jakarta Metropolitan Region (2012).

- What basically influences a property developer when setting the cost of a housing product in Jakarta Metropolitan Region?
- What are the factors that influence a customer's preferences for housing product in Jakarta Metropolitan Region?
- What are the external factors that influence the cost of housing product in Jakarta Metropolitan Region?

Once the main questions were answered, the respondents were required to answer the sub-questions about the underlying factors that influences a real estate developer when setting a price for housing product in Jakarta Metropolitan Region.

- Design plays a significant part in price setting for property developer when dealing with housing product.
- Developer brand plays a significant part in price setting for property developer when dealing with housing product.
- Facilities provided play a significant part in price setting for property developer when dealing with housing product.
- Accessibility plays a significant part in price setting for property developer when dealing with housing product.

Also, it was found that proximity, affordability and livability plays a significant role in influencing the cost of housing product in Jakarta Metropolitan Region.

Subsequently, Aswin *et al.*(2013) classified the findings into several points such as attributes influencing price, the customers' preferences attributes influencing price, real estate developers' perspectives attributes influencing price, external attributes influencing price. Figures 3-6 clearly depicts these findings.

From Figures 3-6, it can be concluded that design is a main factor influencing price from the perspective of both attributes and customer preferences. Additionally, segmentation is a point of view from real estate developer perspective to decide housing price. This finding can either predict buyers' or targets' profile. This is important because banks or other parties would have to ensure they know their clients before lending. Moreover, speculative behavior is identified as an external attributes influencing price. It can be mean, for example, that a consumer purchased a number of bulk real estate in order to get enormous profit immediately.

However, this action caused an increase in the cost of housing due to the instability between supply and demand

Furthermore, in the subsequent research, Aswin *et al.*(2002) addressed the factors or attributes that would influence the price of residential products in Jakarta Metropolitan Region. They successfully formulated factors influencing price and they are depicted as in Figure 7.

From the findings based on Figure 7, it can be observed that the design concept of green building is

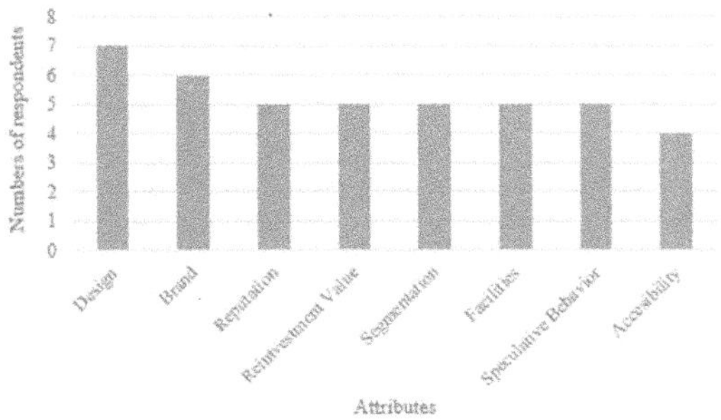


Figure 3. Attributes influencing price (Aswin *et al.*, 2013).

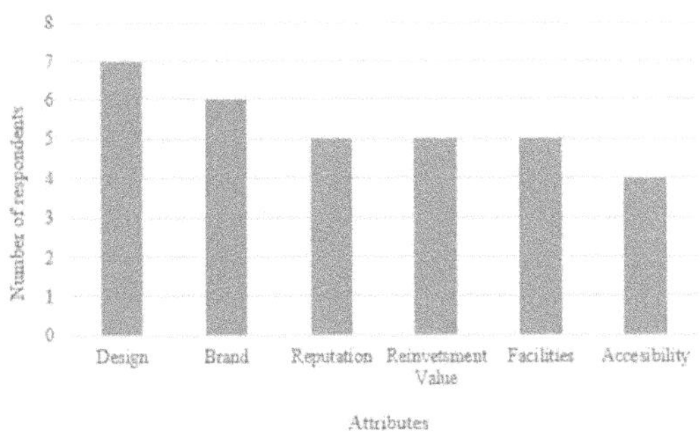


Figure 4. The customer preferences attributes influencing price (Aswin *et al.*, 2013).

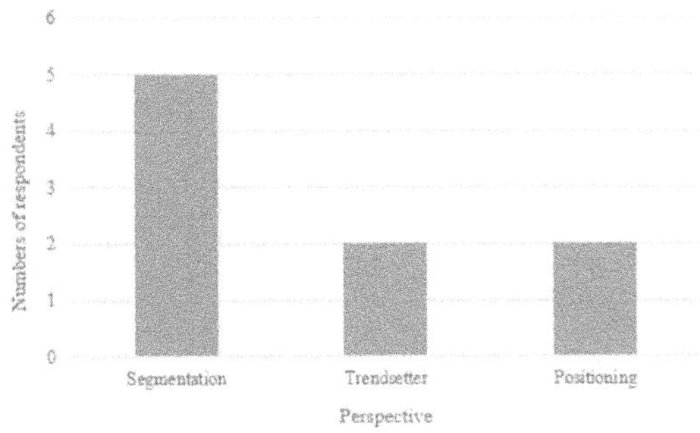


Figure 5. Real estate developer perspectives attributes influencing price (Aswin *et al.*, 2013).

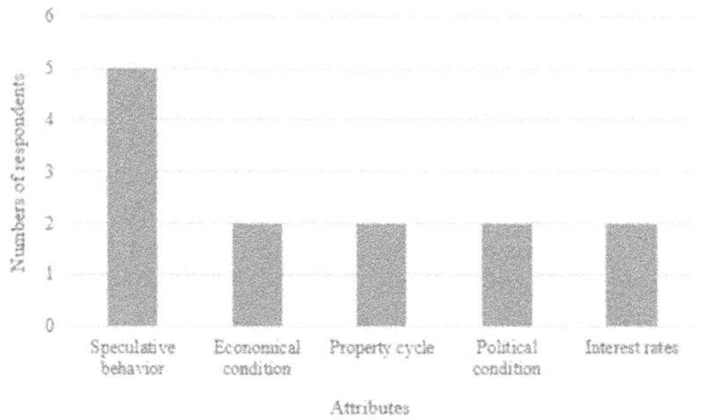


Figure 6. External attributes influencing price (Aswin *et al.*, 2013).

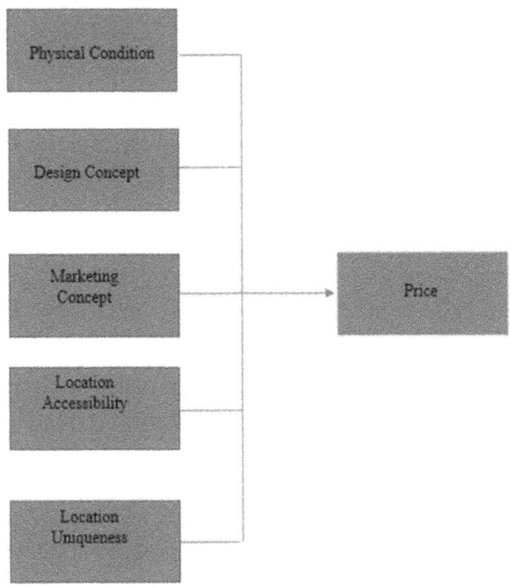


Figure 7. Factors grouping based on research findings (Rahadi et al., 2015).

costlier than the conventional design. This is due to the good indoor environmental quality (IEQ) (W. Fisk, 2002). Indoor air quality (IAQ) is one of IEQ parameters. Building or housing with good IAQ can diminish the indoor air particle by optimizing its ventilation design (J. C. P. Putra *et al*, 2017) and lead to the energy saving as well by using thermal comfort approach (J. C. P. Putra, 2017). Hence, it will create a healthy housing.

Based on the findings above, the complexity of housing problems are located in providing a new house need. It is assumed that the needs are specific to newlyweds or people in the productive age group.

Also, Jakarta is a hub of economic development attracting different kinds of people. Most people work and live there. It is on this basis that the study on the driving factors of low cost housing in Jakarta was conducted.

## 2 EXPERIMENTAL DETAILS

This study will act as an initial comprehension regarding driving factors of low cost housing in Jakarta. The survey included an interview with three people who were in charge of housing development both in private or public sector. All the respondents reside in Jakarta. The results were collected using interview approach and subsequently analyzing the collection of interview answer manuscript.

### 2.1 *Interview question*

The respondents are required to answer the following question during the interview session

- The definition of low cost housing
- The common constraint of low cost housing
- The requirement to purchase low cost housing
- The design of low cost housing
- The supporting factor of low cost housing

Additionally, respondents were asked to give their opinion regarding low cost housing. This session was intended to open discussion between respondents and interviewer. Hence, other valuable inputs were expected to be obtained as well.

## 3 RESULTS AND DISCUSSION

The results from the study are based on the interview question and it presented as follows:

### 3.1 *Interview result - The definition of low cost housing*

This sub-section is really important prior to further investigation in order to synchronize the definition of low cost housing. Subsequently, the result of this sub-section is depicted as in Figure 8.

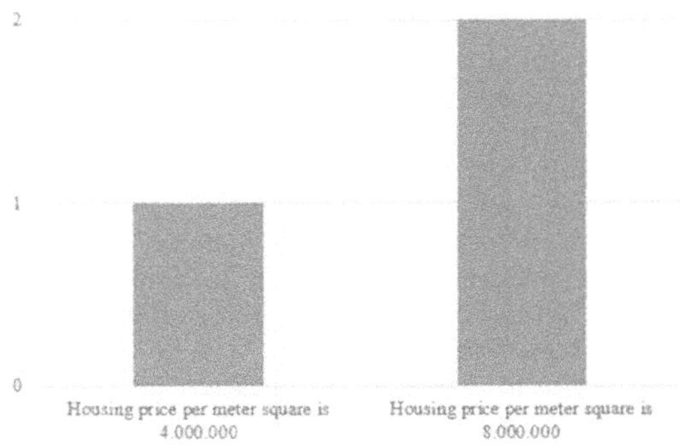


Figure 8. The definition of low cost housing.

The definition of low cost housing is limited to the monthly payroll of housing buyers. The main factor for that determines the housing cost which was within the range of Rp. 8.000.000 is the price of cut and fill, material, and built construction. However, this action impacted to the increment of pricing due to the instability between supply and demand.

Low cost housing has five common parameters. Buyers' profile, purchasing power and maintenance cost are three critical constraints in low cost housing while buyers' psychology and cultural definition is limited to the payroll of buyers.

### 3.2 *Interview result - The common constrain of low cost housing*

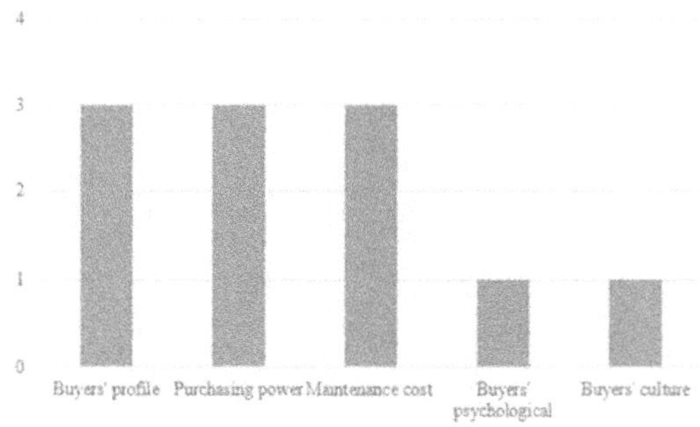


Figure 9. The common constrain of low cost housing.

### 3.3 *Interview result - The requirement to purchase low cost housing*

The major challenge faced as regarding the purchase of low cost housing was workers that had neither a fixed income nor a bank account.

The solution to this problem is it requires assistance from the authority parties should ensure the bank give loans to buy a housing.

In order to solve these problems, it requires assistance from the authority parties to ensure the bank to give credit for buyers to buy housing.

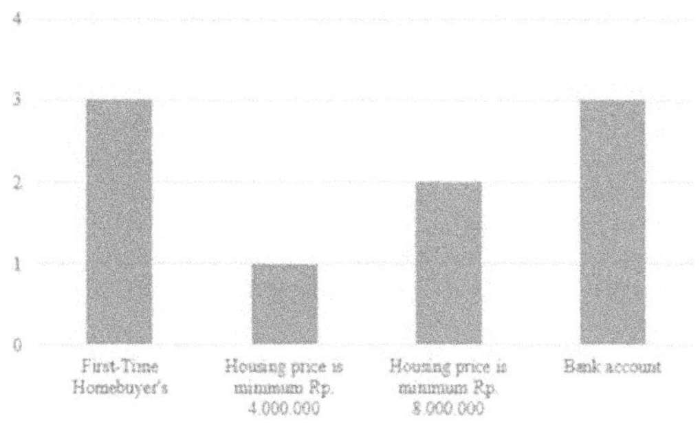


Figure 10. The requirement to purchase low cost housing.

### 3.4 *Interview result – The design of low cost housing*

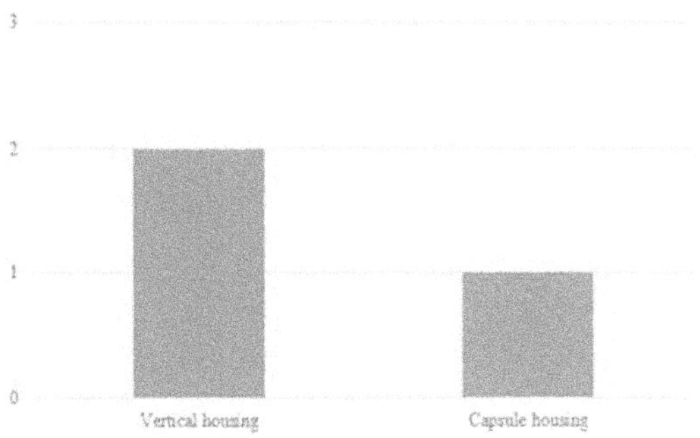


Figure 11. The design of low cost housing.

The design of low cost housing is a main factor that gives effect to the housing occupants. It is expected that the design should also generate a healthy housing.

Besides, by considering the price offered from the low cost housing, the approach of capsule housing is logically raised. However, there is a need for another evaluation for determine the proper design of capsule housing.

### 3.5 *Interview result – The trigger factors of low cost housing*

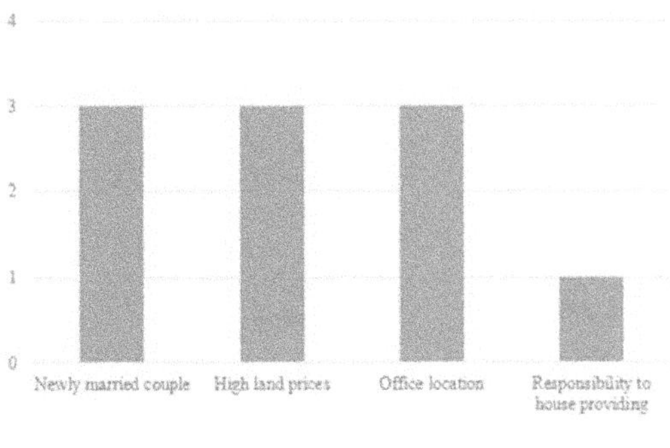


Figure 11. The trigger factors of low cost housing.

Psychologically, the motivation for having a new house for newlyweds is really huge. Nevertheless, the purchasing power of a new house is lower than the increasing cost of lands. Also, the selection of housing location influences cost of housing. Since the working area is located in a strategic region, the cost of housing became relatively high.

For low cost housing to thrive effectively, there is a truly a need for support from the government. Cooperation between the private and government sector on matters such as construction work and provision of land is one of the methods of facilitating the low cost housing program. The cost of construction is most times considered unprofitable so developers tend to seek other projects. Therefore, a subsidy for low cost housing project from the government will help facilitate the program.

Finally, from those findings as represented in the figures above, it can be stated that the essence of this research is as follows:

1. Low cost housing is restricted based on the cost of land, in this initial study its price is restricted to Rp. 8.000.000.
2. The common constraint on low cost housing includes buyer's profile, purchasing power, and maintenance cost.
3. The requirement of purchasing low cost housing is determined by first-time home buyers and bank account.
4. The design of low cost housing in Jakarta is vertical housing.
5. Variables such as newlyweds, high land prices, office location are considered as trigger factors for buying a low cost housing.

## 4 CONCLUSIONS

The study has successfully described factors of low cost housing by reflecting the buyers' behavior as an indicator to assess buyers' profile, buyers' psychological, and buyers' culture. These three aspects are important either prior to approval in purchasing low cost housing or to locate the occupants of low cost housing. Subsequently, the buyers' profile of low cost housing can be predicted for those newlyweds, first-time homebuyers', and workers. It is also important for government to support the low cost housing so they can maintain their selling price.

Subsequently, capsule housing is perceived as a solution to decrease the housing price that is, either land or construction costs. However, the design of capsule housing needs to be further developed in order to create a proper healthy design for occupants.

## 5 FUTURE RESEARCH

The findings from this paper are useful for the further research currently conducted by the authors. This study serves as a preliminary research in order to get valuable input regarding driving factors of low cost housing in Jakarta.

## REFERENCES


B. Hammer, A. Prskawetz, and I. Freund. 2015. Production activities and economic dependency by age and gender in Europe : A cross-country comparison. *The Journal of the Economics of Ageing* 5: 86–97.

Badan Perencanaan Pembangunan Nasional. Badan Pusat Statistik and United Nations Population Fund. 2013. *Indonesia Population Projection 2010-2035*. Jakarta: Badan Pusat Statistik.

Fisk, W.J. 2002. How IEQ Affects Health, Productivity. *ASHRAE Journal* 44(5): 56–60.

H. Winarso and T. Firman. 2002. Residential land development in Jabotabek, Indonesia: triggering economic crisis?. *Habitat International* 26: 487–506.

J. C. P. Putra, and I. A. Rahman. 2017. Effects of inlet air supply on particle deposition in an office building. *Procedia Engineering* 170:189–194.

J. C. P. Putra. 2017. Engineering Physics International Conference, EPIC 2016 A study of thermal comfort and occupant satisfaction in office room. *Procedia Engineering* 170: 240–247.

Justiniano, A., Primiceri, G. and A. Tambalotti. 2015. Credit Supply and the housing boom. *NBER Working Papers* 20874.

M. C. Hoek-smit. 2005. The Housing Finance Sector in Indonesia. *The Housing Finance Business Group Financial Sector.*

R. Aswin, S. Kaderi, and D. P. Koesrindartoto. 2013. Attributes Influencing Housing Product Value and Price in Jakarta Metropolitan Region. *Procedia - Social and Behavioral Sciences* 101: 368–378.

R. A. Rahadi, S. Kaderi, and D. P. Koesrindartoto. 2012. Relationship between Consumer Preferences and Value Propositions: A Study of Residential Product. *Procedia - Social and Behavioral Sciences* 50: 865–874.

R A. Rahadi, Sudarso Kaderi Wiryono. 2015. Factors Influencing The Price of Housing in Indonesia. *International Journal Housing Markets and Analysis* 8(2): 169–188.

*The Social Sciences Empowered – Lumban Gaol, Hutagalung & Fong Peng (eds)*
*© 2020 Taylor & Francis Group, London, ISBN 978-0-367-00090-5*

# The sensitivity analysis of virtualization for Apache Spark

C. Hung & S.Y. Hsu
*Chung Yuan Christian University, Taoyuan, Taiwan, R.O.C*

P.F. Hu & C.Y. Yen
*SYSCOM Computer Engineering Co., Taipei, Taiwan, R.O.C*

T.H. Lin & L.H. Chiang
*Institute for Information Industry, Taipei, Taiwan, R.O.C*

ABSTRACT: This study applies an experimental method to analyze the sensitivity of parameters for the design of Spark's distributed application to the results in a virtualized environment. In the era of big data, a novel computing architecture is required as the size of data is beyond the processing capability of one stand-alone computer. We use the Apache Spark resilient distributed datasets (RDD) in-memory computing framework for data analysis and a Hadoop distributed file system (HDFS) to deal with distributed data storage. Two tasks – secondary sorting and word counting – are applied to test three important parameters: the number of nodes, the number of virtual CPU, and the size of RAM. This sensitivity analysis evaluates the relationship between the targeted parameter and efficiency in order to provide a reference for constructing a Spark distributed system.

## 1 INTRODUCTION

This study proposes an approach of sensitivity analysis for designing an Apache Spark application to deal with big data in a distributed environment. The term, big data, usually indicates the size of data beyond the processing ability of one stand-alone computer. In the era of big data, the Internet is flooded with information such as word of mouth, posts on the social media, Internet of Thing (IoT) sensor signals, mobile phone transmissions, multimedia images, historical stock trading data, online shopping transactions, web browsing logs, etc. The vast amount of data may contain valuable information regarding undiscovered knowledge and applications, which may be used to detect future market trends, the spread of disease predictions, human behavioral predictions, etc. That is the reason why the big data analysis is popular.

There is no commonly acceptable standard for the size of big data. It is generally believed that the size of big data has exceeded the processing capability of traditional tools (Snijders et al., 2012). That is, existing tools are inadequate to deal with such big data. Various data types, such as structured, semi-structured, or unstructured data on the Internet further contribute to the difficulty of analyzing big data. However, big data analysis mainly focuses on unstructured data, for example, text, image, audio, video, signal, etc. (Dedić & Stanier, 2016). Thus, a new computing or storing computer-based structure is required for revealing insights from big data that are diverse and of a massive scale (Hashem et al., 2015).

Hadoop (Ghemawat et al., 2003, Dean & Ghemawat, 2004) is a software framework that provides distributed storing and computing operations. The base Hadoop framework includes Hadoop common, HDFS (Hadoop distributed file system), and Hadoop MapReduce, which is considered one of the ideal solutions for processing big data (Lam, 2010). Besides Google, Yahoo!, Amazon, Microsoft, and Facebook that need to go through terabytes and petabytes of data, many other businesses are also facing the scaling challenge. The Hadoop techniques have been used for various fields, such as image analysis (Wang et al., 2018), scheduling (Varga et al., 2018), frequent item mining (Kusumakumari et al., 2017), opinion mining (Uzunkaya et al., 2015), and network analysis (Bakratsas et al., 2018), etc.

However, there are limitations in the Hadoop framework. For example, the MapReduce cluster computing paradigm uses a linear data-flow structure on distributed programs, causing waiting for processing. Apache Spark deals with the issue and uses the resilient distributed dataset (RDD) as a working set for distributed programs (Zaharia et al., 2012). More specifically, Spark uses in-memory computing technology to analyze and store data in the memory when it has not been written to the disk. Spark not only has fast processing performance, but also has a flexible construction method that can be run separately or built on the existing Hadoop architecture. Due to the maturity of virtualization technology, we use the VMware virtualization technology to analyze the processing efficiency of Spark by adjusting the hardware resource configuration of each node. This paper provides a preliminary assessment of the initial introduction of a Spark distributed system to provide a reference for constructing a Spark distributed system.

## 2 APPROACH OF STUDY

The Spark framework adopts a slave–master architecture in a distributed computing platform. The master is a node that acts as an administrator in the cluster and is responsible for management of the cluster. The slave is a node that acts as a worker in the cluster. A worker is only responsible for analysis and calculation. Spark supports three kinds of distributed architecture: Spark stand-alone, Hadoop YARN, and Apache Mesos cluster management architectures. This paper uses the stand-alone architecture and cooperates with Hadoop HDFS.

### 2.1 *System structure*

In the experimental environment, we use six computers with VMware vSphere installed. VMware vSphere provides a virtualized environment, which can be installed directly on a computer with no operating system pre-installed, and virtualizes hardware resources such as CPU, RAM, a hard drive, and a network interface card. Each computer is considered as a VM host. We assign one VM host running one VM guest, which is an administrator or a master in the distributed system. Each of the other five VM hosts is running two VM guests. Thus, we have ten VM guests, which are workers in the distributed system. The diagram of proposed hardware structure is shown as Figure 1. The system architecture of this study can be divided into three layers, namely a physical layer, a hypervisor layer, and an application layer (Figure 2). We have six computers in the physical layer. VMware vSphere is in the hypervisor layer, which is responsible for the communication between the physical layer and the

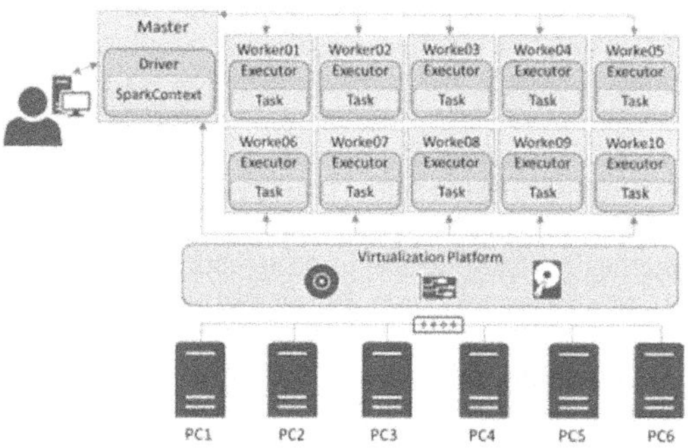


Figure 1. The diagram of proposed hardware structure.

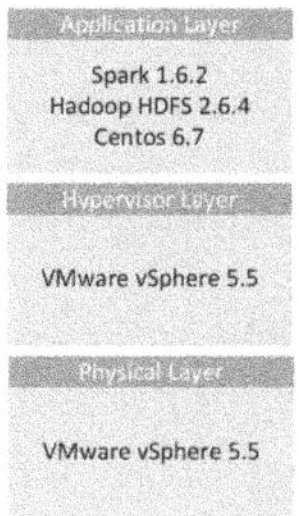


Figure 2. The conceptual diagram for system stack.

application layer. In the application layer, we install CentOS on each VM guest. We then install Hadoop on CentOS. Finally, Scala and Spark are installed. Please note that we use the Spark RDD in-memory computing framework for data analysis and use Hadoop HDFS to deal with distributed data storage. In terms of the hardware structure, we use six computers of the same specification. The CPU is Intel i5-2320-3GHz, the RAM is a pair of Kingston DDR3-1600-4GB, the hard drive is Seagate-5400RPM-16MBCache-500GB, and the network interface card is Intel 82579LM-Gigabit. We assign 8GB RAM to a master node due to its heavy loading and assign 4GB RAM to the other ten worker nodes. To mimic the scenario of big data, whose size is beyond the processing capability of a single computer, we assign each node with a 150GB virtualized hard drive.

### 2.2 *Comparison approach*

This study uses Spark's native programming language (i.e. Scala) as the coding language and uses two tasks – secondary sorting and word counting – for sensitivity analysis of Spark in virtualization.

#### 2.2.1 *Secondary sorting*

Secondary sorting is one of the external and stable sorting evaluation tasks. The targeted data contains two fields: a primary field and a secondary field. First, rows of the primary field are sorted by ascending order. Second, rows of the secondary field are sorted by ascending order if their values in the primary field are the same. The task is used by the Sort Benchmark Competition in order to compare the performance of different sorting algorithms. This study also uses the secondary sorting as a performance test for Spark's sensitivity analysis. All experiments use the same sorting approach; namely, the Spark's sortByKey function.

#### 2.2.2 *Word counting*

The task of word counting calculates the number of occurrences for each word that appears in an article or a document collection. The procedure of word counting is one of the fundamental steps in the field of text mining. Generally, stop words, such as *the*, *of*, *that*, *which*, *it*, etc. are removed in the preprocessing stage. When finishing the procedure of word counting, the most frequent *k* words, known as top-*k*, imply these words are more important than the others. The top-*k* word counting approach can be used for consumer behavior analysis, popular consumer analysis, user browsing topic analysis, user searching popular topics, etc. This study uses the word counting approach to find top-*k* words as a performance test for Spark's sensitivity analysis.

### 2.3 *Data set generation*

We use two methods to generate experimental data for secondary sorting (S-Sorting) and word counting (W-Counting) respectively. For the task of secondary sorting, the primary field contains seven bits of English letters and the secondary field contains four bits of numbers. All English letters and numbers are chosen randomly. Some 6,306 files are generated, each file size is 50MB on average, and the total data size is 300GB. For the task of word counting, we search for novels on the Internet. The file format is text. Each file has an average size of 40MB. There are 7,522 files. The total data size is also 300GB.

### 2.4 *Sensitivity analysis*

Sensitivity analysis is used to analyze the influence of parameters on a model. In order to show the influence caused by the change of parameters, this study only adjusts one parameter at a time. In this study, we have two tasks: secondary sorting and word counting. We adjust three important parameters: the number of nodes, the number of virtual CPU (vCPU), and the size of RAM in a Spark distributed environment. We show the conceptual diagram of sensitivity analysis in Figure 3. We use ten nodes, two vCPUs, and 4GB of RAM as the default values. We then test one of the parameters to evaluate the influence of this parameter on performance.

On the other hand, in order to test a wider range of parameter values under the existing hardware constraints, we use five nodes as workers in the Spark distributed environment. Thus, we modify the default parameters to five nodes, two vCPUs and 8GB of RAM (Figure 4). Then, we analyze the influence of the number of vCPU and the size of RAM on performance.

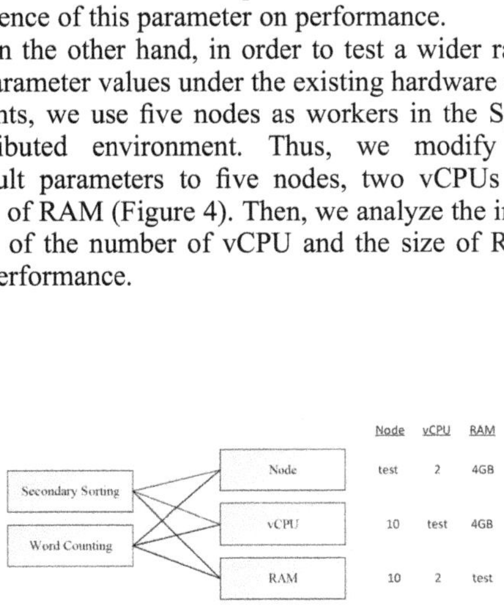


Figure 3. The conceptual diagram of sensitivity analysis for node, vCPU, and RAM.

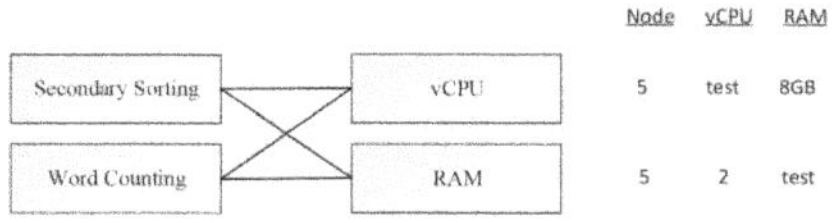


Figure 4. The conceptual diagram of sensitivity analysis for vCPU, and RAM based on five nodes.

## 3 EXPERIMENTS

In this study, we use two tasks to test three important parameters: the number of nodes, the number of virtual CPU, and the size of RAM, for the Spark distributed application in the virtualization environment. In order to present generalized results, each experimental result is the average of three experiments.

First, we test the influence of the node number from four to ten and show the results in Figure 5. This study takes 17.2 hours to complete the task of word counting when using four nodes, and 2.2 hours when using ten nodes. For the task of secondary sorting, it takes 11.6 hours when using four nodes and 4.3 hours when using ten nodes. In these experiments, the more nodes, the less time to complete the work. However, for the secondary sorting task, the curve in Figure 5 has a tendency to become smoother, indicating that the number of nodes has been increased to a certain amount, and the contribution to execution efficiency is getting less.

Second, we test the influence of the virtual CPU number from one to two for one node. According to experimental results in Figure 6, the model using two vCPUs needs more time to complete the tasks than that using one vCPU. Due to hardware limitations, one computer in the experiment is configured with only one physical CPU. Although each computer is virtualized into two nodes, and each node is configured with two vCPUs, the calculation for the task is performed by the physical CPU. Therefore, the increase of the vCPU cannot improve the operation efficiency.

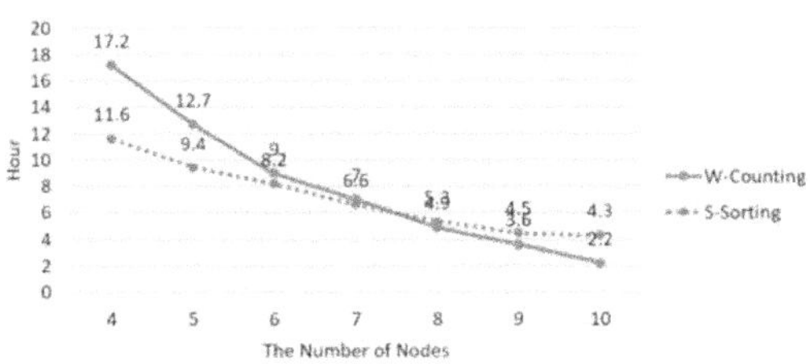


Figure 5. Sensitivity analysis of the number of nodes.

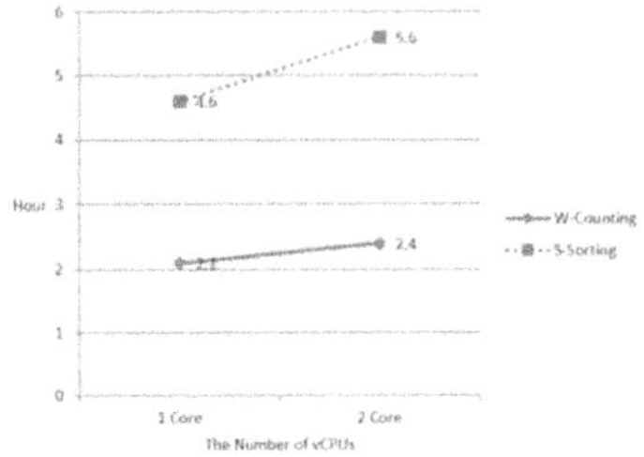


Figure 6. Sensitivity analysis of the number of vCPU.

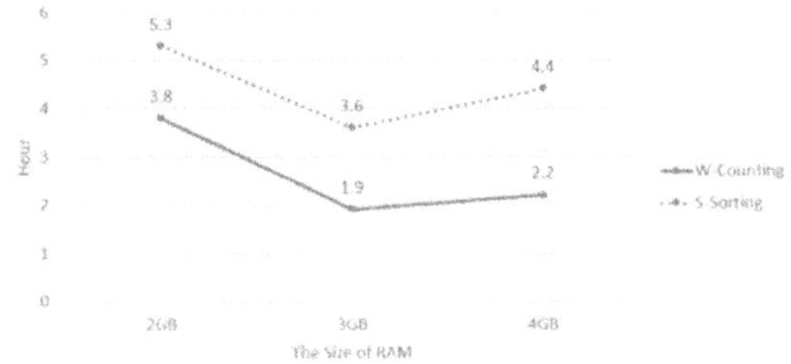


Figure 7. Sensitivity analysis of the size of RAM.

Third, we test the influence of the size of RAM from 2GB to 4GB for one node. We show the experimental results in Figure 7. When the RAM size of each node is increased from 2GB to 3GB, the experimental results for both tasks show that efficiency can be improved. However, when the RAM size of each node is increased to 4GB, the efficiency of the two tasks is decreased. This situation implies that the memory for the operation system is not enough. In this case, the operation system will store and retrieve data from the secondary storage, such as the hard drive, for use in main memory. We check the usage of SWAP in the CentOS operation system (Table 1).

Table 1. CentOS SWAP usage status in KB based on 10 nodes.

| Node | 2GB | 3GB | 4GB |
|---|---|---|---|
| node1 | 25,356 | 12,315 | 19,872 |
| node2 | 34,287 | 19,872 | 29,726 |
| node3 | 45,556 | 13,654 | 15,982 |
| node4 | 33,412 | 12,111 | 25,443 |
| node5 | 64,332 | 14,211 | 15,662 |
| node6 | 63,566 | 12,981 | 26,753 |
| node7 | 54,148 | 11,189 | 30,192 |
| node8 | 42,123 | 12,716 | 51,256 |
| node9 | 54,125 | 19,999 | 43,125 |
| node10 | 71,237 | 15,621 | 22,651 |
| average | 48,814 | 14,467 | 28,066 |

Generally, the more memory a node has, the higher the computational efficiency. We use 4G RAM for a node and its computational efficiency cannot be improved with memory, which is also limited by hardware.

Fourth, we use one node as the master and five nodes as workers in the Spark distributed environment to test a wider range of parameter values under the existing hardware constraints. We test the influence of the number of virtual CPUs from one to four for one node. According to experimental results in Figure 8, a model using two vCPUs performs better than other models.

Fifth, based on the five-worker-node distributed environment, we test the influence of the RAM size from 4GB to 8GB. We show the experimental results in Figure 9. When the RAM size of each node is increased from 4GB to 6GB, the operation performances for both tasks are also improved. However, when the RAM size exceeds 6GB, the execution efficiency is getting worse. We check the usage of SWAP in the CentOS operation system (Table 2). When a node has 6G RAM, its operating system requires almost no SWAP, which indicates that the RAM size is sufficient when the Spark application is executed. Conversely, when a node is equipped with 8G RAM, the SWAP usage is very high, which results in poor performance.

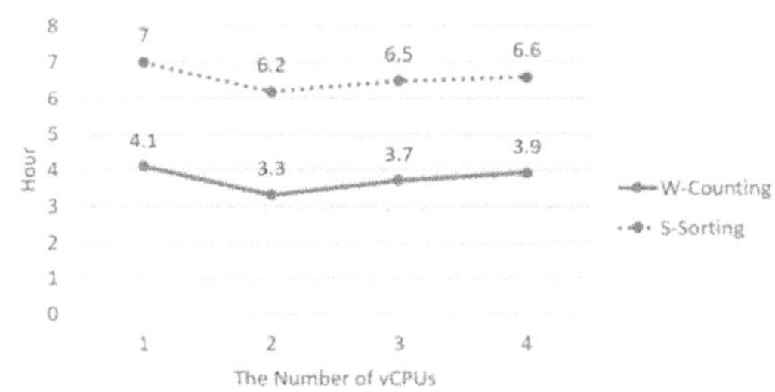


Figure 8. Sensitivity analysis of the number of vCPU based on 5 nodes.

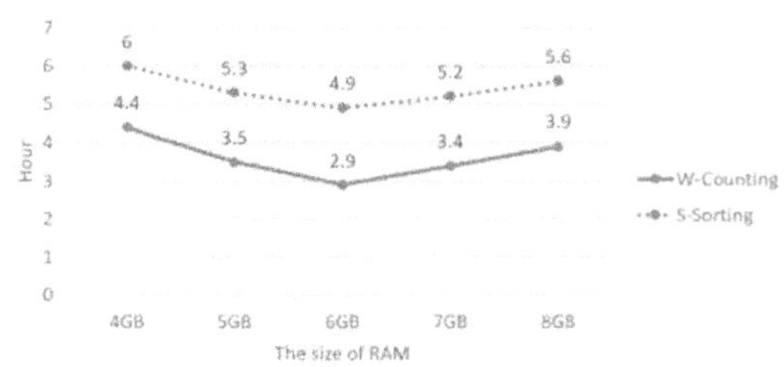


Figure 9. Sensitivity analysis of the size of RAM based on 5 nodes.

Table 2. CentOS SWAP usage status in KB based on five nodes.

| Node | 4GB | 5GB | 6GB | 7GB | 8GB |
|---|---|---|---|---|---|
| node1 | 11,212 | 16,404 | 0 | 3,640 | 55,763 |
| node2 | 19,832 | 14,804 | 0 | 2,228 | 36,160 |
| node3 | 18,288 | 16,360 | 0 | 5,320 | 45,564 |
| node4 | 15,948 | 18,880 | 260 | 1,044 | 47,460 |
| node5 | 18,152 | 12,289 | 380 | 876 | 31,209 |
| average | 16,686 | 15,747 | 128 | 2,622 | 43,231 |

Thus, it is inefficient to assign all the memory to the node in a virtualized environment.

## 4 CONCLUSION AND FURTHER WORK

This study builds the Spark distributed system in a virtualized environment. We analyze three important parameters: number of nodes, number of vCPU, and RAM size, for two common big data analysis tasks, namely, secondary sorting and word counting. This sensitivity analysis evaluates the relationship between the targeted parameter and efficiency. Under the hardware limitations of this study, the more nodes the model uses, the less time it takes to complete the tasks, generally. However, this situation becomes unclear when the node number gets greater. For the number of virtual CPU of a node, we fail to prove that more vCPU is able to produce higher performance. In terms of RAM usage, generally speaking, more RAM can achieve higher performance.

For further work, some possible directions are introduced. For example, machine learning and deep learning have shown extraordinary results in various fields (Zhang et al., 2018). Hadoop 3 (https://hadoop.apache.org/docs/r3.0.0/index.html) is able to use GPU within the cluster to execute deep learning algorithms on a Hadoop cluster. Thus, sensitivity analysis for GPU-installed hardware is a possibility for further work. There are many different virtualized techniques and products. Integrating Spark with different virtualized platforms is another avenue for further work. In the era of big data, the amount of data is getting larger and larger. The network bandwidth can be included in the further experimental study to verify whether the network bandwidth will affect the data analysis time.

## ACKNOWLEDGMENTS

This work was supported in part by the Ministry of Science and Technology of Taiwan under Grant MOST 106-2410-H-033-014-MY2.

## REFERENCES

Bakratsas, M., Basaras, P., Katsaros, D., & Tassiulas, L. 2018. Hadoop MapReduce performance on SSDs for analyzing social networks. *Big Data Research* 11:1–10.

Dean, J,. & Ghemawat, S. 2004. MapReduce: Simplified data processing on large clusters. *Proceedings of Sixth Symposium on Operating System Design and Implementation*, pp. 137–150.

Dedić, N., & Stanier, C. 2016. Towards differentiating business intelligence, big data, data analytics and knowledge discovery. *Proceedings of Innovations in Enterprise Information Systems Management and Engineering*, pp 114–122.

Ghemawat, S., Gobioff, H., & Leung, S.-T. 2003. The Google file system. *Proceedings of 19th Symposium on Operating Systems Principles*, pp. 29–43.

Hashem, I. A. T., Yaqoob, I., Anuar, N. B., Mokhtar, S., Gani, A., & Khanb, S. U. 2015. The rise of "big data" on cloud computing: Review and open research issues. *Information Systems* 47: 98–115.

Kusumakumari, V., Sherigar, D., Chandran, R., & Patil, N. 2017. Frequent pattern mining on stream data using Hadoop CanTree-Gtree. *Proceedings of 7th International Conference on Advances in Computing & Communications.*

Lam, C. 2010. *Hadoop in Action*, (1st ed.), Manning Publications.

Snijders, C., Matzat, U., & Reips, U.-D. 2012. Big data: Big gaps of knowledge in the field of Internet. *International Journal of Internet Science* 7: 1–5.

Uzunkaya, C., Ensari, T., & Kavurucu, Y. 2015. Hadoop ecosystem and its analysis on Tweets. *Procedia-Social and Behavioral Sciences* 195: 1890–1897.

Varga, M., Petrescu-Nita, A., & Pop, F. 2018. Deadline scheduling algorithm for sustainable computing in Hadoop environment. *Computers & Security* 76: 354–366.

Wang, W., Zhao, W., Cai, C., Huang, J., Xu, X., & Li, L. 2018. An efficient image aesthetic analysis system using Hadoop. *Signal Processing: Image Communication* 39: 599–508.

Zaharia, M., Chowdhury, M., Das, T., Dave, A., Ma, J., McCauley, M., Franklin, M. J., Shenker, S., & Stoica, I. 2012. Resilient distributed datasets: A fault-tolerant abstraction for in-memory cluster computing. *Proceedings of the 9th USENIX Conference on Networked Systems Design and Implementation.*

Zhang, Q., Yang, L.T., Chen, Z. & Li, P. 2018. A survey on deep learning for big data. *Information Fusion* 42: 146–157.

*The Social Sciences Empowered – Lumban Gaol, Hutagalung & Fong Peng (eds)*
*© 2020 Taylor & Francis Group, London, ISBN 978-0-367-00090-5*

# Preschool teachers' views on education of children with a physical disability in Czech kindergartens

V. Pacholík
*Faculty of Humanities, Tomas Bata University in Zlín, Zlín, Czech Republic*

ABSTRACT: The question of integration of children and pupils with special educational needs has recently been an important topic in the Czech education system. However, MEYS (Ministry of Education, Youth and Sports) attempts to integrate these children into ordinary schools and common education raises concerns of teachers and many pedagogy and psychology experts. The aim of this study was to cover attitudes of kindergarten's teachers in the Zlín region toward the problems of integration of children with physical disabilities into common kindergarten classes. We conducted a questionnaire survey among teachers. The key parts of questionnaire consisted of scale items expressing respondent's attitudes toward selected aspects of integration of children with a physical disability into ordinary kindergarten. The results imply a prevailing positive attitude of teachers toward integration of children with a physical disability. They demonstrated a positive relationship between teacher's length of teaching practice and openness to children's integration. On the other hand, a relationship between teacher's work experience with a child with a physical disability and openness to integration of these children into ordinary kindergartens has not been proved. The research proved that the size of school does not correlate with quality of conditions for integration of children with a physical disability. Based on the results, it is recommended to diversify the offer of further education of pedagogical staff in the field of work with children with special educational needs. Nevertheless, health and didactic issues are not the only concern. It is also necessary to support teachers' understanding of social and psychological aspects of a physical disability and possible difficulties with child's performance in a peer group. General perception of people with a physical disability in Czech society also plays an important role.

## 1 INTRODUCTION

Kindergarten is a place where children adopt a range of skills, knowledge, habits, and also attitudes. In social contact, children adopt different social roles, create their own self-perception, and form their attitudes towards the world that surrounds them. A part of their life are other people with all their uniqueness. Some individual differences are hidden and children discover them in a more intensive contact (various psychological dispositions, attitudes, etc.); others are visible at first sight. Physical disability is one of these obvious differences. An effort to integrate children with a physical disability into common kindergartens thus respects this variety and meets the needs of the children and also facilitates natural social development of the majority population.

A handicap, disorder, or defect are disruptions of psychological, physiological, or anatomic structure or function (Slowík, 2007). According to Gruber and Lendl (In Vítková, 2006), physical disorder is a limitation in motor abilities as a result of damage in supporting apparatus or musculoskeletal system or of a different organic damage. Renotiér (2003) adds that it results from an injury, illness, or heredity. Hartl and Hartlová (2000, 442) define a physical disability as a disability obstructing mobility and work activity and is divided into temporary and permanent, congenital and acquired. According to Ješina (2014), the term physical disability is very wide; some authors therefore prefer the term movement disorder which better reflects the functional limitation. The term "physical disability" includes the following groups: paralysis, deformation, malformation, and amputation. Cerebral palsy, a type of paralysis, is the most common physical disability in children (up to 80%).

We didn't focus on specific types of physical disabilities in our research. It's why we used the term "physical disabilities" only.

A disabled individual accepts, to a great extent, attitudes of the society towards his handicap. The view of a disabled individual thus more or less reflects the relation of society to the handicap. This results in various defenses that help the disabled individual maintain inner balance and, if possible, not disrupt his own identity. Thus there may be reactions, such as trying to "do it by oneself" to avoid negative reactions from people (either negative reactions or on the contrary too-compassionate ones related to protective tendencies), external presentation of oneself as a strong self-confident individual in order to hide inner insecurity and low self-esteem, etc. Such behavior is naturally reflected in satisfying the need for security and safety, which demands unconditional acceptance and adequate empathy from people surrounding the individual, respecting principles of Rogers' approach (Nykl, 2012). According to Hadj-Moussová (Vágnerová, Hadj-Moussová, Štech, 2000), the relationship toward disabled people requires human society develop tolerance and the ability to accept differences, variety, and individuality.

Erikson defines the preschool period as a stage of play, the typical feature of which is the initiative. In this stage, activity and self-enforcement plays an important role (Erikson, 2015). Physical movement, which is in its spontaneous form one of the basic means of self-realization of a child, plays an important role in complex development, so not only in the motor and physical development. Renate Zimmer (2015) claims that a child needs movement to support health, physical and psychological well-being and satisfaction, self-realization and presentation of motor skills to strengthen social contacts with peers and to establish new contacts. It is the area of activity where a child with a physical disability is considerably limited on the primary and secondary level. Vágnerová et al. (2000) also mention disturbances of the socialization process, namely (1) in characteristics of society influence, especially of family influence and (2) in the sphere of an individual's psychological and somatic disposition. We must realize there is a very close relationship between the two areas. Socialization must be understood as a system of mutual interactions between an individual and the environment (Nakonečný, 2009; Výrost, Slaměník et al., 2008; Řezáč, 1998). The above-mentioned division therefore represents only an artificial, formal classification facilitating orientation to the problem.

## 2 METHODOLOGY

Our research focused particularly on views of teachers in preschool education and of students of related programs in Zlín on the problems of integration of children with a physical disability. Using a questionnaire research method, we tried to detect some characteristics of the views and their relation to the length of practice, education level achieved, and other teachers' characteristics.

### 2.1 *Research aims*

The research aim was to identify views of preschool teachers, and students of programs focused on preschool education, on integration and inclusion of children with a physical disability into common kindergarten classes. This study presents only partial research findings.

In order to accomplish our goals, we set the following research questions (RQ) and related hypotheses (H):

RQ1: How do preschool teachers in the Zlín region perceive integration of children with a physical disability into common kindergartens?

RQ2: How do views of preschool teachers in the Zlín region on integration of children with a physical disability into common kindergartens vary according to the characteristics of teachers?

- H1: Teacher's openness towards integration of children with a physical disability into a common kindergarten increases depending on education level achieved in the field of education.
- H2: Teacher's openness towards integration of children with a physical disability into a common kindergarten depends on teacher's age.
- H3: Teacher's openness towards integration of children with a physical disability into a common kindergarten increases depending on the length of educational practice.
- H4: Teachers having experienced work with a child with a physical disability integrated into a common kindergarten will exhibit a higher level of openness towards integration than teachers without this experience.

### 2.2 *Research sample*

Research sample (n = 183) consisted of pedagogical staff working in preschool education in the Zlín region (n = 99) and of students of Tomas Bata university in Zlín (from now on TBU) in a bachelor's study Preschool Teacher Training program, and a follow-up master's Pedagogy of Preschool Age program (n = 84). More detailed characteristics are shown in the Table 1.

Table 1. Composition of research sample.

| | AGE (years) | | | PRACTICE (years) | | |
|---|---|---|---|---|---|---|
| | min. | max. | x̄ | min. | max. | x̄ |
| TOTAL *n = 183* | 19 | 60 | 27.7 | 0 | 41 | 4.7 |
| TEACHERS *n = 99* | 19 | 60 | 32.9 | 0 | 41 | 8.7 |
| STUDENTS *n = 84* | 19 | 45 | 21.9 | 0 | 0 | 0 |

### 2.3 *Methods*

We used the questionnaire method for data collection. As a follow-up to the goals set in the research, we focused the questionnaire items on teacher's views of selected aspects of integration of children with a physical disability into common kindergarten classes. In the initial questions, we inquired basic demographic data about respondents, such as gender, age, length of practice, and highest teacher training education achieved. The crucial part of the questionnaire included 41 mostly scale items in which the respondents expressed their agreement with a given statement on a scale of 0 (completely disagree) to 6 (completely agree). In compliance with Gavora (2012), we formed the items so that the designated scale value expressed level of positive attitude towards the problem of integration. In order to prevent respondents' tendency to answer mechanically, we included some reverse items.

In the pre-research we verified comprehensibility of the questionnaire items and, based on discussion with the educational community, we added some extra statements. With a factor analysis, we searched for suitable structuration of items so that the individual categories were filled with following categories:

- **factor 1:** Teacher's openness towards integration/inclusion
- **factor 2:** Teacher's reservations towards integration/inclusion
- **factor 3:** Individualization of approach and using physical activities for personality development
- **factor 4:** School conditions for integration/inclusion
- **factor 5:** Subjectively perceived teacher's readiness for work with an integrated child

We also determined the reliability of individual factors (Table 2) and used a normality test (Table 3).

Table 2. Reliability of the questionnaire (Cronbach alpha).

| Questionnaire factors | F1 | F2 | F3 | F4 | F5 | Total |
|---|---|---|---|---|---|---|
| *Cronbach alpha* | 0.95 | 0.91 | 0.88 | 0.86 | 0.80 | 0.94 |

Table 3. Normality test.

| | Kolmogorov-Smirnov | | | Shapiro-Wilk | | |
|---|---|---|---|---|---|---|
| | Statistic | df | Sig. | Statistic | df | Sig. |
| Total | 0.052 | 183 | 0.200 | 0.988 | 183 | 0.116 |

## 3 RESEARCH FINDINGS

Results analysis indicates a rather positive perception of integration of children with a physical disability into common kindergartens. Table 4 shows average values of answers (on scale 0 – 6).

We designated the first factor as *Teacher's openness towards integration/inclusion.* The answers in all items of this area are on average around positive evaluation. This is reflected also in the average value of answers to items of factor No. 1. In general, we can say preschool teachers are quite positive towards integration of children with SEN or with only a physical disability into common kindergartens. The second factor was designated as *Teacher's reservations towards integration/inclusion* and it represented a kind of an antipole to factor No. 1. Our aim in this part was to observe what negative impacts of integration on an integrated child teachers view as significant. Comparison of factor 1 and 2 again indicates rather positive attitudes towards integration. Factor 3 focuses on *Individualization of approach and using physical activities for personality development.* As shown in Table 4, it is an area with the best evaluation. Average values of answers to individual items are above value 4.5. However, the results must be again analyzed with more detail. Items related to using physical activities were evaluated better than items related to individual approach. In case of using physical activities for social development of children, no respondent answered completely disagree and determinative deviations of individual items show low values (0.701–1.059). The fourth factor was focused on *School conditions for integration/inclusion.* In this area there was the largest difference in answers of the respondents (SD = 1.48). According to views of the teachers, the addressed kindergarten supports integration and inclusive efforts (the average value of answer to item 1 is 4.07; SD = 1.639, which is the lowest determinative deviation in this area), but evaluation of readiness of the school for integration is paradoxically evaluated as insufficient. The weakest point seems to be the adaptation of interior spaces of the institution for movement of people with a physical disability. But we must add that the answers of individual respondents varied greatly (SD = 2.212). In view of the fact that respective adaptations were financially demanding (construction adjustment, equipping the institution with

Table 4. Average scores according to factors.

| Questionnaire factors | $\bar{x}$ | SD |
|---|---|---|
| F1 Openness | 4.72 | 0.99 |
| F2 Reservations | 2.81 | 1.09 |
| F3 Individualization | 5.00 | 1.10 |
| F4 Conditions | 2.84 | 1.48 |
| F5 Preparedness | 2.51 | 1.12 |

devices for transfer of wheelchair users on the stairs, etc.), and that new buildings should respect the valid regulations of wheelchair accessibility policy, we assume that the problem concerns primarily older buildings. However, most of the Czech kindergartens are probably in old buildings. Factor 5 was designated *Subjectively perceived teacher's readiness for work with an integrated child.* Taking into account the aims of the study, we focused exclusively on work with children with a physical disability. The factor is internally divided into two areas: perceived competences and willingness to educate in the field of work with children with SEN. Both areas are evaluated quite positively. Teachers do not perceive presence of a child with a physical disability as a source of psychological or work strain. The second area is also evaluated relatively well: teachers show willingness to educate themselves in respective areas, but mainly during standard working time.

**Age of the respondents** correlated with overall results very weakly ($r = 0.214$; $\alpha = 0.01$). The correlation of individual factors with the age of the respondents shows that with age, openness towards integration rises and that older teachers are more likely to realize the benefits of integration and less likely to stress possible risks (Table 5).

In addition, we considered the relationship between **the highest teacher training education achieved** and views of respondents of integration. It shows that the level of achieved education is not related to teachers' attitude to integration of children into kindergartens. The opinion of the wider social environment is more important than the level of education. The only factor that demonstrated a statistically significant difference between groups with different levels of education was factor 3 *Individualization of approach and using physical activities for personality development.* Subsequently, a post-hoc test (Tukey HSD) was applied and revealed that respective teacher training study supports using physical activities as a means of socialization, while the level of achieved education is irrelevant. Analysis of the relationship between **length of teaching practice** and teachers' attitudes towards integration reveals a statistically significant relationship of these quantities; length of teaching practice thus positively, but very weakly, correlates with a positive attitude of teachers towards integration of children with a physical disability into common classes of kindergartens ($\rho = 0.148$). This relationship is however manifested mainly in factor *3 Individualization of approach and using physical activities for personality development.* Therefore it cannot be definitely said that the length of teaching practice influences teachers' attitudes to integration. Relationship analysis of other variables (student/teacher with practice, size of school, integration of children at school, and experience of a teacher with an integrated child) did not demonstrate statistically significant differences in teachers' views of integration of children with a physical disability into common kindergarten classes. Neither did a detailed analysis of answers to individual questionnaire items provide more fundamental findings.

Table 5. Correlation of respondents' age with questionnaire factors *(*significance with $\alpha = 0.01$).*

| | Age | F1 | F2 | F3 | F4 | F5 |
|---|---|---|---|---|---|---|
| Spearman's Correlation Coefficient | 1.00 | 0.23* | 0.23* | 0.34* | –0.05 | –0.05 |
| Sig. (2-tailed) | | 0.00 | 0.00 | 0.00 | 0.62 | 0.48 |
| N | 183 | 183 | 182 | 133 | 108 | 183 |

## 4 CONCLUSIONS AND RECOMMENDATIONS

The research of teachers working in the field of preschool education and university students of related programs demonstrates rather positive attitudes towards integration of children with a physical disability. They are aware of the benefits of integrating a child with a physical disability into a common class, especially regarding building appropriate social relationships of a child with healthy population. The same is true vice versa – kindergarten teachers in Zlín stress the importance of the presence of a child with a physical disability in a common kindergarten class for the rest of the children in the class.

Although the participants of our research did not experience a stronger deficit in the area of their own readiness for work with an integrated child, we believe it is appropriate to include at least some recommendations which are to be respected when working with a child with a physical disability:

**Development of motor activity**. It is necessary to pay attention to development and practice of specific skills. A physical disability may prevent a child from performing normal movements a healthy child performs automatically. However, unlike a healthy child, he or she naturally adopts skills which are indispensable for locomotion and which are hard to master for a healthy child. Depending on type of a physical disability, some body parts are unilaterally loaded. Therefore, it is convenient to include more frequent relaxation exercises and put emphasis on relaxation of muscles of overloaded body parts.

**Emotional development and socialization.** A child with a physical disability may experience frustration in satisfying some needs and accomplishing his goals, particularly when estimating his possibilities unrealistically. This may result in unusual reactions which may seem inadequate in a given situation. Nevertheless, we must realize that the child has a strongly frustrated noticeable need for natural, spontaneous activities. A teacher should try to create an atmosphere of acceptance, respect, and partnership among children without highlighting the differences.

**Communication.** Although there may not appear to be objective difficulties in communication with a child with a physical disability, we must pay cinsiderable attention to this area. The child with a disability may have worsened articulation due to disrupted motor activity of the speech organs or there may appear problems with gesticulation and facial expressions, for example, in the form of a lower occurrence or involuntary movements. Communication should be with the child, not his assistant. The assistant is not a child's "spokesperson," so every message should be addresses directly to the child. Apart from communication with an integrated child, attention must be paid also to specific topics in communication with the other children. Especially observation and continuous evaluation of relationships between children and, depending on the situation, a clear explanation for the other children of, for instance, physical differences of a child with a physical disability, unusual behavioral manifestations, but also of specific demands a teacher has for a child with a disability and of different criteria in the evaluation of the child.

**Evaluation.** Individualization of approach is present in all areas of a teacher's work. Therefore it is not realistic to use the same evaluation criteria for all children. With a child with a physical disability, a teacher must take into consideration all specifics he or she did not have a chance to encounter before and which are therefore something new or unknown. Evaluation of a child and his progress must be based on a thorough knowledge of the situation of a particular child.

**Assistant.** Although it is important for a child with a physical disability to develop independence and self-sufficiency, in some situations he or she cannot get along without help of others. Therefore it is appropriate to use the services of an assistant in the kindergarten who helps the child with more difficult tasks and at the same time may focus on the child when the teacher has to work with other children.

This text is focused on the integration of preschool children, although experience has shown that the same problems also arise in primary school. A detailed description of the situation requires future research on the primary school teacher.

## REFERENCES


Erikson, E. H. 2015. *Životní cyklus rozšířený a dokončený: devět věků člověka.* Praha: Portál.

Gavora, P. 2012. *Tvorba výskumného nástroja pre pedagogické bádanie.* Bratislava: Slovenské pedagogické nakladateľstvo - Mladé letá.

Hartl, P., & Hartlová, H. 2000. *Psychologický slovník.* Praha: Portál, s.r.o.

Ješina, O. 2014. Tělesné postižení. In Dagmar Trávníková (Ed.), *Vybrané aplikované pohybové aktivity: Teorie a praxe* (pp. 185). Brno: Masarykova univerzita.

Nakonečný, M. 2009. *Sociální psychologie.* Praha: Academia.

Nykl, L. 2012. *Carl Ransom Rogers a jeho teorie: Přístup zaměřený na člověka.* Praha: Grada Publishing.

Renotiérová, M. 2003. *Somatopedické minimum.* Olomouc: Univerzita Palackého.

Řezáč, J. 1998. *Sociální psychologie.* Brno: Paido.

Slowík, J. 2007. *Speciální pedagogika: prevence a diagnostika, terapie a poradenství, vzdělávání osob s různým postižením, člověk s handicapem a společnost.* Praha: Grada Publishing.

Vágnerová, M., Hadj-Moussová, Z., & Štech, S. 2000. *Psychologie handicapu.* Praha: Karolonum.

Výrost, J., & Slaměník, I. 2008. *Sociální psychologie.* Praha: Grada Publishing, a.s.

Zimmer, R. 2015. *Sport und Spiel im Kindergarten.* Aachen: Meyer & Meyer Verlag.

*The Social Sciences Empowered – Lumban Gaol, Hutagalung & Fong Peng (eds)*
*© 2020 Taylor & Francis Group, London, ISBN 978-0-367-00090-5*

# Literature review on E-commerce adoption stage. Proposing a new integrative E-commerce adoption mode

Dionysia Kowanda
*Department of Economics, University of Gunadarma, Jakarta, Indonesia*

Muhammad Firdaus*
*Department of Computer Science, University of Gunadarma, Jakarta, Indonesia*

Guyub Nuryanto & Rowland Bismark Fernando Pasaribu
*Department of Economics, University of Gunadarma, Jakarta, Indonesia*

ABSTRACT: The purpose of this study is to conduct a literature review on the E-commerce Adoption Stage and to propose a New Integrative E-Commerce Adoption Model using a content analysis research technique which involves employing data sources from the Science Direct and ProQuest electronic databases. This is achieved by using four search terms related to the e-commerce adoption model without any date limitations. Of 18 scholarly journals reviewed, 4 articles showed relevance to the scope of the present study. Indeed, it was found that Lewis and Cockrill (2002), Teo and Pian, (2004), Lefebvre et al. (2005), and Maswera et al. (2008) all illustrates a connection to e-commerce adoption stages by defining criteria in order to identify stages of e-commerce adoption in corporations or on websites. This study proposes an integrative research model that encompasses the four models of the e-commerce adoption stage and creates a theoretically-grounded basis for evaluation, which will be used to analyse the e-commerce adoption stage of a country.

*Keywords*: e-commerce adoption stage, e-commerce adoption model, literature review, decision support

## 1 INTRODUCTION

The wave of globalisation has given rise to major changes in the business environment and a familiarity with these changes is necessary if companies are to remain competitive. The development and promotion of technologies, especially the Internet, has led to e-commerce development, which makes the business environment more complex (Bakhtiari, Daneshpajooh, & Masoumi, 2014).

The exponential growth of the Internet has changed the rules of competition in many industry sectors. Corporations including SMEs are integrating their business into the Internet as they can reach a larger market scope online (Lane, Wafa, Hassan, & Belkhamza, 2014). Moreover, with advancements in technology, there have been changes in the methodology of business transactions. The rapid Information Technology progress along with its widespread diffusion have led to 'frictionless' economies in which transaction costs are nearly zero, barriers to entry and contestability disappear, and markets clear instantly (Sharma, 2016).

The Internet has facilitated a new wave of economic growth and development through e-commerce (Bieron & Ahmed, 2012). The e-commerce phenomenon is one of the most important developments in commercial and industrial practice since the Industrial Revolution and growth resulting from the impact of the Internet (Kolzow & Pinero, 2001). E-commerce offers the most exciting business opportunities in the marketplace, as the Internet gives rise to the possibility of replacing the traditional method of communications and provides an unprecedented infrastructure for moving information (Soliman & Youssef, 2003).

The introduction of the Internet as a shopping channel offers a new opportunity for both manufacturers and retailers to reach potential consumers in a more efficient way (Almousa, 2011). The critical edge for businesses today is e-commerce. E-commerce has been blooming in the business cycle, where traditional brick-and-mortar businesses have undergone a transition to leading dotcom or click-only businesses (Lane, Wafa, Hassan, & Belkhamza, 2014). E-commerce stands for electronic commerce. It means dealing in goods and services through electronic media and the Internet. E-commerce refers to the movement of business onto the World Wide Web (Johri, 2016). On the Internet, the e-commerce concept relates to a website of the vendor, who sells products or services directly to the customer through the portal using a digital shopping cart or digital shopping basket system, and allows payment via credit card, debit card, or EFT (electronic fund transfer) payments. E-commerce, or e-business, involves carrying on a business with the help of the Internet and by using information (Johri, 2016). The Internet and e-commerce as the main wings not only trigger major changes in the business, but everyday life is also affected (Fakhri, Hajighafori, & Jafarzadeh, 2014).

The early years of e-commerce were a period of explosive growth and extraordinary innovation, beginning in 1995 with the widespread use of the Web to advertise products. Even with the collapse in 2000, a sobering period of reassessment occurred, followed by strong double-digit growth through the current period. The e-commerce sector saw unprecedented growth in 2014 (Sharma, 2016). Said growth was driven by rapid technology adoption led by the increasing use of devices such as smartphones and tablets, and access to the Internet through broadband, 3G, etc., which in turn led to an increased online consumer base. Companies are trying to maximise their revenue by providing not only quality products but also making it easy to buy such products online (Kim, Yim, Sugumaran, & Rao, 2016). Furthermore, favoured demographics and a growing Internet user base have helped this growth (Sharma, 2016).

As e-commerce is becoming a mainstay in the global service management economy (Kim, Yim, Sugumaran, & Rao, 2016), it brings about enormous possibilities and creates easier access to the markets of developed countries, especially the B2C markets, as well as higher potential revenues resulting from the development of new trading opportunities (Ezzi, 2016).

Global trends indicate that e-commerce has been an important field of advanced countries in the last century. Developing countries are also moving rapidly towards implementing its use. However, all countries in proportion to their existing infrastructure and capabilities have invested in this matter (Fakhri, Hajighafori, & Jafarzadeh, 2014). E-commerce has a significant positive impact on GDP as an indicator of economic growth; indeed, the GDP is positively and significantly affected by e-commerce sales (Johri, 2016). The adoption of e-commerce is something which has been strategically used by organisations, regardless of the size of such organisations, as it provides opportunities for various types of businesses to flourish in the context of the maturing e-commerce (Lane et al., 2014).

This literature review aims to explore the e-commerce adoption phase in existing research. The secondary aim is to create a new model for e-commerce adoption for future research in a specific country.

## 2 E-COMMERCE

E-commerce stands for electronic commerce, or e-business, and involves carrying on a business with the help of the Internet and by using information technology, such as Electronic Data Interchange (EDI) (Johri, 2016); the latter is a methodology used by modern businesses which addresses the need of business organisations, vendors and customers to ease the business, reduce cost and improve the quality of services while increasing the speed of delivery and facilitating the paperless exchange of business information (Sharma, 2016). E-commerce is seen as an activity in which consumers obtain information and purchase products (Korsakiene, 2006), deal in goods and services (Johri, 2016; Gandhi, 2016), and conduct business transactions or managerial activities (Sharma & Lijuan, 2015) involving consumer goods to industrial goods, professional consultation to social networking, physical and digital products (Kiang, Ye, Hao, Chen, & Li, 2011) through electronic media and the Internet.

Over the past decade, the growth of e-commerce has been quite impressive (Liu, Chen, & Lu, 2015) and the concept has become an important part of everyday life for consumers during the 21st century. The variety of services in e-commerce has broadened in recent years and consumers have adopted those services as part of their everyday lives (Sharma & Lijuan, 2014). The tremendous popularity and acceptance of e-commerce in global business are due to the relative ease, speed and flexibility when it comes to completing a transaction online (Anyu & Ofong, 2015). According to studies and surveys conducted world-wide, the Internet is an efficient medium for buying goods and services and is driving e-commerce growth (Xanthidis & Nicholas, 2004). E-commerce has an influence on cost saving (Liu, Chen, & Lu, 2015) by eliminating existing intermediaries and drastically reducing transaction costs. These lower production costs will encourage the entry of new businesses and thus increase competition and pressure to pass lower costs on to consumers as lower prices, ease the business, reduce cost and improve the quality of services while increasing the speed of delivery. Enhancing international trade and boosting the global economy (Sharma, 2016) increase the productivity of business organisations, government organisations, and society (Deshmukh, Chhangani, & Thampi, 2016); these measures also lead to a shift in market power from the producer to the consumer, providing the latter with the ability to search among thousands of merchants for the lowest prices; indeed, this significantly improves the efficiency of economies, enhances competitiveness, and increases long-term growth (Sharma, 2016).

## 3 METHODOLOGY

The inception step in this literature review commences with the identification of research articles on e-commerce adoption using the following keywords: "e-commerce adoption model", "e-commerce adoption phase" and "e-commerce adoption stage". Accessions in the search process were applied in the screening process in the Science Direct and ProQuest computerised databases which were subscribed to. Moreover, consideration was given to hosting a large and reputable scholarly journal and easily differentiating between scholarly journals and proceeding papers.

Below is a review of the selected articles in the reference list.

### 3.1 *Inclusion criteria*

The inclusion criteria for this review are defined as follows: studies published without using year's limitation up until 2018, studies which are considered scholarly journals, studies which were published in English, and studies which focused on e-commerce adoption stages using three of the above keywords.

In the first step, the search terms "e-commerce adoption model", "e-commerce adoption phase" and "e-commerce adoption stage" were used to look for relevant studies. Reference lists from these studies were then used to add more literature pertaining to e-commerce adoption stages.

### 3.2 *Exclusion criteria*

During the second step, the references of the selected studies were inspected and relevant studies were included in the review. All irrelevant and non-staging approach work was excluded. Thus, a list of 18 articles was assessed; those which showed relevance to the scope of the study were included in the review, and the most relevant were then assessed. As a result, this study focuses on 4 articles relating to the topic.

### 3.3 *Findings*

#### 3.3.1 *Sampling demography*

The selected articles consisted of 12 scholarly journal search results in Science Direct and 5 scholarly journal search results in ProQuest using the "e-commerce adoption model" keywords. Furthermore, use of the "e-commerce adoption phase" keywords led to 0 scholarly journal search results in Science Direct and 0 scholarly journal search results in ProQuest. Finally, use of the keywords "e-commerce adoption stage" led to 1 scholarly journal search result in Science Direct and 0 scholarly journal search results in ProQuest.

All results were carefully reviewed and, based on the methodology approach, it was either deemed that the 14 articles were providing a model for e-commerce adoption based on, e.g., factors which influence e-commerce adoption, or there were no models related to e-commerce adoption.

In total there are four relevant articles which provide a model based on e-commerce adoption on a "stages basis". Indeed, these articles provide characteristics or features of each stage of the e-commerce adoption.

These e-commerce adoption stages are still relevant when it comes to addressing and assessing the e-commerce adoption of a certain country or company; said stages are also useful in describing and further expanding companies' capabilities at certain stages, and in using a mix of quantitative and qualitative measures in a country-wide decision-making process.

#### 3.3.2 *E-commerce adoption stages*

##### 3.3.2.1 *The six steps of the Ladder*

Lewis and Cockrill's (2002) research focused on companies' level of e-commerce usage, in accordance with the Connectivity Indicator and E-commerce Adoption Model, developed by the UK Department of Trade and Industry. According to Lewis and Cockrill (2002), in order to engage in any form of e-commerce, a business must have a basic level of connectivity, i.e. a basic connection to the Internet either by having a website, making frequent use of external e-mail or using electronic data interchange, irrespective of the technology deployed. Once a business is connected, this six-step model measures the progression of any business from relatively simple to more complex e-commerce activities. The six steps of the Ladder, leading to full e-business participation include:

1. Messaging: The use of e-mail to send text messages.
2. Online marketing: Creation of a website or e-mail to publish information about products and services, so that customers can access this information online.
3. Online ordering: Online interaction between a business and its customers, or a business and its suppliers, for the placement of an order.
4. Online payment: A transaction which can be split into two separate online activities, namely issuing or receiving an invoice and completing the transaction online by an electronic payment.
5. Order progress/online sales support: The use of e-commerce to support the business relationship between a customer and a supplier, e.g. through online post-sales support such as courier information.
6. E-business: The theoretical end point when a business has fully integrated ICT into its operations, potentially redesigning its business processes around ICT or completely reinventing its business model

##### 3.3.2.2 *Web adoption model*

Teo and Pian's (2004) research focused on a model for Web adoption and examined the characteristics of different levels of websites in terms of their features, which are:

1. Level 0 Web adopters – no website, only e-mail account; this level is classified as non-adopters.

2. Level 1 – Web presence: The first level is Web presence, where websites at this stage provide information and brochures and tend to be non-strategic in nature.
3. Level 2 – prospecting: at this level websites provide customers with product information, news, details on events, interactive content, personalised content, e-mail support, etc.
4. Level 3 – business integration: Web adoption is incorporated into the business model and there is also the integration of business processes; for example, there are cross-functional links between customers and suppliers and a Web strategy is integrated into the firm's business strategy.
5. Level 4 – business transformation: The highest level of Web adoption and there is an overall business transformation model throughout the organisation

3.3.2.3 *E-commerce adoption model*
Lefebvrea, Lefebvrea, Eliaa, and Boecka's (2005) research focused on capturing the gradual unfolding of e-commerce adoption. This model differentiates between nonadopters (stage 00 and stage 0) and adopters (stages 1, 2, 3 and 4).

1. Stage 00 refers to non-adopters with no intention of becoming involved in any e-commerce initiatives; in contrast,
2. Firms at stage 0 are not presently conducting any e-commerce activity but are planning to do so within the next 12 months.

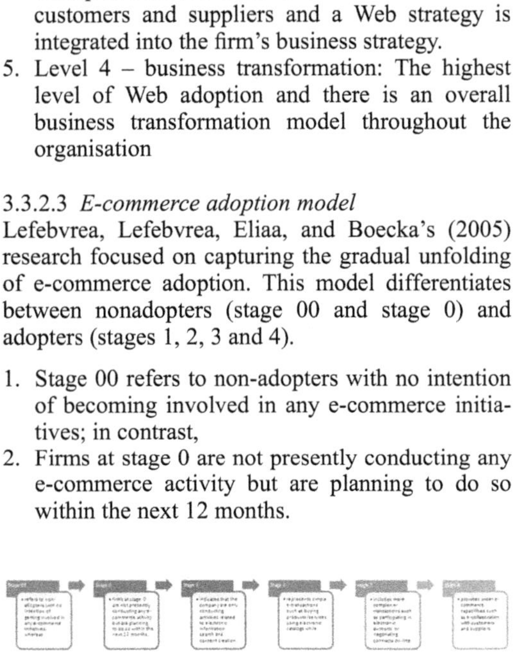

Figure 1. Lefebvrea et al.'s E-commerce adoption model. Source: Lefebvrea et al. (2005).

This model further captures the different stages of e-commerce penetration among adopters:

1. Stage 1 indicates that the company is only conducting activities related to electronic information searches and content creation.
2. Stage 2 represents simple e-transactions such as buying products/services using electronic catalogues.
3. Stage 3 includes more complex e-transactions such as participating in e-auctions or negotiating contracts online.
4. Stage 4 provides wider e-commerce capabilities such as e-collaboration with customers and suppliers.

3.3.2.4 *E-commerce adoption features*
Maswera, Dawson, and Edwards' (2008) research focused on discovering the nature and extent of e-commerce adoption by accessing examples and then evaluating them against a list of e-commerce features such as:

1. Corporate information
2. Product information
3. Non-product information
4. CRM Reservation
5. Payment

## 4 PROPOSED MODEL

The stage model of e-commerce adoption taken from previous research provides criteria which make it possible to identify stages of e-commerce adoption in corporations or on websites. This study proposes an integrative research model (shown in Figure 3) that encompasses the four models of the e-commerce adoption stage and creates a theoretically-grounded basis for evaluation that will be used to analyse the e-commerce adoption stage of a country (shown in Figure 2 and Table 1).

1. Stage 0 – In the e-commerce stage model from Lefebvre et al. (2005), the adoption stage is split into two major conditions, which are nonadopters and adopters. In this stage, the common characteristic is no website present. In addition to that, the nonadopters stage further describes by Teo and Pian, (2004), using the same name which are nonadopters and the same characteristic which is no website present but e-mail. Further according to Teo and Pian (2004), even firms that do not have independent domain names and websites, they normally have an e-mail account that they use to establish connectivity with customers and business partners. This common criterion was also found by Lewis and Cockrill (2002), although the terminology used was the 1st ladder, with the criteria of the use of e-mail to send text messages.
2. Since the common characteristic of stage 0 or non-adopters is that there is no website present, stage 1 – on an e-commerce adoption stage – will possess an early stage of website presentation characteristic. According to Teo and Pian (2004), the first level of web presence is where the implementation is still in progress, with the characteristic of occupying a domain name or simply having a presence. The websites at this stage provide information and brochures and tend to be non-strategic in nature. This is also aligned with Maswera et al.'s (2008) criteria at level 1, where the websites provide corporate information. According to Lewis and Cockrill (2002), the 2nd phase of e-commerce adoption involves the website publishing information about products and services; as such, we can consider this phase as being relevant in stage 1, because the publication of information is possible in the form of simple

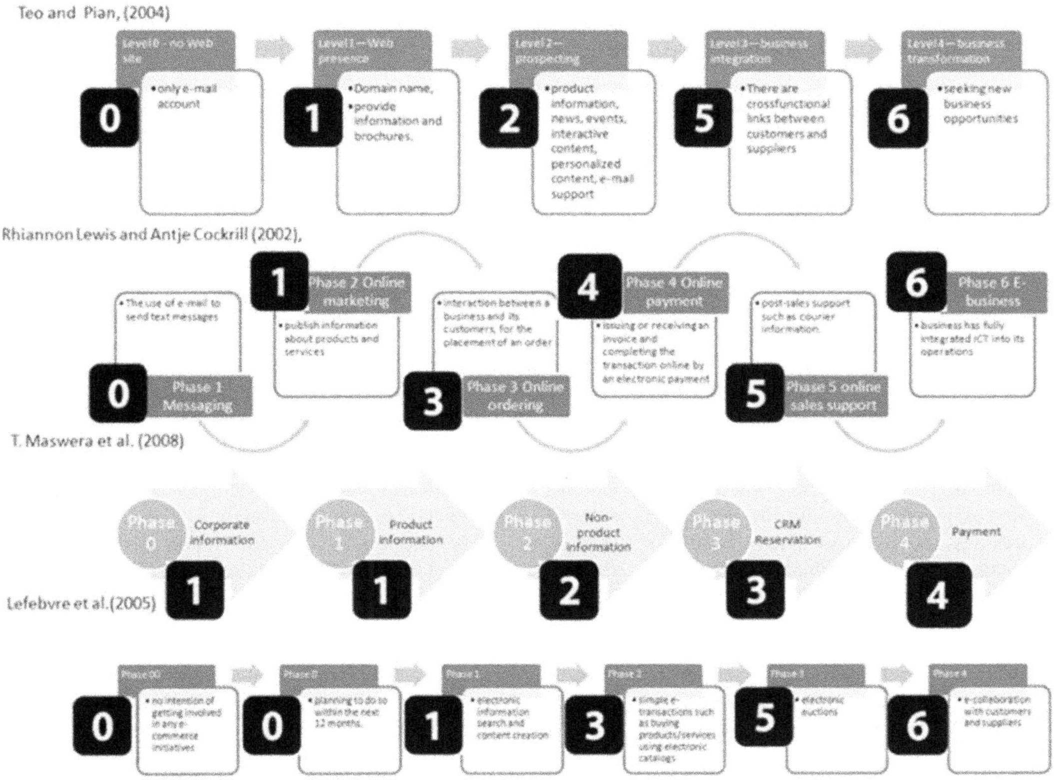


Figure 2. Integrated E-commerce adoption model research.
Source: Lewis and Cockrill (2002), Teo and Pian (2004), Lefebvrea et al. (2005), Maswera et al. (2008).

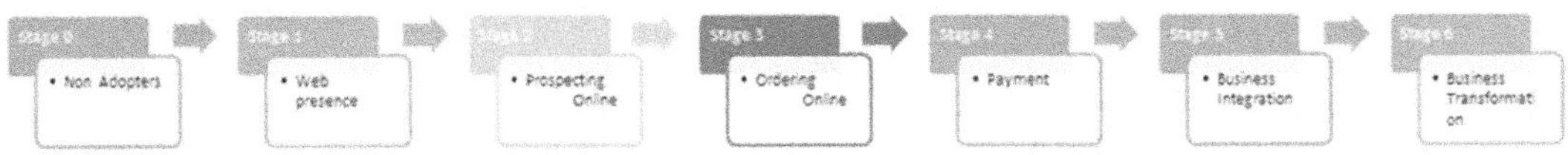


Figure 3. Proposed E-commerce adoption model.

information and brochures, but also relevant to the more advanced stage.

3. After the establishment of websites, the next stage is to focus on a more prospecting direction or the direction of conducting commerce. In stage 3, the common characteristic is providing information on products and services, as well as further information about news and events, and non-product information. In line with Teo and Pian (2004), at level 2 the websites provide customers with product information, news, details on events, interactive content, personalised content, and e-mail support for their potential customers. In the view of Lewis and Cockrill (2002), this phase is called online marketing, where the website publishes information about products and services for its online customers. In Maswera et al. (2008), the commonalities are described in level 2 and 3 where the websites provide product and non-product information. Lefebvre et al. (2005) suggested that, at stage 1, SMEs are only conducting activities related to electronic information searches and content creation.
4. The next stage of this model is the ability to complete the prospecting stage, which involves proceeding with the Online Payment. In stage 4 the common criterion is completing the transaction via online payment. In Lewis and Cockrill (2002), the online payment stage is split into two separate online activities, namely invoicing and completing the transaction online via an electronic payment. Maswera et al. (2008) described it by providing Payment in level 5; on the other hand, with Lefebvre et al.'s (2005) model, the ability to carry out transactions electronically is described in stage 2.
5. In stage 5, the Web's capability is expanded further to Business Integration, where the common function is providing cross functionality between

Table 1. Integrative e-commerce adoption model.

| Adoption Model | Stage 0 | Stage 1 | Stage 2 | Stage 3 | Stage 4 | Stage 5 | Stage 6 |
|---|---|---|---|---|---|---|---|
| | Non Adopters | Web presence | Prospecting | Online Ordering | Online Payment | Business Integration | Business Transformation |
| | • No website<br>• Using email for business | • Domain name<br>• Information & Brochures | • Product & Service Information<br>• News, events & non product Information | • Business Interaction for order placement<br>• Reservation | • Completing transaction via online payment | • Crossfunctional between customers & suppliers<br>• Post sales support – such as courier Info<br>• Electronic auction | • IT Integration into its operation<br>• Seeking new business opportunity<br>• E-collaboration |
| Lewis and Cockrill (2002) | Step 1 | Step 2 | Step 2 | Step 3 | Step 4 | Step 5 | Step 6 |
| Teo and Pian, (2004) | Level 0 | Level 1 | Level 2 | | | Level 3 | Level 4 |
| Lefebvre et al.(2005) | Stage 00, Stage 0 | | Stage 1 | Stage 2 | Stage 2 | Stage 3 | Stage 4 |
| T. Maswera et al. (2008) | | | Phase 0, Phase 1, Phase 2 | Phase 3 | Phase 4 | | |

customers and suppliers such as e-auction and also post-sales support. This stage is described by Teo and Pian (2004) in level 3, where business integration is incorporated into the business model and business processes are also integrated; in addition, there are cross-functional links between customers and suppliers. Similar to that, Lewis and Cockrill (2002) described their phase 5 as providing online sales support, such as the business relationship between a customer and a supplier, e.g. through online post-sales support such as courier information. On the other hand, Lefebvre et al. (2005) described their phase 3 as capability in providing more complex e-transactions such as participating in e-auctions or negotiating contracts online; they also described the business relationship between a customer and a supplier.

6. With regards the final stage, stage 6, the proposed model is described in this stage as Business Transformation, where there is Web adoption which leads into IT Integration and then into its operation and provides the ability to seek new business opportunities and e-collaboration. According to Teo and Pian (2004), their level 4 model is described as business transformation. This is the highest level of Web adoption. It will transform the overall business model throughout the organisation. The focus is on building relationships and seeking new business opportunities. Moreover, Lewis and Cockrill (2002) described the phase 6 model as the end point, when a business has integrated ICT into its operations, potentially redesigning and/or reinventing its business processes around ICT. In Lefebvre et al. (2005), phase 4 is also described as the transformation of the overall business model throughout the organisation.

## 5 CONCLUSION AND FUTURE WORKS

This study concludes that, based on the underlying research on e-commerce adoption, there is a correlation between these past research studies in terms of defining the stages and characteristics of e-commerce adoption. Thus, we can propose to integrate the stages based on the characteristics of the e-commerce adoption stages provided by each research study.

The exploration of the e-commerce adoption model is possible thanks to sources selected from the Science Direct and ProQuest electronic databases. Indeed, ScienceDirect hosts over 12 million articles from 3,500 academic journals and 34,000 e-books (ScienceDirect), while ProQuest is the largest multidisciplinary, full-text database (ProQuest), providing a variety of content types across over 175 subjects; this includes millions of much-coveted full-text articles from thousands of scholarly journals, over 100,000 full-text dissertations and more than 455,000 working papers. Despite this, however, there is a chance to improve the research by acquiring more sources for this study.

The present literature review study was conducted using a qualitative method, in order to explore and understand a phenomenon, rather than the statistical approach; this also limits the generalisability of its findings.

In the future, the proposed model can be used to further examine e-commerce adoption in a certain country, provide condition analysis, apply the statistical approach and further understand the topic; said model can also be used for decision support for e-commerce initiatives by government and, furthermore, we can use the approach to make comparisons between known models of e-commerce adoption.

## REFERENCES


Almousa, M. 2011, "THE INFLUENCE OF RISK PERCEPTION IN ONLINE PURCHASING BEHAVIOR: A MULTI DIMENSIONAL PERSPECTIVE", International Journal of Arts & Sciences; Cumberland Vol. 4, Iss. 12, pp. 373–382.

Anyu, J.N. & Ofong, C. 2015, "TEETHING PROBLEMS IN LITIGATION AND REGULATION OF E-COMMERCE", International Journal of Organizational Innovation (Online), Vol. 7, No. 3, pp. 47–64.

Bakhtiari, M., Daneshpajooh, A.A., & Masoumi, S.S. 2014, "QUANTITATIVE APPROACH TO THE

CONCEPT OF E-COMMERCE IN THE THIRD MILLENNIUM", Kuwait Chapter of the Arabian Journal of Business and Management Review, Vol. 3, No. 12, pp. 143–148.
Bieron, B. & Ahmed, U. 2012, "Regulating E-commerce through International Policy: Understanding the International Trade Law Issues of E-commerce", Journal of World Trade, Vol. 46, No. 3, pp. 545–570.
Ezzi, S.W. 2016, "Exploring the Characteristics of the E-commerce Marketplace in Saudi Arabia", International Journal of Economic Perspectives, Vol. 10, No. 4, pp. 5–20.
Fakhri, P., Hajighafori, F., & Jafarzadeh, J. 2014, "THE ROLE OF ELECTRONIC INSURANCE AND ELECTRONIC COMMERCE IN EXPANSION OF INSURANCE COMPANIES", Kuwait Chapter of the Arabian Journal of Business and Management Review, Vol. 3, No. 10, pp. 70–79.
Gandhi, S.K. 2016, "India's Jumbo Jump from E-Commerce to Mobile Enabled Services (MES): A Review", Productivity, Vol. 56, No. 4, pp. 326–331.
Johri, B. 2016, "Econometric Analysis of the Relationship between Indian E-Commerce Transaction and GDP", Productivity, Vol. 56, No. 4, pp. 319–325.
Kiang, M.Y., Ye, Q., Hao, Y., Chen, M., & Li, Y. 2011, "A service-oriented analysis of online product classification methods", Decision Support Systems, Vol. 52, pp. 28–39.
Kim, D.J., Yim, M., Sugumaran, V., & Rao, H.R. 2016, "Web assurance seal services, trust and consumers' concerns: an investigation of e-commerce transaction intentions across two nations", European Journal of Information Systems, Vol. 25, No. 3, pp. 252–273.
Kolzow, D.R. & Pinero, E. 2001, "EDRI White Paper: The Internet economy and its impact on local economic development", Economic Development Review, Vol. 17, No. 3, pp. 82–99.
Korsakiene, R. 2006, "CHALLENGES OF E-BUSINESS DEVELOPMENT IN LITHUANIA", Studia Universitatis Babes-Bolyai.Studia Europaea, Vol. 51, No. 3, pp. 103–120.
Lane, P.P.M., Wafa, S.A., Hassan, R.A., & Belkhamza, Z. 2014, "PERCEIVED USEFULNESS AND PERCEIVED EASE OF USE OF E-COMMERCE ADOPTION AMONG ENTREPRENEURS IN SABAH", Kuwait Chapter of the Arabian Journal of Business and Management Review, Vol. 3, No. 9, pp. 94–103.
Lefebvre, L., Lefebvre, E., Elia, E., & Boeck, H. 2005, "Exploring B-to-B e-commerce adoption trajectories in manufacturing SMEs", Technovation, Vol. 25, pp. 1443–1456.
Liu, T., Chen, C., & Lu, W. 2015, "The Impact of E-Commerce Adoption on Operating Cost: An Empirical Study of Taiwanese Manufacturing Firms", International Journal of Economic Perspectives, Vol. 9, No. 2, pp. 71–84.
Liu, T., Chen, C., & Lu, W. 2015, "The Impact of E-Commerce Adoption on Operating Cost: An Empirical Study of Taiwanese Manufacturing Firms", International Journal of Economic Perspectives, Vol. 9, No. 2, pp. 71–84.
Maswera, T., Dawson, R., & Edwards, J., 2008 "E-commerce adoption of travel and tourism organisations in South Africa, Kenya, Zimbabwe and Uganda", Telematics and Informatics, Vol. 25, pp. 187–200.
ProQuest, 2018, "What is ProQuest Central?", https://proquest.libguides.com/pqc. [Online] Accessed 23 Aug 2018.
ScienceDirect, 2018, https://www.elsevier.com/solutions/sciencedirect. [Online] Accessed 23 Aug 2018.
Sharma, G. & Lijuan, W. 2014, "Ethical perspectives on e-commerce: an empirical investigation", Internet Research, Vol. 24, No. 4, pp. 414–435.
Sharma, G. & Lijuan, W. 2015, "The effects of online service quality of e-commerce Websites on user satisfaction", The Electronic Library, Vol. 33, No. 3, pp. 468–485.
Sharma, R. 2016, "E-Commerce: An Analytical Study on Its Spread and Impact on Indian Economy", Productivity, Vol. 56, No. 4, pp. 338–344.
Soliman, F. & Youssef, M.A. 2003, "Internet-based e-commerce and its impact on manufacturing and business operations", Industrial Management & Data Systems, Vol. 103, No. 8, pp. 546–552.
Teo, T & Pian, Y. 2004, "A model for Web adoption", Information & Management, Vol. 41, pp. 457–468.
Xanthidis, D. & Nicholas, D. 2004, "Evaluating internet usage and ecommerce growth in Greece", Aslib Proceedings, Vol. 56, No. 6, pp. 356–366.

*The Social Sciences Empowered – Lumban Gaol, Hutagalung & Fong Peng (eds)*
*© 2020 Taylor & Francis Group, London, ISBN 978-0-367-00090-5*

# The academic worker's training and preferences in the area of professional growth

Jana Vasikova & Marie Pavelkova
*Faculty of Humanities, Tomas Bata University in Zlín, Zlín, Czech Republic*

ABSTRACT: The article focuses on the professional training of postgraduate students who are also academics at a university. The aim is to present the preferences of this specific group in the field of professional growth with a focus on research activities. The study presents selected results of qualitative research realized via semi-structured interviews with five participants. Categories have been created that are crucial to the research of the novice researcher. It was a finding of certainty in the field of research methodology and determinants influencing procedures in research projects, as well as the ability to cooperate and work in a team, for example, during project activities. The analysis shows that postgraduate students who are simultaneously employed as academics prefer a contextual orderliness of their specialization, which leads to deepening of their expertize. However, in a real environment, they must adapt flexibly to the actual needs of the workplace.

*Keywords*: postgraduate students, preference, research methodology, member of academic staff

## 1 INTRODUCTION

The doctoral student who is enrolled in a doctoral study program and has the option of commencing work in the university as an academic worker during their studies undoubtedly finds themselves in a very interesting situation. From one perspective, this situation may seem perfect in the sense that the student has the opportunity to apply their freshly acquired knowledge in practice and in an environment which represents to them both work and school, which they yearn to complete successfully. From another perspective, however, the doctoral student is employed within the scope of a certain position, which brings with it responsibilities that are very similar to the responsibilities of other academic workers, and therefore tasks are placed on their shoulders which they must solve and resolve, if possible at the same prestigious level as their colleagues. Academic workers who are simultaneously doctoral students experience their professional start in a very specific way. Therefore, their preferences in the area of professional growth are understandably also specific and original.

Every student approaches their studies differently, and also realizes pedagogic research differently.

Research activity is the alpha and omega of both the doctoral student's work and the academic's. Everyone assumes a different attitude to this area of their development, and their preferences in the given development area mature. Undoubtedly, however, both roles are interconnected and influence one another. This interconnection is primarily in the area of research. The first-stage researcher searches for their place in the area of science, and begins to specialize in a research area. Certain preferences in the area of professional growth can be typical for this specific group, being doctoral students simultaneously employed as academic workers, and can fundamentally influence their search for themselves in this profession.

This study aims to (1) present facts about the preferences that this specific group has in the area of their professional growth, whereby we find the area of research activity fundamental, and (2) describe how a first-stage researcher is influenced by working in a team, and how this specific group manages joint work on projects.

## 2 CURRENT SITUATION REGARDING THE ISSUE OF DOCTORAL STUDENTS WHO ARE SIMULTANEOUSLY ACADEMICS

The respondents of the presented research are academics, first-stage researchers. They are postgraduate students, who simultaneously work in the university as assistants. For these persons, it applies that they belong to a group of pedagogic workers who are in the position of both teacher and scientist.

As stated by Kalenda (2013), a master's student, doctoral student, or postdoctoral student can find themselves in the position of first-stage researcher. These first-stage researchers are influenced considerably by the educational conditions of individual universities, which are fundamental for their professional growth. The position of a first-stage researcher is also defined by the European

Framework for Research Careers. As for the definition of research workers, there are four categories (as per the European Framework for Research Careers): the First-Stage Researcher is in the lowest position, followed by a researcher recognized by the community (Recognized Researcher), followed by an Established Researcher, while a so-called Leading Researcher finds themselves at the top of the imaginary ladder (Kalenda, 2013).

The definition of the position of university academic worker, and the definition of the term itself, is governed by Act no. 111/1998 Coll. (Act on Higher Education Institutions and on Amendments and Supplements to some other Acts) specifically in §70, as follows: "*Academic workers are university employees who perform both pedagogic and scientific, research, developmental and innovative artistic or other creative activity.*" These include professors, associate professors, extraordinary professors, professional assistants, assistants, lecturers, and scientific workers. In our research, we devote ourselves to university employees in the position of assistant who are simultaneously doctoral program students.

The discourse of the university teacher's self-concept is also defined by Horká, Kratochvílová, and Grůzová (2014), when they speak of how the work of a university pedagogue consists of teaching and research, and these activities cannot be separated from one another: "*The university teacher mediates the students' scientific/technical/artistic knowledge cognition in teaching.*" (Horká, Kratochvílová, Grůzová in Szimethová et al, 2014). The doctoral student studies a doctoral program in their chosen field. This is the third level of university preparation and, as per the Universities Act, "*it is focused on scientific research and independent creative activity in the area of research or development, or on independent theoretical and creative activity in the area of art.*" The studies last three to four years, and are realized on the basis of an individual study plan under the guidance of an instructor (§ 47, Act no. 111/1998 Coll.). The instructor is an important participant in the doctoral student/academic worker's professional training. Gavora (2013) states that the instructor's duties are defined in the specific university's directive; the aforementioned author describes and evaluates the specific requirements of both a Czech university and a foreign one.

The professional public does not decisively deal with the issue of doctoral students who are simultaneously academics, because an exact classification cannot be found here. Hattie and Marsh (1996) describe a first-stage academic (i.e. a person who has already acquired the title of Ph.D. at a university), specifically in connection with their concept of teaching, and how it can be influenced by the incipient teacher simultaneously also being a researcher. Here we find the need to specify the role of a first-stage researcher. The concept of a teacher's schooling in university is also fundamentally influenced by their research plans. However, research has shown that just because an academic is an excellent researcher, does not mean that they are also an excellent teacher. Akerlind (2007) also researched these academics.

As part of the study, a sample of academics (18 men and 10 women) were interviewed for the purpose of ascertaining their method of cultivating and understanding themselves as teachers and simultaneously researchers in a university. Three qualitatively different ways of thinking about the concept of development as a teacher were discovered:

a) concept of teaching – teaching is becoming less sophisticated
b) teacher's knowledge and skills – expansion of their repertoire in teaching strategies
c) results of teaching – students' learning and development improves.

Here, we can reason that if an academic is also an active researcher, they will find benefits in participating in research activity which this activity brings. Interesting results in this area are also presented in a study by Navrátilová (2013), who basically claims that the academic worker/doctoral student tends towards the position of pedagogue or researcher on the basis of which of these areas they regard as cardinal from their perspective: "*Every academic worker consciously or intuitively regards themselves more as a scientist, or on the contrary as a pedagogue*" (Navrátilová, 2013, p. 51). Information from a different perspective is presented in a study by Coromina, Capó, Coenders, and Guia (2011), which examines doctoral students specifically by showing the most important aspects of their successful integration into the academic worker community. The study identifies 115 discovered aspects, and therefore crucial elements, of which 92 relate to their supervisors (i.e. persons who guide them during doctoral studies). Furthermore, many aspects relate to their research groups or their community of doctoral students as a whole. Similarly, it addresses what they find to be obstacles to their progress, of which 20 relate to the network of doctoral students or relationships. The most frequently presented topics relating to the network are members of a group of research workers who call for doctoral students to publish, meet up with research workers outside the research group, and have the opportunity to discuss the topic of their Ph.D. work with experts, as well as frequent contact with their supervisors and members of the research group.

It is also important to draw attention to a summarizing study by Mareš (2013), who clearly sums up here the numbers of graduates of Pedagogy doctoral study programs in the Czech Republic in the years 1999–2011, and characterizes this postgraduate study in that it "*fulfilled and fulfils the function of some kind of incubator*" (Mareš, 2013, p. 6) in connection with the growing independence of professional didactics and with the focus of every postgraduate student. In one of his essays in this research probe, Professor Mareš focuses on characterizing a group of doctoral students who are already in actual pedagogic practice: *"The largest group of applicants is formed by teachers who are already in the practice, and work in universities. They are pushed by their institution to increase their qualifications, and the workplace should have a corresponding qualification structure. Only some of them commence their studies out of an interest in examining pedagogic problems; the others see their studies more as a formal step which however has existential impacts.*" This crucial motivating factor undoubtedly influences the work of the simultaneous doctoral student and academic worker, but it does not have to mean merely a necessary evil for everyone; it also creates room for combining the pedagogue's academic path with study, and more space for practice. So how is it in the situation of the respondents of the presented research, if the simultaneous academic worker and doctoral student is placed in a position where they are to strategically and systematically realize teaching, bring research results, and at the same time also fulfil the duties of a simultaneous academic worker and doctoral student?

## 3 RESEARCH METHODOLOGY

On the basis of the methodology of the chosen research, the following basic research questions were stipulated:

1) What preferences do academic workers, who are simultaneously (academic worker with master's degree) in the position of doctoral student in the area of research activity, have?
2) What certainty is there in the area of research methodology?
3) Do projects realized in the faculty bring positive or negative aspects for doctoral students who are simultaneously academic workers?

Interviews were conducted in order to stipulate the research questions. Basic questions, which determined the framework of the realized semi-structured interviews, were stipulated. The interviews were recorded in written form, and then coded. An analysis of the information led to the selection of codes, and their subsequent transformation into the form of results led to the discovery of fundamental information that was crucial for the researchers.

### 3.1 *Research sample*

The participants in our research were selected deliberately. An available research sample was involved – persons who are in full-time employment as academic workers in the university, in the position of assistant, and are concurrently also doctoral students. These were five workers from two universities in the South Moravian Region.

### 3.2 *Interpretation of the obtained data*

On the basis of a detailed analysis of the realized interviews, which were converted into written form on the basis of literal transcripts, we identify the following semantic categories.

Table 1.

| | |
|---|---|
| 1 | First-stage researcher |
| 2 | Adapting worker |
| 3 | Duties versus salary |
| 5 | Advantageous position – benefits |
| 6 | Professional preferences |

<u>First-stage researcher</u>

From the analyzed interviews, it is evident that all the workers except one perceive their position as that of a researcher who is at the start of their career path in this area. One of the participants referred to this position as "*constantly learning.*" These are persons who, over the years, acquire both theoretical and practical experience from the area of methodology, on the basis of teaching or participation in all types of research projects. This position of a doctoral student and first-stage researcher seemed to mean, in the perception of some of the participants, that certain shortcomings in knowledge are accepted in this phase of their career path; in other words, after defending their title this will no longer be possible. This is demonstrated by the following statement: *"You always have a chance to learn, and it is legitimate for you to learn constantly. Everyone knows that you will still have methodology lectures, and that you are trying out some contributions. After I complete it (the title), nobody will be holding my hand."*

They see a great benefit in opportunities to become involved in research projects, on whose basis they can attend conferences and therefore acquire more professional experience.

Another positive aspect and significant moment for the first-stage researchers we contacted is teamwork: "*It's a matter of even a beginner having enormous support, data making its way to them with the assistance of other people, and they know that it is not a dead topic; moreover, it automatically applies that everyone will accept the topic.*"

Here, in addition to teaching, a first-stage researcher can acquire experiences from the area of methodology, where uncertainty in methodological issues is evident from all the interviews.

"*A person must constantly learn, search and attend all kinds of workshops*" or "*it's best to try things out in practice, then a person understands what and how.*"

They also see publishing activity, which they perceive as a "*springboard,*" as very demanding: "*In doctoral studies, emphasis is placed on publications; they are a kind of springboard for academic activity..., but it probably takes some people longer – for example, me.*"

Adapting worker

According to the statements, the position of an academic worker is initially very demanding. The person is simultaneously in the role of both student and pedagogue, and they must distinguish between these two roles.

"*It can often even be unpleasant for the person, when they are expected to be an academic worker while in certain situations they are a mere student who must fulfil their assignments, and is constantly tested and evaluated ... it's basically quite hard work.*"

It's multiform work, which includes lots of varied activities. The person must manage teaching, as well as organizational, publishing, and other affairs. The fact that the person works and studies in the same department can be a certain benefit. Initially, respondents state that it is difficult to adapt to their new position, as well as "*master*" all the information and skills connected with the realization of teaching and operation of the department.

Duties versus salary

An assistant's job description is teaching, research, and publishing activity all in one. Doctoral students describe their obligations as broad-spectrum, and the authors view that positively. "*The work is not stereotypical, but a person must be flexible to manage it.*"

The interviewed doctoral student describes her obligations as follows: "*teaching of subjects, research, work on studies, professional articles, involvement in other faculty events and activities, effort to participate in projects, activities relating to promoting the faculty, and of course the worker's self-development such as English courses.*"

For them, managing doctoral studies is a "*level*" of professional qualification, within whose scope the worker can perform other work activities, such as guided master's theses, etc. As for the definition of the position of assistant, one of the interviewed parties replies: "*Assistant means that they are not a qualified professional assistant; they have not yet completed their doctoral studies, and are ranked in a salary class accordingly.*" The participants only mention an academic worker's salary marginally; nevertheless, they are mostly sceptical.

Advantageous position – benefits

The interviewed doctoral students agree on the fact that the position of a doctoral student (first-stage researcher) and academic worker may be very demanding, but in a certain way it can also be advantageous, in that the position of a doctoral student involves acquisition of information which they can utilize in the position of academic worker.

"*A person who takes up the position of assistant from practice is cut off from the academic world, so the findings and knowledge from the subjects helped me in my work as assistant.*"

Participation in a wide range of projects opens up a range of other experiences for the person, and offers them the opportunity to verify their existing theoretical findings in practice.

Systematic work on their own project allows a person to acquire in-depth knowledge of the given issue, and they can then apply this knowledge in teaching. The following statement discusses the perception of a personal shift in the academic worker's career path on the basis of the position of first-stage researcher.

"*I'm moving forward. It's a fact that this progress is visible. I find that I now have the courage to write something in the project, and formulate things which I would not have managed before. As for education, in the past I taught, and explained, what was in the books. Suddenly I've noticed that I'm teaching differently; that I can present things which a student won't read.*"

Professional preferences

The area of the pedagogic worker's preferences in the area of professional growth is a crucial element for everyone who is trying to find their way in this area. Contending with a situation in which the academic worker teaches subjects that fit their professional specialization, but their research activity implicitly leads them in a different direction within the scope of their doctoral studies, can be completely unmanageable – or, on the contrary, beneficial. The contacted persons are in diverse situations. Some of them find that their orientations intertwine, while for others they are fundamentally different. The respondents in the presented research probe agree on the fact that: "*If I do not work on doctoral research within the scope of teaching at university, then I will definitely not complete my doctoral studies.*"

Preparation for teaching at a high theoretical level, combined with theoretical specialization in a doctoral study program, is exhausting, and a risk exists here that they will not be able to manage the required area at an acceptable level. The respondents in the presented research plan perceive the disparity between the specialization of simultaneous academic and doctoral student as an obstacle and, as Mareš (2016, 20) states, the "polarity between scientist and practitioner" must be systematically directed; on the contrary, as the same author cites: "*do not stagnate in the field.*"

## 4 CONCLUSIONS

This research probe analyzed semi-structured in-depth interviews with doctoral students, who are simultaneously also in the position of academic workers. This position was implicitly perceived as a clearly intertwining role of academics who enrich their status with another desired title.

This qualitatively oriented research probe clearly shows that this situation is enormously demanding, but if the doctoral student chooses it and decides to go down this path, then they are internally determined to battle with it and extract what's best from it. It was shown that the respondents are very strongly aware of the position of a first-stage researcher, and perceive it as a space in which they are entitled to "make mistakes." They see their doctoral studies as the beginning of their research. Key places that the respondents find most demanding were identified. They expressed themselves most fundamentally in relation to the expectations of their environment, which expects quality teaching, organizational management of an academic's work, and simultaneously also fulfillment of doctoral seminars at the required level. The most critical focal point was found to be the ability to transform oneself from the position of doctoral student into the position of academic worker. The respondents agreed on the fact that the salary with which they are remunerated does not correspond to the energy they expend. According to our conclusions, the same demands are placed on the respondents as on academics who have completed their doctoral studies, and they are required to present results of the same quality. However, we can also state that the research respondents do not see doctoral program studies as a formal obligation for taking up the position of academic worker. The doctoral students evaluated the opportunity to participate in research projects very positively, because this option is a way of participating in excellent topics which they would probably not have addressed as individuals, and they see this chance as room for learning from older, experienced professionals.

To conclude, we can state that the respondents find the situation of a university academic worker who decides to complete a doctoral study program simultaneously with their employment as highly demanding, but at the same time see the benefits that this work brings.

## REFERENCES


Akerlind, G. S. (2007). *Growing and Developing as a University Researcher.* Springer: 55(2), 241–254.

Coromina, L., Capó, A., Coenders, G. & Guia, J. (2011). PhD Students' Research Group Networks. A Qualitative Approach. *Metodološki zvezki.* 8(2), 173–189.

Gavora, P. (2013). The instructor's role in educating the doctoral student. In Wiegerová, A., Szimethpvá, M., Gavora, P., Kalenda, J., Navrátilová, H. & Kočicová, S. (Eds.), *First Stage Researcher* (pp. 35–49). Zlín: Tomáš Bata University.

Hattie, John & Marsh, H. W. (March 2004). *The Relationship Between Research.* Marwell Conference Centre.

Horká, H., Kratochvílová, J. & Grůzová, L. (2014). The university teacher as a scientist/researcher and educator of teachers. In Szimethová, M. et al. (Eds.), *The University Teacher: A Position between Teaching and Research* (pp. 8–21). Zlín: Tomáš Bata University.

Kalenda, J. (2013). From a "skholá [?] situation" to an "irruere situation": The researcher's changing environment. In Wiegerová, A., Szimethová, M., Gavora, P., Kalenda, J., Navrátilová, H. & Kočicová, S. (Eds.), *First Stage Researcher* (pp. 11–18). Zlín: Tomáš Bata University. *Springer.* 2(2). 241–254.

Mareš, J. (2013). Neviditelná skupina aneb Co s postdoktorandy? *Pedagogická orientace*, 23(1), 5–26.

Mareš, J. (2013). The invisible group or What to do with postdoctoral students? *Pedagogic Orientation*, 23(1), 5–26.

*Act no. 111/1998 Coll., on Higher Education Institutions. Available at: http://www.msmt.cz/vyzkum-a-vyvoj-2/zakon-c-111-1998-sb-o-vysokych-skolach*

*The Social Sciences Empowered – Lumban Gaol, Hutagalung & Fong Peng (eds)*
 *ISBN 978-0-367-00090-5*

# Self-regulation in peer rejected students

K. Hrbackova & I.Balaban Cakirpaloglu
*Department of Pedagogical Sciences, Faculty of Humanities, Tomas Bata University in Zlín, Czech Republic*

ABSTRACT: Social rejection and exclusion, which is experienced as painful, are a major risk factor for physical and mental illness. Unfortunately, social rejection and exclusion become a common part of school life. It is rather difficult to reintegrate rejected students to classroom, because it is necessary to apply mechanisms for regulating one's own behavior. These mechanisms may be undermined by social rejection. The aim of the research was to find out the level of self-regulated skills related to interpersonal cognitive problem solving in socially rejected students. We wanted to reveal the relationship between peer rejection and failure in self-regulation among socially rejected students compared to other (accepted) classmates. For this purpose, we used *sociometric nomination method, the Failure of Self-Regulation Questionnaire* and *The Means-Ends Problem Solving* (MEPS) procedure that included 5 stories related to problem situations that students can experience in a school environment. 1625 lower secondary school students (aged 11 - 16) from 20 schools took part in the research. The results imply that the level of use of students' self-regulated skills is related to peer preference. Socially rejected students achieve lower levels of self-regulation than other classmates. We found deficiency in the regulation of behaviour, emotions and the regulation of will in the peer rejected students with the exception of cognitive regulation in which no differences were detected between socially rejected students and other classmates. The results have implications for early prevention and intervention efforts to foster adaptive self-regulation and reduce the risk of later social rejection.

## 1 INTRODUCTION

### 1.1 *Social rejection*

We consider social rejection to be a significant threat which results in deliberate exclusion of an individual from social relations or social interaction (by a person or group of people). Social exclusion can be collectively referred to as a number of specific mechanisms (and the consequences of their actions), which threaten the integrity and social cohesion of a given collectivity and question the identity of their members (Mareš & Sirovátka, 2008). We can perceive social exclusion as "restriction of access to opportunities and limitation of the capabilities required to capitalise on these opportunities." (Hayes, Gray & Edwards, 2008). In a broader sense, we can perceive rejection as a term which expresses the fact that people perceive that their relational value is lower than they desire. Specific forms of rejection (such as peer rejection or romantic rejection, or else familial estrangement) can be considered as instances of rejection which occur within specific types of relationship (Leary, 2005).

### 1.2 *Social rejection within the school environment*

Researches show that social rejection and exclusion, which is experienced as painful, are a major risk factor for physical and mental illness. In this context, the phenomenon of social pain is discussed. MacDonald & Leary (2005) define social pain as, "a specific emotional reaction to the perception that one is being excluded from desired relationships or being devalued by desired relationship partners or group." In general, it can be said that social pain occurs in reaction to a socially disconnecting event. Many studies have confirmed (Borsook & MacDonald, 2013; Eisenberger, 2013) that social pain (rejection) is experienced biologically in a similar manner as physical pain, as it shares with physical pain not just affective centres, but also some sensory centres of a brain. Unlike physical pain, however, they differ in the intensity of re-use, which is significantly higher in the experience of social pain.

Social exclusion within the school environment is more and more frequently considered to be a hidden form of aggression or bullying in which a child is subjected to a certain level of exclusion caused by manipulation with his or her social relationships and social status (Cross, Shaw, Hearn, Epstein, Monks, Lester & Thomas, 2009). Similarly to Juvonen and Gross (2005), we include bullying into a broad category of peer rejection. Although the manifestations of rejection and bullying vary, the action of exclusion serves similar functions for the group. We understand rejection and bullying as two conceptually related phenomena that are important to understanding social outcast.

Research shows that approximately 1 in 6 children experience social exclusion in the school environment, but the actual number of socially excluded children may be even higher, as revealing social exclusion is difficult and often occurs in a hidden way (Cross, Shaw, Hearn, Epstein, Monks, Lester & Thomas, 2009). Social exclusion can occur both in active form (e.g. swearing, mockery) and passive (e.g. ignorance) and can go from physical and verbal aggression to exclusion from joint activities, lack of interest, defamation, intrigues (Williams, Forgas & Hippel, 2005). Research on peer rejection in the school environment unfortunately shows that social rejection is an integral part of children's lives (Bacete, Planes, Perrin & Ochoa, 2017). Rather, social exclusion is fundamentally about a lack of connectedness and participation from a peer group.

One of the fundamental human needs is defined as the need for social interaction and relationships with other people. Baumeister & Leary (1995) perceive the need to belong as the fundamental human need to "create and maintain at least a minimum number of longer-term positive and significant interpersonal relationships." This need is manifested by the fact that one has a strong tendency to avoid social rejection and exclusion and seeking acceptance by others (Leary & Kelly, 2009). The mere awareness of the existence of a relationship does not fulfil the need to belong. People need close personal contacts or interaction which are emotionally positive or pleasant and which also meet the condition of being a stable, emotionally advantageous and prospective relationship, which means that it results in satisfying interactions with the people in those relationships. This means that in order to fulfill the need to belong, relationships with others should be seen as meaningful.

The need to belong somewhere and to identify with a specific group is most strongly perceived during the preadolescent period (Svoboda, 2011; Rawlins, 1992). The period of preadolescence is a period of independence from parents and family and increasing dependence on one's peer group. The strongest social group which a preadolescent is a member of during this period is his or her school class. This is given by its institutional boundaries (that is to say, it is a formal group formed on the basis of a different motivation than the need for mutual relations) and it creates a structure of relations (including informal) that is a result of the process of identification with peers. Significant dynamics of relationships not only pose positive emotions, but is accompanied by a number of negative manifestations associated with the experience of social rejection (for example, as a result of the "fight for position" in a given class). (Leets & Wolf, 2005). In this environment, there is a risk of exclusion not only in risk groups (at risk of social exclusion) but also in non-risk groups because of the difference from others perceived by the group as a disadvantageous characteristic (Harrist & Bradley, 2002). The latest research shows that rejection is not a property of the rejected student or a characteristic of her/his behaviour (Dirks, Treat & Weersing 2007), i.e. what the rejected child says or does, does not lead directly or inevitably to rejection. Although studies have confirmed that socially excluded students display a lower level of prosocial behaviour and a higher level of aggressive behaviour (Bierman, 2004; Twenge, Ciaracco, Cuervo & Baumeister, 2003), it is possible that these behavioural manifestations are not the cause, but rather the result, of social rejection. Manifestations of aggressive behavior may occur amongst children who are popular within the class group. Bacete, Planes, Perrin & Ochoa (2017) have come to the conclusion that the behavior of a rejected student is not the actual reason for exclusion, but rather the rejectors' interpretations of a child´s behavior and whether they think this behavior will have a negative impact on themselves or on their social group.

Attempts at reintegrating excluded individuals within a social group are fairly difficult, as social rejection tends to be stable over time (Cillesen, Bukowski & Haselager, 2000). Although there is an assumption that experiencing social rejection leads an individual to make greater attempts at becoming integrated within the group, paradoxically the opposite tends to occur, i.e. difficulties in adjustment (Caldwell, 2003; Wood, Cowan & Baker, 2002) and an increase in selfish and self-defeating behaviors (Blackhart, Baumeister & Twenge, 2006). Experience of social rejection may lead to a number of adverse psychological consequences such as loneliness, low self-esteem, depression and increased manifestations of aggressive behavior (McDougal, Hymel, Vaillancourt & Mercer, 2001). There is an assumption that the link between social exclusion and subsequent behavior manifestations is caused by a hidden inner process, which is influenced by social exclusion and which also has a subsequent impact on increased manifestations of negative behavior.

The question arises as to whether this relationship can affect an individual's emotions, in particular his or her negative emotions. This assumption, however, has been rejected by many studies (Baumeister, Twenge & Ciarocco, 2014); Twenge, Baumestier & Stucke, 2001). It was found out that excluded individuals often deny negative experience of exclusion (Leary, 2015). Furthermore, social exclusion does not exhibit a deficit in mood either. Self-assessment scales socially excluded individuals who deny rejection and show a more positive mood, show higher levels of aggressive behavior (Twenge & Baumeister, 2013).

Besides emotions, other major inner processes include cognitive information processing and self-regulation. Self-regulation is another trait that seems much more powerful in human beings than in others (Baumeister & DeWall, 2005).

### 1.3 *Self-regulation*

Self-regulation, defined as the capacity to control or alter one's responses, is a vital mechanism for producing adaptive and socially desirable behavior (Baumeister, DeWall, Ciarocco & Twenge, 2005). Self-regulation is manifested in two possible ways. In the first case, this situation occurs when one checks his or her impulses and some behavior is prevented (one stops doing something). In the latter case, it is the fulfillment of the primary need to realize your own goals and the activation of a certain behavior (a person starts to do something). In this regard, self-determination theory distinguishes autonomous self-regulation and controlled self-regulation (Sokol, Grouzet & Müller, 2013). Self-controlled self-regulation plays a central role when there is a dissonance in the interaction between the inner processes of man and the social environment (e.g. as a result of social rejection or exclusion).

Self-regulation is the ability to manage emotions, behavior, or thoughts in accordance with the demands of the situation. It is a set of skills that enable peoples to direct their own behavior towards a goal, despite the unpredictability of the world and their own feelings. (Child Mind Institute, 2017)

Current research shows that when people are socially rejected, they exhibit decrements in self-regulation (Blackhart, Baumesiter & Twenge, 2006). Self-regulation may act as a possible mediator between social exclusion or rejection and self-defeating behaviors, because self-regulation failure is implicated in antisocial behavior, aggression, a lack of prosocial behavior, and self-defeating behaviors. Impeding the need to belong leads to significant impairments in self-regulation and intelligent performance. In sum, a series of studies found a broad pattern of impaired self-regulation amongst people who experienced social exclusion (Baumeister, DeWall, Cirocco & Twenge, 2005). Social exclusion and rejection produce inner disruption sufficient enough to produce self-regulation failure (Baumeister & DeWall, 2015). The question is how and why does rejection impair self-regulation? The limited resource model of self-control (Baumister, Vohs & Tice, 2007) assumes that self-regulation works in a similar way to a muscle. It shows that anti-social and aggressive behaviour occurs very often when an individual does not see any prospects to acceptance. It is likely that when social acceptance is prevented, an individual loses the motivation or will for self-regulation, which is a mechanism which could provide assistance in reintegration to a social group. Self-regulation or will to self-regulate have emerged from most recent data as the most important inner processes to change in response to social exclusion. (Baker & Baumeister, 2017)

In our research, we aim to verify the relationship between the degree of self-regulation skills of students in classes and the social rejection from a class group.

## 2 METHOD

### 2.1 *Research goals*

The aim of the research is to find out the level of use of self-regulation skills amongst socially rejected students within the class group. Our main objective was to find out the connection between students' social preference (level of social rejection by peers) and their level of self-regulation associated with interpersonal cognitive problem-solving. At the same time, we study what is the degree of failure in self-regulation among socially rejected students compared to other (accepted) classmates.

### 2.2 *Participants*

1625 lower-secondary school students (Year 6 – Year 9) from 20 schools took part in the research; the number is comprised of 849 boys and 776 girls of between 11 and 16 years of age (M = 13.17, SD = 1.287).

Schools were chosen by random selection (random number generator) from amongst all the lower-secondary schools in the Czech Republic[1].

### 2.3 *Measures*

Self-regulation was evaluated using the *Means-Ends Problem Solving technique* (MEPS) and the *Failure of Self-Regulation Questionnaire* (FSRQ). The measure of social rejection was determined using *peer nominations* (sociometric-rating questionnaire).

The MEPS (Platt & Spivack, 1989) measures the degree to which children who were and were not self-regulated possessed skills related to interpersonal cognitive problem-solving. The MEPS uses a story-based format where students are provided the beginning and ending of a story. The beginning poses a problem (for example "You and your classmate had agreed on working on a collective task together, but s/he chose to work with someone else in the end."), the ending reports the outcome (for example "The end of your story is that you work with the classmate you agreed to work with in the end."). Students were instructed to say what happened in the middle of the story that connects the two. In essence, students are asked to generate the means by which the outcome was achieved, given the problem. Students write this middle section, which can be as long or short as necessary. Because students require additional time for the stories and responses only 5 of the 10 scenarios were selected for administration. The MEPS allows students to generate as many means as they possibly can, and these are scored as being relevant or irrelevant. To provide some standardization in the process, we

1 Ministry of Education, Youth and Sport of the Czech Republic register of school and school facilities, situation as on 26/04/2017 (incomplete, mixed grade, special and practical schools were eliminated from the list).

asked students to generate only the best answer for the middle of the story. This answer is then evaluated on a scale of 0 to 3, with 0 being no means or completely irrelevant means and 3 being a relevant means. The students were able to obtain a maximum of 15 points, whereby a higher score corresponded to a greater degree of use of self-regulation skills during interpersonal cognitive problem-solving. On the basis of a factor analysis (analysis of the main components), we ascertained that the questionnaire is one-dimensional and this explains 56.64% of the variance. The internal consistency of the questionnaire in all five items, measured using Cronbach's coefficient, attains a value of $\alpha = .81$, which represents an acceptable degree of reliability.

The FSRQ was created on the basis of the team of author's previous studies in self-regulation amongst children and young people (Hrbáčková & Hladík, 2018). This is a self-judging scale, consisting of 17 statements, and examines the extent to which students experience a deficit in selected areas of self-regulation – in the area of behavioral regulation (4 items, e.g. "I have problems submitting a task in time."), emotional regulation (4 items, e.g. "I get upset easily."), cognitive regulation (4 items, e.g. "I start things without thinking.") and for each statement, students answer true or false depending on whether they are often experiencing the situation. For every True answer they get one point (they can get 0-17 points). Higher points indicate a higher deficit in selected areas of self-regulation of students. The perceived failure rate in individual areas of self-regulation is expressed for the sake of clarity by the perceived failure rate percentage score (the maximum score is 100% failure in self-regulation).

In order to identify socially rejected students, a *sociometric-rating questionnaire* was used. The most widely used method to measure sociometric status is peer nomination, in which the participant is asked to nominate peers he or she likes most or likes least. The measurement of sociometric peer status is based on liking (e.g. acceptance) and disliking (e.g. rejection) peer nomination items.

There were six sociometric questions in the questionnaire to measure peer status. Unlimited nominations were used. Self-nomination was not allowed. Students were asked, for example, "Who do you like the most?" (LM) and "Who do you like the least?" (LL). The LM and LL items were used to calculate a social preference index for each student, according to the procedure of Coie, Dodge & Cappotelli (1982). The raw nominations for LM and LL ratings were tallied, standardized, and transformed into a social preference score (a continuous social preference score was calculated by taking the difference between standardized LM and LL ratings). The rejected group consisted of all of those students who received a social preference standardized score of less than -1.0, a LL standardized score of greater than 0, and a LM standardized score of less than 0.

### 2.4 *Procedure*

The students filled in the questionnaires using the "paper - pencil" method. The data was processed using the IBM SPSS program version 24. The Independent-samples t-test and Pearson's correlation coefficient were used for the data analysis. At the same time, the prerequisites for use of the chosen test were examined; i.e. normality and homoscedasticity (Levene's test) were verified.

## 3 RESULTS AND DISCUSSION

### 3.1 *Level of self-regulation amongst socially-rejected students*

In the research sample of lower-secondary students (aged 6 – 9 years, N = 1625) we identified 13.5 % socially rejected students (N = 219). This implies that every seventh to eighth student in our sample is rejected by the classroom.

We ascertained that these students (rejected from the class group) achieve a significantly lower level of self-regulation (M = 5.95; SD = 3.573) linked to interpersonal cognitive problem solving (Table 1) compared to other classmates who are not rejected from the class group (M = 7.10; SD = 3.451, $p < .001$). The analysis implies that socially rejected students use self-regulation skills to solve problem situations to a lesser extent. The difference between average values measured within both groups is statistically significant, although substantively both groups achieve quite a low score in level of self-regulation linked to interpersonal cognitive problem solving (from a maximum of 15 points).

The perceived level of failure of self-regulation is significantly higher amongst socially rejected students (M = 7.57; SD = 4.087) compared to the other classmates (M = 6.32; SD = 3.620, $p < .001$). Socially rejected students perceive their failure in self-regulation more intensively (from a maximum of 17 points) than other classmates.

The research results imply (Table 2) that the social preference level (level of rejection by classmates) relates to the level of students' self-regulation in dealing with interpersonal cognitive problem-solving ($r = .130$, $p < .001$). The lower the social preference (i.e. the higher the level of social rejection), the

Table 1. Comparison of levels of self-regulation between rejected and other students.

| | | Mean | SD |
|---|---|---|---|
| Self-Regulation (MEPS) | Rejected | 5.95** | 3.573 |
| | Others | 7.10 | 3.451 |
| Failure of Self-Regulation | Rejected | 7.57** | 4.087 |
| | Others | 6.32 | 3.620 |

** The results are significant at a significance level of .01.

Table 2. Connection between social preference (level of rejection) and students' self-regulation.

| | Social preference | Mean | SD |
|---|---|---|---|
| Self-Regulation (MEPS) | .130** | 6.94 | 3.489 |
| Failure of Self-Regulation | -.145** | 6.50 | 3.713 |

** Correlation is significant at the .01 level (2-tailed).

lower the level of students' self-regulation in terms of interpersonal cognitive problem-solving and vice-versa. Similarly, the perceived level of failure in self-regulation relates to the level of social preference (level of rejection by classmates) ($r = -.145$, $p < .001$). The higher the perceived level of failure in self-regulation, the lower the social preference, i.e. the higher the level of social rejection by classmates.

### 3.2 *Level of failure in self-regulation in socially-rejected students compared to other students*

In order to find out how the level of failure in selected areas of self-regulation differs (Table 3), we compared the perceived rate of failure in self-regulation of behavior, self-regulation of emotions, cognitive self-regulation and self-regulation of will and observance in students of socially refused and other classmates manifestations of social rejection from the classroom). We found out that socially rejected students perceive their failure in self-regulation more intensively than other classmates in behavioral self-regulation (32 % failure, $p < .001$), in self-regulating their emotions (49 % failure, $p < .001$) and in self-regulation of will and attention (45 % failure, $p = .005$).

In cognitive regulation, we did not notice significant differences in perceived failure between socially rejected students and other classmates ($p = .140$). Students perceive failure in this area (cognitive self-regulation)similarly (regardless of the social

Table 3. Level of failure in selected areas of self-regulation amongst rejected and other students.

| SRL Failure | | Mean | % failure |
|---|---|---|---|
| Behavioural self-regulation | Rejected | 1.30 | 32 % |
| | Others | .99** | 25 % |
| Emotional self-regulation | Rejected | 1.94 | 49 % |
| | Others | 1.43** | 36 % |
| Cognitive self-regulation | Rejected | 2.10 | 53 % |
| | Others | 1.97 | 49 % |
| Will, attention self-regulation | Rejected | 2.24 | 45 % |
| | Others | 1.94** | 39 % |

** The results are significant at a significance level of .01.

preference), but the perceived level of this failure is the greatest for both groups. According to their responses, socially rejected students achieve 53 % failure in this area, while we noticed 49 % failure amongst other classmates. Socially rejected students also achieve a significant deficit in self-regulation of emotions and self-regulation of will.

While studying what situations socially rejected students most commonly failed in, we analyzed in more detail the frequency of responses to individual statements (statements with over 50 % frequency of failure). The analysis implies that socially rejected students very often procrastinate (70 % of students), get angry quite easily (55 % of students), are easily distracted (54 %), have concentration problems (53 % of students) and often start things without thinking first (50 % of students). Compared to other students (who do not have experience of social rejection), they admit to a greater extent to arguing with others (47 % of socially rejected students compared to 27 % of other students), to having difficulties in handing in assignments in time (37 % of socially rejected students compared to 21 % of other students), and to often having difficulties in calming down when something makes them angry (44 % of socially rejected students compared to 28 % of other students).

According to Bierman (2004), most children who are rejected by peers display some behavioral patterns, such as disruptive behavior, impulsive behavior or social anxiety. The research is largely correlational, but there is evidence of reciprocal effects. This means that children with problems are more likely to be rejected, and this rejection then leads to even greater problems for them (Coie, 1990).

## 4 CONCLUSIONS

The results of the research suggest that on average we can label every 7th – 8th student in the sample set as a socially rejected student from the class group. Current studies similarly confirm that social exclusion can be encountered in almost every class and that it is an everyday part of children's lives (Bacete, Planes, Perrin & Ochoa, 2017).

We found out that socially rejected students achieve a relatively low score in the level of their own self-regulation. This means that in dealing with interpersonal cognitive problem-solving situations, students use self-regulation skills to a very limited extent and are not particularly good at regulating themselves such so that they can deal with problem situations relating to interpersonal relationships. At the same time, we verified the connection between level of self-regulation and level of peer preference. Our results show that a perceived deficit in self-regulation and a low level of use of self-regulation skills in dealing with interpersonal problem situations is linked to students' social preference. The lower the students' social preference is (i.e. the higher their level of social rejection), the lower the level of use of self-regulation skills and also the higher the level of perceived failure in self-regulation.

In line with current studies which confirm the connection between social exclusion and self-regulation (Baker & Baumeister, 2017; Baumeister, Twenge & Ciarocco, 2014) we can conclude that self-regulation plays an important role when a child is subjected to the experience of social rejection. In our research, we did not study the focus orientation, and as such as cannot determine in what way this works as a result of social exclusion.

The results also imply that socially rejected students are aware of their deficit in self-regulation. The perceived level of failure in self-regulation is significantly higher amongst socially rejected students compared to other classmates.

Socially rejected students most strongly perceive a deficit in the area of cognitive self-regulation. In this area, however, they are not significantly different from other classmates. Both groups (socially rejected and other students) state that they are not particularly good at regulating their own thoughts (they often, for example, start things without thinking first, or procrastinate).

In contrast to other classmates, socially rejected students feel a significant deficit in regulating their emotions, regulating their will and in regulating behavior. This deficit is significantly higher amongst socially rejected students compared to other classmates in the class group. We believe that the perceived failure of socially rejected students may relate to the stigmatisation of individuals – stereotyped perception of mutual differences (Bauman, 1999) and also to the fact that students are subjected to significant pressure from classmates and their criticisms, through which they may get into a so-called social spiral (Baumeister & DeWall, 2005) which reduces their capacity for self-regulation and reintegration into the class group. We shall investigate the dynamic of how self-regulation skills act in future studies in which we want to verify the effect of social rejection on the level of students' self-regulation, amongst other focuses. In so doing, we want to contribute to more effective prevention of self-regulation failure and the risk of students' social exclusion.

## ACKNOWLEDGEMENTS

The research was supported by Czech Science Foundation (GACR) based on the research project 17-04816S *The Dynamics of Self-Regulation in Socially Excluded Students).*

## REFERENCES

Bacete, F. J., Planes, V. E., Perrin, G. M., & Ochoa, G. M. (2017). Understanding Rejection between First-and-Second-Grade Elementary Students through Reasons Expressed by Rejecters. *Frontiers in psychology*, 8, 462. doi: 10.3389/fpsyg.2017.00462

Baker, L. R., & Baumeister, R. F. (2017). Alone and impulsive: self-regulatory capacity mediates and moderates the implications of exclusion. In K. D. Williams. & S. A. Nida (Ed.), *Ostracism, exclusion, and rejection* (pp. 29–45). Abingdon, Oxon, United Kingdom: Routledge.

Bauman, Z. (1999). *Globalizace/Globalization*. Praha: Mladá fronta.

Baumeister, R. F. & DeWall, C. N. (2005). The inner dimension of social exclusion: Intelligent thought and self-regulation among rejected persons. In Williams, K. D., Forgas, J. P., von Hippel, W. (Eds.), *The Social Outcast* (pp. 1–15). New York, NY: Taylor & Francis Group, LLC.

Baumeister, R. F., DeWall, C. N., Ciarocco, N. J., & Twenge, J. M. (2005). Social Exclusion Impairs Self-Regulation. *Journal of Personality and Social Psychology*, 88(4), 589–604.

Baumeister, R. F., Twenge, J. M., & Ciarocco, N. (2014). The inner world of rejection: Effects of social exclusion on emotion, cognition, and self-regulation. In J. P. Forgas & K. D. Williams (Ed.), *The social self: cognitive, interpersonal, and intergroup perspectives* (pp. 161–174). New York, NY, United States: Psychology Press. doi:10.4324/9781315800516

Baumeister, R. F., & Leary, M. R. (1995). The Need to Belong: Desire for Interpersonal Attachments as a Fundamental Human Motivation. *Psychological Bulletin*, 117(3), 497–529.

Baumeister, R.F., Vohs, K.D., & Tice, D.M. (2007). The strength model of self-control. *Current Directions in Psychological Science*, 16, 351–355.

Bierman, K. L. (2004). *Peer rejection: Developmental processes and intervention strategies*. New York: NY: Guilford Press.

Blackhart, G. C., Baumeister, R. F., & Twenge, J. M. (2006). Rejection's Impact on Self-Defeating, Prosocial, Antisocial, and Self-Regulatory Behaviors. In K. D. Vohs & E. J. Finkel (Eds.), *Self and Relationships: Connecting Intrapersonal and Interpersonal Processes* (pp. 237–253). New York: Guilford Press.

Borsook, T. K., & MacDonald, G. (2013) Social pain. In C. N. DeWall, (Ed.), *The Oxford Handbook of Social Exclusion*. New York, NY: Oxford University Press. pp. 1–41.

Caldwell, M. S. (2003). *The effect of peer victimization on social behavior and children's ability to negotiate conflict*. Dissertation Abstracts International: Section B: The Sciences and Engineering, 64(3-B).

Cillessen, A. N., Bukowski, W. M., & Haselager, G. T. (2000). Stability of sociometric categories. In A. N. Cillessen, & W. M. Bukowski (Eds.), *Recent advances in the measurement of acceptance and rejection in the peer system* (pp. 75–93). San Francisco, CA: Jossey-Bass.

Coie, J. D. (1990). Toward a theory of peer rejection. In S. R. Asher & J. D. Coie (Eds). Peer rejection in childhood (pp. 365–401). Cambridge, England: Cambridge University Press.

Coie, J. D., Dodge, K. A., & Coppotelli, H. (1982). Dimensions and types of social status: A cross-age perspective. *Developmental Psychology*, 18(4), 557–570.

Cross, D., Shaw, T., Hearn, L., Epstein, M., Monks, H., Lester, L., & Thomas, L. (2009). *Australian Covert Bullying Prevalence Study (ACBPS)*. Cowan University, Perth: Child Health Promotion Research Centre.

Dirks, M. A., Treat, T. A., & Weersing, V. R. (2007). The situation specificity of youth responses to peer provocation. *Journal of Clinical Child & Adolescent Psychology*, 36, 621–628. doi:10.1080/15374410701662758

Eisenberger, N. I. (2013). Why rejection hurts: the neuroscience of social pain. In C. N. DeWall (Ed.), *The Oxford Handbook of Social Exclusion*. (pp. 152–162). New York, NY: Oxford University Press.

Harrist, A. W. & Bradley, K. D. (2002). Social exclusion in the classroom: Teacher and student as agents of change. In J. Aronson (Ed.), *Improving academic achievement: Impact of psychological factors on education* (pp. 364–383). New York: Academic Press.

Hayes, A., Gray, M., & Edwards, B. (2008). *Social inclusion: Origins, concepts and key themes*. Canberra: Australian Government.

Hrbáčková, K., & Hladík, J. (2018). Autoregulace žáků v závislosti na vnímaných projevech rizikového chování/ Self-regulation of pupils depending on the perceived manifestations of risk behavior. *Studia Paedagogica*, 23 (1),81–104. doi:http://dx.doi.org/10.5817/SP2018-1-5

Juvonen, J., & Gross, E. F. (2005). The Rejected and the Bullied: Lessons About Social Misfits from Developmental Psychology. In K. D. Williams, J. P. Forgas, & W. von Hippel (Eds.), Sydney Symposium of Social Psychology series. The social outcast: Ostracism, social exclusion, rejection, and bullying (pp. 155–170). New York, NY, US: Psychology Press.

Leary, M. R. (2015). Emotional responses to interpersonal rejection. *Dialogues in Clinical Neuroscience*, 17(4), 435–441.

Leary, M. R. (2005). *Varieties of interpersonal rejection* (pp. 35–51). New York: Psychology Press.

Leary, M. R., & Kelly, K. M. (2009). Belonging motivation. In M. R. Leary & R. H. Hoyle (Eds.), *Handbook of individual differences in social behavior* (pp. 400–409). New York, NY: Guilford Press.

Leets, L. & Wolf, S. (2005). Adolescent rules for social exclusion: when is it fair to exclude someone else? *Journal of Moral Education*, 34(3), 343–362.

MacDonald G., & Leary M. R. (2005). Why does social exclusion hurt? The relationship between social and physical pain. *Psychological Bulletin*, 131(2), 202–223.

Mareš, J., & Sirovátka, T. (2008). Sociální vyloučení (exkluze) a sociální začleňování (inkluze) – koncepty, diskurz, agenda. *Sociologický časopis/Czech Sociological Review*, 44(2), 271–294.

McDougall, P., Hymel, S., Vaillancourt, T., & Mercer, L. (2001). The consequences of childhood rejection. In M. R. Leary (Ed.), *Interpersonal rejection* (pp. 213–247). New York, NY: Oxford University Press.

Platt, J. J., & Spivack, G. (1989). *The MEPS procedure manual*. Center for Addiction Research, University of Medicine & Dentistry of New Jersey: Camden, NJ.

Rawlins, W. K. (1992). *Friendship matters: Communication, dialectics, and the life course*. Hawthorne, NY: de Gruyter.

*Register of schools and school facilities*, Ministry od Education, Youth and Sports. (2015). Retrieved from http://www.msmt.cz/vzdelavani/skolstvi-v-cr/statistika-skolstvi/adresare-skolskychinstituci

Sokol, B., Grouzet, F., & Müller, U. (2013). *Self-regulation and autonomy: Social developmental dimensions of human conduct*. New York: Cambridge University Press.

Svoboda, J. (2011). *Zdroje, důsledky a návrhy řešení agresivního chování v současné škole/Resources, Consequences and Suggestions for Aggressive Behavior in the Recent School*. Unpublished manuscript.

Twenge, J. M., Baumeister, R. F., Tice, D. M., & Stucke, T. S. (2001). If you can't join them, beat them: Effects of social exclusion on aggressive behavior. *Journal of Personality and Social Psychology*, 81, 1058–1069.

Twenge, J. M., Ciarocco, N. J., Cuervo, D., & Baumeister, R. F. (2003). *Social exclusion reduces prosocial behavior*. Unpublished manuscript.

Williams, K. D., Forgas, J. P., & Hippel, W. (2005). *The social outcast: Ostracism, social exclusion, rejection, and bullying*. New York: Psychology Press.

Wood, J. J., Cowan, P. A., & Baker, B. L. (2002). Behavior problems and peer rejection in preschool boys and girls. *Journal of Genetic Psychology*, 163, 72–88.

*The Social Sciences Empowered – Lumban Gaol, Hutagalung & Fong Peng (eds)*
 *ISBN 978-0-367-00090-5*

# A prototype DWIPA trip planner: When e-tourism collides with Ant Colony Optimization (ACO) algorithm

Guson Prasamuarso Kuntarto, Khalish Ramadhansyah & Irwan Prasetya Gunawan
*Computer Science, Faculty of Engineering and Computer Science, Universitas Bakrie, Jakarta*

ABSTRACT: Planning travel tours, or what is commonly called the Tourist Trip Design Problem (TTDP), is one of the problems often faced by tourists when traveling in an area. The difficulties faced by tourists are to determine the tourist location to be visited, the distance between destinations, and travel time during tourist trips. Therefore, each tourist chooses the fastest and shortest tourist route that can be achieved so that they can maximize their limited time. The application of e-tourism is now growing very rapidly: one of the fastest-growing domains is ontology-based travel applications. In Indonesia, the application of ontology-based tourism support engines has been widely developed, but supporting features such as trip planners are still very minimal. One of the search engine applications in the field of tourism is DWIPA search engine. This study aims to develop the DWPA search engine architecture by adding a prototype trip planner feature. The Borobudur Temple destination area was chosen as the object of research where this object is one of the 10 National Priority Tourism Destinations as stated in Presidential Regulation No. 93/2017. DWIPA's trip planner feature utilizes DWIPA Ontology and implements the Ant Colony Optimization (ACO) algorithm to determine the best tourist travel routes from destinations or points-of-interest located around the Borobudur Temple tourist area. The prototype is implemented using the Java programming language. The prototype shows that the Ant Colony Optimization algorithm successfully provides travel solutions and produces optimum routes for destinations around the tourist area of Borobudur Temple.

*Keywords*: Tourist trip design problem, travel planning, Ant Colony Optimization, ACO, e-tourism, ontology, DWIPA trip planner

## 1 INTRODUCTION

Planning travel tours, commonly called the Tourist Trip Design Problem (TTDP) is one of the problems often faced by tourists when traveling in an area. The TTDP here refers to tourists scheduling to visit some points-of-interest (POI) with limited time owned by tourists (Vathis & Zaroliagis, 2016). The TTDP problem can be solved with the application of travel recommendations with a web or mobile interface-based approach. The use of web-based interfaces is the most commonly used by e-tourism recommenders because it allows tourists to get information in a user-friendly manner. However, web-based applications are not designed to be used during stays, because most tourists don't get easy access to Internet connections. Although many tourists already have mobile or tablet devices with an Internet connection, the information is not easy to read or manipulate on small screens (Borràs, Moreno, & Valls, 2014).

To overcome web interface problems, especially in the trip-planning function currently, ontology-based travel applications have been developed. Ontology is very useful researchers and organizations likewise engaged tourism, but each entity uses its own data model. If the entities need to communicate with one another, then it is necessary to represent a common concept that can be mutually understood (Jakkilinki, 2008).

At present, many trip-planner applications have been developed. One of them is applying ontology as a tourism knowledge base, namely the trip planner application developed by the Australian Sustainable Tourism Ontology (AuSTO), built by Roopa Jakkilinki and Nalin Sharda, which includesall general concepts used in the field of tourism (Jakkilinki, 2008). In Indonesia there are already ontology-based search engine applications in the field of e-tourism, but still not equipped with a trip-planner feature. The search engine application is called DWIPA, which is a search engine built to solve the problem of e-tourism in Bali by implementing the semantic web. Semantic web technology is used because it makes it easy to get information about accommodation, attractions, cultural events, etc., by utilizing ontology (Kuntarto & Gunawan, 2012).

Thus there are still opportunities to develop DWIPA search engines by adopting the Australian Sustainable Tourism Ontology (AuSTO) into a DWIPA trip-planner application based on the Ant Colony Optimization (ACO) algorithm to overcome Tourist Trip Design Problem (TTDP).

## 2 LITERATURE REVIEW

### 2.1 *Related research*

A study entitled "Ontology of Information for Tourism Development Planning in Thailand," aims to create an ontology that contains information on tourism in Thailand. The research and development method used is first to do documentation analysis and study the user to identify information needed for tourism development planning. Second is the classification and development of information structures that will be used for the development of the ontology. Third is to build and evaluate the ontology. The results of the study were an ontology containing information for tourism development planning consisting of 15 main classes (Chinnapatjeerat, 2017).

In Indonesia, research with the theme of tourist design problems has attracted the attention of researchers. One article, entitled "Designing application of Ant Colony System Algorithm for the shortest route of Banda Aceh and Aceh Regency Tourism by using graphical user interface Matlab," aims to implement the Ant Colony Algorithm used in calculating the shortest route and applied to tourism Banda Aceh destination and Aceh Besar district. In this study, the algorithm was used to determine the shortest route from the tourism destination of Banda Aceh and the district of Aceh Besar. The results of this study are the shortest routes that can be obtained with a minimum distance in one trip (Rusdiana, 2017). However, the result has not shown the optimum calculation, therefore there are still opportunities to optimize these results.

Outside Indonesia, the algorithm to overcome the problems of the TTDP has been tested. One study is entitled "Using Ant Colony Optimization for tourist route construction automation", which aims to study how the Ant Colony Optimization (ACO) algorithm can be used to build attractive tourist routes. This study uses a tourist route model described as a hierarchy of concepts that describe information relating to tourists such as speed and energy costs; information relating to attractions such as name, type, cost of visit, travel time, location, and value of attraction; and information relating to visits such as the list of attractions provided, and time and duration of visits (Skripal, 2016). However, this research has not used the domain of knowledge about ontology-based tourism.

The study entitled "The eCOMPASS multimodal tourist tour planer", aims to solve the problem of the TTDP by creating a web/mobile-based context-aware application called eCOMPASS where the application is useful as a tour trip recommendation planner to tourists in the destination of Athens (Greece) and the destination of Berlin (Germany) by using public transportation to reach tourist sites and schedule breakfast or lunch at restaurants along the route. The algorithm used is based on the new cluster-based heuristic approach, SlackRoutes (Gavalas, Kasapakis, Konstantopoulos, & Pantziou, 2015).

### 2.2 *Ontology*

Ontology is a conceptualization of a domain consisting of entities, attributes, relationships, and axioms. Ontology has an important role in knowledge management and is used as a basic understanding of the semantic web (Tang, 2010). Ontology is very useful for entities such as researchers and organizations engaged in the same domain, but each entity uses a separate data model. For example, in the tourism domain, if every tourism entity needs to communicate with the others, then a common data representation is needed. The representation must represent both concepts in the domain, and the relationship between the two concepts. Creating ontologies for tourism makes it possible to share knowledge between different tourism organizations and also allows for the creation of e-tourism applications such as search engines and tour planners (Jakkilinki, 2008).

In Australia, the development of tourism ontology has been grown rapidly. One of the ontologies on the tourism domain is the Australian Sustainable Tourism Ontology (AuSTO). It covers all general concepts used in tourism, both from a customer's perspective and from a company perspective, especially for tourism in Australia. AuSTO ontology reuses knowledge from several ontologies that already exist. This AuSTO ontology also developed an intelligent travel application generator. The framework of the application generator makes it possible to produce a variety of intelligent travel guidelines such as trip planning, and recommendation systems. The model used to build this ontology is based on the application of e-tourism called the E-Tourism Application Generator Architecture (e-TAGA). The e-TAGA model consists of three layers: ontology layer (OL), the business logic layer (BLL), and graphical user interface layer (GUIL) (Jakkilinki, 2008). The business logic layer (BLL) section uses the inference engine and custom logic, and is responsible for producing results, namely returning results based on user interaction. Users determine their travel requirements on ontology and tourism vendors advertise their offers on ontology. The offer and requirements are loaded into the Jena subsystem ontology model. Pellet reasoners match travel requirements to vendor offers and send them to the travel application manager. Tour trip planners can submit requests to the travel application manager and produce travel solutions (Jakkilinki, 2008).

In Indonesia, there is some research that focuses on the development of ontology in the tourism domain. One of the ontologies is the DWIPA Ontology. The DWIPA Ontology has been transformed several times since 2012. The current development of DWIPA Ontology III has successfully enriched the initial DWIPA Ontology II with new classes, subclasses, and instances using ontology-enrichment methodology. This study succeeded in summarizing some of the results of research related to the development of enrichment ontology in the tourism domain from 2014 to 2017. The study also succeeded in enriching DWIPA Ontology II, which was successfully carried out by the population to the object/instance (Kuntarto & Moechtar, 2017). The method used in this study is the ontology enrichment method. This method is used to expand the background of ontology by adding new concepts and relationships through the information extraction process. The enrichment process is carried out on semi-automatic web documents (corpus). The process of selecting an instance is done by a statistical and linguistic approach. By applying logical tests using query logic (DL) and data-driven evaluation techniques, DWIPA Ontology III produces 4 main classes, 15 subclasses and 199 instances/objects with new concept extensions: cultural park, artist, and monument (Kuntarto, Gunawan, Moechtar, & Ahmadin, 2017).

### 2.3 *Ant Colony System (ACS) and Ant Colony Optimization (ACO) algorithms*

The Ant Colony System (ACS) algorithm consists of three main aspects: (1) the state transition rule provides a direct way to balance the exploration of new parts and the exploitation of the main and accumulated knowledge about the problem; (2) global updating rules are applied only to the best part of the path; and (3) when an ant creates a solution, a local updating rule is applied. The first step is to calculate the state transition rule is to calculate the inverse distance using the equation:

$$\eta_{ru} = \frac{1}{d_{ru}} \tag{2.1}$$

Next, calculate the position of an ant at point r, choose the Destination s to move by applying the equation:

$$s = \begin{cases} \arg max_u \in J_k(r)\left\{[\tau(r,u)].[n(r,u)]^{\beta}\right\}, \\ \quad if\ q \leq q_0(exploitation) \\ S, otherwise\ (biased\ exploration) \end{cases} \tag{2.3}$$

Where q is a random number distributed evenly in [0 … 1], $q_0$ is a parameter ($0 \leq q_0 \leq 1$), and S is a random variable. The equation of the global updating rule is as follows:

$$\tau(r,s) = (1-\alpha).\tau(r,s) + \alpha.\Delta\tau(r,s) \tag{2.4}$$

Where

$$\Delta\tau(r,s) = \begin{cases} (L_{gb})^{-1}, if(r,s) \in global\ best\ tour \\ 0, otherwise \end{cases} \tag{2.4}$$

0 <α <1 is the pheromone decay parameter and $L_{gb}$ is the overall length of the best path.

The local updating rule equation is as follows:

$$\tau(r,s) = (1-p).\tau(r,s) + p.\Delta\tau(r,s)$$

Where

$$\Delta\tau_{r,s} = \frac{1}{L_{nn}.C} \tag{2.5}$$

where 0 <r <1 is the parameter (Dorigo & Gambardella, 1997; Huang, 2013).

Then, after many modifications, the Ant Colony (AC) algorithm is optimized and is known as the Ant Colony Optimization (ACO) algorithm. In foraging activities, ant colonies transport food and return to the nest using the same pathway. Ant colonies find pathways between nests and food sources by releasing a substance called a pheromone along the path to communicate. Ant colonies usually choose pathways with high levels of pheromone concentration. If there are obstacles, the ant colony tries several directions, then the next ant determines the route to be taken according to the highest number of pheromone concentrations (Huang, 2013).

The Ant Colony Optimization algorithm is based on the possibility of ant movement from point x to point y:

$$p_{xy}^{k} = \frac{(\tau_{xy})^{\alpha}(\tau_{xy})^{\beta}}{\sum z \in\ allowed_z(\tau_{xy})^{\alpha}(\tau_{xy})^{\beta}} \tag{2.6}$$

In equation (2.6), α and β are the coefficients used to manage $\tau_{xy}$ and $n_{xy}$, $\tau_{xy}$ represent the number of pheromones stored for the transition from point x to point y and for $n_{xy}$ transition xy.

The pheromone conditions that are updated by transitions are as follows:

$$\tau_{xy}(t+1) = (1-p)\tau_{xy}(t) + \sum_k \Delta\tau_{xy}^{k}(t) \tag{2.7}$$

Where

In equation (2.5), 0 <p <1 is the evaporation of the pheromone trace, $\Delta\tau_{xy}^{k}(t)$ is the number of pheromones given by the ant (k) to the arc it encounters as in equation (2.6), where $L^k$ is the length of the path that the ant passes (Skripal, 2016; Yousef, Shuqeir, Amjad, & Qublan, 2014)

## 3 RESEARCH METHODS

### 3.1 *Output model*

Based on Figure 1, the research is divided into four main activities. The first stage is the business requirement, which gathers the requirements needed in a travel business.

The next step is to design an architectural model of DWIPA trip planner that combines the DWIPA search engine architecture and AuSTO architecture. The third activity is selecting the reasoner that applies logic to the knowledge contained in the ontology to get some conclusions. The last step is the trip-planner function, which is to produce a function from the trip planner including recommending tourist attractions and travel routes.

### 3.2 *A proposed model of the Ant Colony Optimization (ACO) algorithm on DWIPA trip planner*

A model of the Ant Colony Optimization (ACO) algorithm is the stage that is carried out in order to produce the optimal path for planning travel trips. Figure 2 shows the process of implementing the ACO algorithm to produce a travel route planning path.

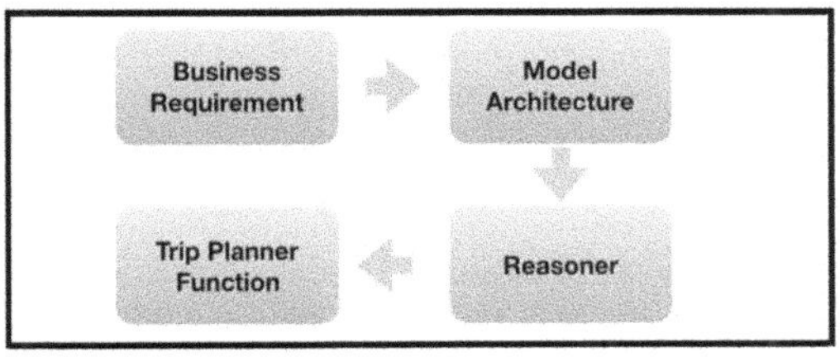


Figure 1. Research methods.

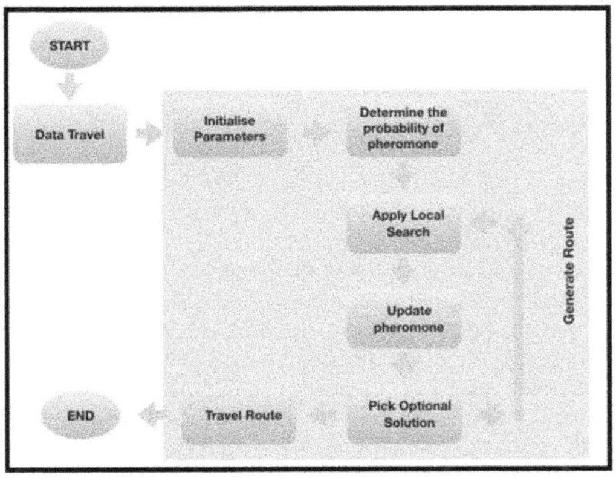


Figure 2. The generate route process of ACO that implemented on the DWIPA trip planner.

## 4 ANALYSIS AND DISCUSSION

### 4.1 *Output model/prototype*

The architectural model of DWIPA trip planner gives users a proposal for planning travel tours through a series of business logic steps. First, the travel needs of the users are queried into the ontology tourism, namely DWIPA Ontology III. DWIPA Ontology III is a knowledge base on tourism in Indonesia that was developed in 2017 (Kuntarto, Gunawan, Moechtar, & Ahmadin, 2017). After that, the query is forwarded to reasoners. The function of reasoner is a trip planner model logic that changes the results of queries from ontology to information that is processed by the travel application manager. Second, this travel application manager then returns the information to the trip planner or search engine.

Figure 3 Shows the proposed model of DWIPA trip planner. The model consists of two main parts: search engine and trip planner. Both of the features use DWIPA Ontology III as a main knowledge-based system in order to provide relevant information about tourism in Indonesia. It is proposed that the tourists are strongly encouraged to query for the information which proceeds by the reasoner and return the information related to user desire. The travel application manager is basically a user interface to facilitate tourists selecting the DWIPA feature whether searching or trip planning.

The tourist destination that used on this research is one of the ten locations of National Priority of Tourism Destinations (DPN) according to Presidential Regulation No. 93/2017. The selected location is the Borobudur Temple Area, located in Central Java Province as shown in Table 1:

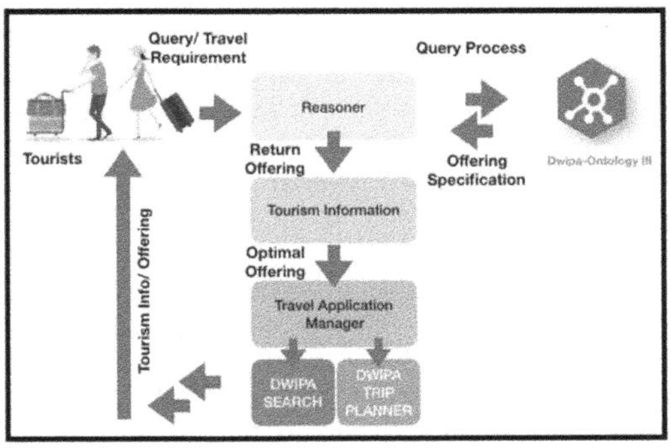


Figure 3. 1 DWIPA search engine architectural model that has developed the trip planner feature.

Table 1. List of the amenities nearby the borobudur temple area.

| Amenities | Numbers |
|---|---|
| Hotel | 18 |
| Rental Cars | 4 |
| ATM | 6 |
| Pharmacy | 5 |
| Restaurants and Cafe | 16 |

### 4.2 *Elaboration of a proposed model of the Ant Colony Optimization (ACO) algorithm on DWIPA trip planner*

The ACO algorithm is divided into several substages which refer to the ACO algorithm stages: (1) initialize parameters, (2) determine the probability of pheromone, (3) apply local search, update pheromone, and (4) pick the optional solution.

The first stage, initialize parameter, is used in the ACO algorithm. There are eight parameters: destination, rho, ant, start destination, alpha, beta, and food parameters. The first parameter is the destination parameters used to determine the number of destinations and also the distance between one destination and another, or what is called a cost. The second parameter is the rho used to determine how much the pheromone evaporates at each iteration. The third parameter is the number of ants used. In this study, ants are represented as tourists. The fourth parameter is the starting point for ants to start the journey. The fifth parameter is the alpha parameters used in supporting variables for pheromone values. The sixth parameter is the beta parameter used as a supporting variable for the value of the visibility of the destination. The seventh parameter is the food parameters used as iterations. The second stage is to determine the probability of pheromone. This is done by determining the initial pheromone value for each route between destinations. Determination of the value of pheromone is intended so that each path has interest for each ant. In this study, the initial pheromone value used is defined as 1. The third stage is to apply the local search and update pheromone. The last stage is to select the optimal solution for planning travel. Figure 4 shows that the model of DWIPA Trip Planner has been visualized using Java Programming language. The dots represent the numbers of tourism destinations and the straight lines represent the optimal routes for tourists to travel from one destination to other destinations.

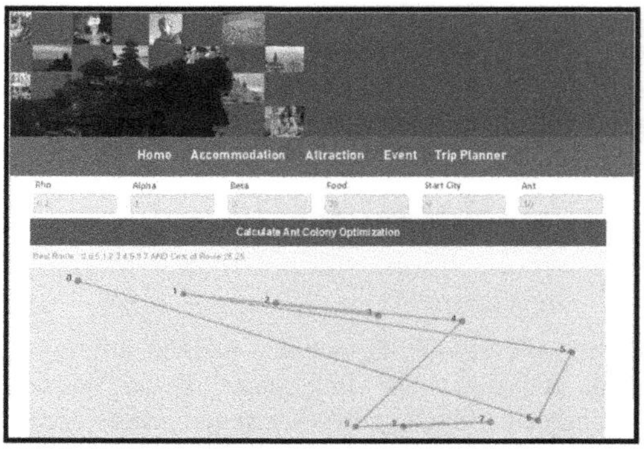


Figure 4. Visualization of ant colony optimization at DWIPA trip planner.

## 5 CONCLUSION

In summary, DWIPA search engine was successfully developed by adding a new feature named DWIPA trip planner. On the trip planner page, there are several inputs – parameters such as rho, alpha value, beta value – determining the initial destination. In addition, there are also buttons for calculating the ACO algorithm. The implementation of the ACO algorithm has been successfully implemented in order to determine the planning of tourist routes on the DWIPA Trip Planner. ACO was successfully visualized by using the Java programming language in order to produce the optimal destination route from tourism locations nearby the Borobudur Temple Region Area.

## ACKNOWLEDGMENT

The authors would like to thank the Ministry of Research and Technology and the Higher Education of the Republic of Indonesia for the research grant Number 107/SP2H/LT/DRPM/IV/2018 and Agreement/Contract No. 017/SPK/LPP-UB/III/2018.

## REFERENCES

Borràs, J., Moreno, A., & Valls, A. (2014) Intelligent tourism recommender systems : A survey, *Expert Syst. With Appl.*, 41, 7370–7389.

Chinnapatjeerat, R. (2017) Ontology of information for tourism development planning in Thailand, *5th Int. Conf. Comput. Sci. Netw. Technol.*, 2, 856–878.

Dorigo, M., & Gambardella, L. M. (1997) Ant Colony System : A cooperative learning approach to the traveling salesman problem, *IEEE Trans. Evol. Comput.*, 1(1), 53–66.

Gavalas, D., Kasapakis, V., Konstantopoulos, C., & Pantziou, G. (2015) The eCOMPASS multimodal tourist tour planner, *Expert Syst. Appl.*, 42(21), 7303–7316.

Huang, H. (2013) The Application of Ant Colony Optimization, *J. Theor. Appl. Inf. Technol.*, 52(3) 343–347.

Jakkilinki, R. (2008) A Framework for ontology-based tourism application generator, *Inf. Commun. Technol. Support Tour. Ind.*, 26–49.

Kuntarto & Gunawan, D. (2012) DWIPA search engine : When e-tourism meets the semantic web, *Int. Conf. Adv. Comput. Sci. Inf. Syst.*, 155–169.

Kuntarto, G. P., & Moechtar, F. L. (2017) DWIPA Ontology II : A semi-automatic ontology population process for Bali Tourism based on the ontology population methodology, *Int. Conf. Smart Cities, Autom. Intell. Comput. Syst.*, 42–47.

Kuntarto, G. P., Gunawan, I. P., Moechtar, F. L., & Ahmadin, Y. (2017) DWIPA ontology III : Implementation of ontology method enrichment on tourism domain, *Int. J. Smart Sens. Intell. Syst.*, 10(4), 903–919.

Rusdiana, S. (2017) Designing application of Ant Colony System algorithm for the shortest route of Banda Aceh and Aceh Besar Regency Tourism by using graphical user interface Matlab, 17(2), 2017.

Skripal, B. (2016) Using Ant Colony Optimization for tourist route construction automation, *Proc. 2nd Int. Conf. Appl. Inf. Technol.*, 103–105.

Tang, S. (2010) Tourism domain ontology construction from the unstructured text documents, *Proc. 9th IEEE Int. Conf. Cogn. Informatics, ICCI*, pp. 297–301.

Vathis, N., & Zaroliagis, C. (2016) Scenic Athens : A personalized scenic route planner for tourists, *IEEE Symp. Comput. Commun*, 1151–1156.

Yousef, S., Shuqeir, A., Amjad, T., & Qublan, A. (2014) Hybrid algorithm based on Ant and genetic algorithms for task allocation, Int. J. Comput. Networks Commun., 6(1). 191–202.

*The Social Sciences Empowered – Lumban Gaol, Hutagalung & Fong Peng (eds)*
*© 2020 Taylor & Francis Group, London, ISBN 978-0-367-00090-5*

# Logistication of Russia as basis for further development of economy

V.A. Shumayev
*Russian Transport University, Moscow, Russia*

L.P. Goncharenko
*Plekhanov Russian University of Economics, Moscow, Russia*

A.A. Odintsov
*National Research University Bauman Moscow State Technical University, Moscow, Russia*

A.A. Sazonov
*PRAMO Ltd. Logistics, Moscow, Russia*

S.A. Sybachin
*Plekhanov Russian University of Economics, Moscow, Russia*

ABSTRACT: Further development of the Russian economy, in particular development of northern, Siberian, and Far East territories, and new mineral and raw material facilities, becomes impossible in conditions of the existing transport and logistical system of the country. There is an acute necessity to developing transport, logistics, and infrastructure as a growth basis for the country's economic potential.

The article shows conceptual directions of the Russian logistical system development, forming a common system of commodities flow, creating logistical infrastructure and, based on this, further development of the whole Russian economy.

Development and creation of the common transport and logistical system in Russia is proposed on the basis of a program approach. It is recommended to develop a Federal target program "Logistication of the Russian Federation till 2030," including comprehensive development of all types of transport, logistical infrastructure, transport routes and corresponding productions, service facilities, social infrastructure, creating a Federal logistical center that, which implements common product flow policy and controls common transport and logistical systems, an International transportation center, as well as a section about development and functioning of the North Sea route and adjacent territories.

According to the authors, implementation of the program will create a basis for further development of energetics, industry, and agriculture, in other words, will be a basis for transformation of the economy and country as a whole.

## 1 INTRODUCTION

It is necessary for Russia to access world economic ties for its integration in the world economic space. Similar to integration in international systems of financial circulation (for example, Internet), Russia is facing the necessity to be integrated in the international logistical network of people and commodities flow with respective infrastructural and institutional mechanisms and information subsystems. However the condition of roads, transport, and logistical centers in Russia is still unsatisfactory. For further development of the economy, northern and Siberian regions, as well as for participation in international transportation and economic ties, we need a developed transport and logistical network, meeting requirements of industry, agriculture, and the population.

## 2 METHODS AND THE RESEARCH BASE

A special place in logistics is held by transport, providing physical motion, and the movement of people and commodities across the country.

This part of Russia, capable of linking countries of Europe and Asia, Caucuses, Middle East, Middle Asia, India, China, Korea, and Japan is quite lucrative and efficient for attracting world capital and the development of production. It is situated between European and Asian countries and using its international transport corridors, can contribute to trade and economic, scientific, and cultural exchange between countries and nations. Transport lines are the logistical basis for providing foreign economic ties to Russia and its integration in the global economic system. The economic situation of the country directly depends upon the functioning and development of the transport system.

Development of the transport infrastructure, as well as of neighboring territories, must be aimed at securing a common economic space for the country. This will facilitate the solution of issues facing the Russian and world economy, as well as consolidate economic integration in the country as a whole and in separate regions (Bandy, DeShields, Cunningham, & Britt, 2017).

Export and import ties, including cargo transit to Asia–Europe and South–North are currently limited due to insufficient capacity of port facilities, which must be developed. The North Sea route can become a self-sustained Euro-Asian corridor, linking European Union countries and Asian Pacific region, because the Russian part of the Arctic will become, in the long term, a rich mineral and raw material basis not only for Russia, but for the whole planet.

Highly developed Euro-Asian transport systems can link all types of transport and international transport corridors. Railway, motor road, and air transport should be brought together by integral hubs with common traffic control systems, transport and logistical centers, and multipurpose automated storage terminals with nearby commodity exchanges and wholesale markets (Shumayev, 2014).

One of main directions of transport system development is improving the technical level of transport vehicles and equipment, using modern transportation, management, and information and telecommunication technologies. Transport systems should aim at using Russian transport machinery and equipment. Cargo and passenger transportation in the Russian Federation shall be performed by Russian carriers. A procedure should be adopted whereby Russian carriers get ready to offer transport services to foreign clients.

To optimize the structure of transport vehicles, we should move along the following lines:

In the railway transport – increasing the share of modern, highly efficient locomotives, cargo cars of high-load capacity and reliability; increasing the share of high-speed passenger trains – both intercity and interregional electric-powered, comfortable trains;

In the sea transport – construction of large tankers of Baltmax and Bosphormax series to increase oil export; construction of special gas carriers to bring hydrocarbons from Yamal, Sakhalin, and other regions; construction of a new generation of ferry boats and a new generation of ice breakers for transportation in the Arctic;

In the river transport – building new, more economical types of ships, designed for operation at small depths; wider use of big ships; creating a new generation of comfortable tourist ships; building high technology ports with fast cargo handling;

In the motor road transport – bigger share of large-load capacity auto trailers, larger variety of semi-trailers for intercity and international trips; more special cargo underframes, equipped with self-loading devices and replaceable bodies for retail trade, small business, farmers, and the communal sector. We should also keep in mind the vector of modern electronic automation of transportation, based on Drive-by-Wire control, using ADAS (Advanced Driver Assistance Systems) (Sarkar, Ullah, & Kim, 2017).

In the air transport – creating a new family of economical planes.

Of special importance in Russia is inner-water transport. In the market economy, the use in the world's longest inner-water ways of efficient Russian-made "river–sea" type of ships can yield a number of extra advantages:

Possibility of direct cargo handling–free transportation from a river berth of the Russian enterprise to a foreign inner port or even to a berth of the consignee;

Serious fuel economy by using large ships, as one mid-size river–sea ship has the same cargo capacity as a whole railway train;

Much better impact on the environment as emissions from river transport are lower than those from railway and motor road;

Possibility to service remote and difficult-to-access areas, also during spring floods;

Growth of production and town-building potential around discovered deposits, forests, and river beds;

Development of water tourism and recreation near water sides and thus raising regional revenues;

Expansion of interregional and international economic ties, which is hard to achieve by land routes.

It is worth noting that in countries with rivers, the water transport has priority in the national economic policy, attracting federal and private investment. Without intensive use of inner-water ways of river–sea fleet, the national transport system of Russia is incomplete.

## 3 STUDY RESULTS

A progressive factor in cargo transportation is integration of production and transport processes on the principles of logistics, developing the supporting infrastructure. A basic management target here is implementation of coordinated state policy in promoting transport services, and industrial development of territories, communications, and supporting infrastructure. First of all, we need to take a political decision on development of the transport system, as a basis of our economic development. To implement it we ought to provide project management mechanisms, properly considering Federal, regional, and private interests. The State must initiate business proposals. It would be correct to set up several international consortiums for development and implementation of joint projects (Liu & Ren, 2016).

Economic sanctions made our leadership turn to Russian producers of food and industrial goods. However, this is not an easy task, as we are still lacking the required infrastructure. We are still short of storages, fridges, grain silos, logistical centers, bases, etc. We need large investment to their construction. But before investing we should develop modern projects on their location close to consumers, storage and processing facilities, as well as possible ways of transportation. Thus we come up the with necessity to solve a comprehensive task of Federal scale, as it will require creation of a common logistical network in Russia, in other words, logistication of the whole space (Vecera & Pribyl, 2017).

Setting up entire market infrastructure should be done on the basis of the Federal target program "Logistication of the Russian economy," which is to be developed. It is worth noting that spontaneous non-coordinated creation of separate infrastructure facilities in each region, ignoring technical standards and organizational practice, will result not only in freezing of large investment, but also in aggravating trade and transport isolation of economic structures at the local and regional level from the rest of the world, which is tantamount to limiting Russian business potential and weakening economic (in particular, food) security in Russia.

In perspective, creation of logistical infrastructure should be aimed at formation of the common national cargo grid. Coordination of the regional logistical infrastructure and cargo flow across the country must be done by a Federal logistical control center, which should be formed (Dumeignil, Paul, & Paul, 2017).

Regions are undoubtedly interested in development of the transport system, as it will give them an opportunity to solve complicated problems requiring considerable resources such as roads and construction of industrial and civil facilities. Development of infrastructure demands involvement of qualified personnel of higher, secondary, and special education to fill 2–3 million work places (Chang, Jung, Ross, & Kim, 2017).

Creation and implementation of the transport and technological infrastructure has the following key directions:

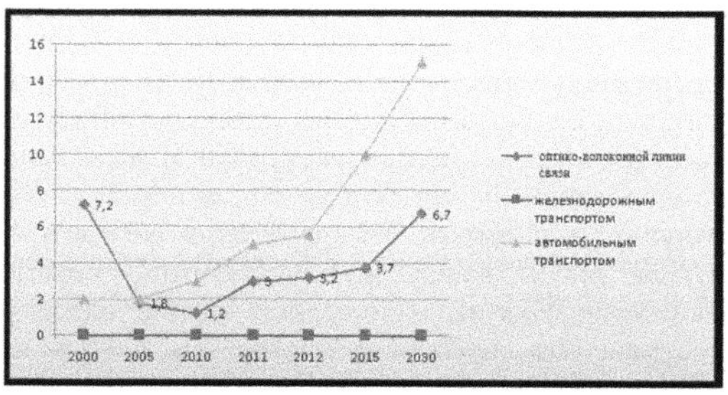


Figure 1. Construction time schedule of railways, motor roads, and fiber-optic lines[1].

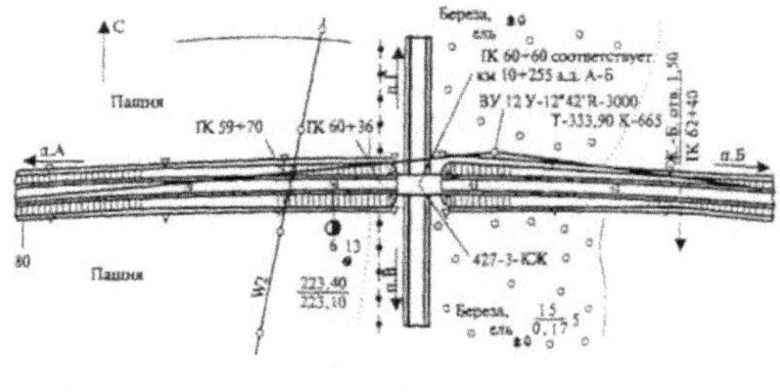

Figure 2. Motor road diagram[2].

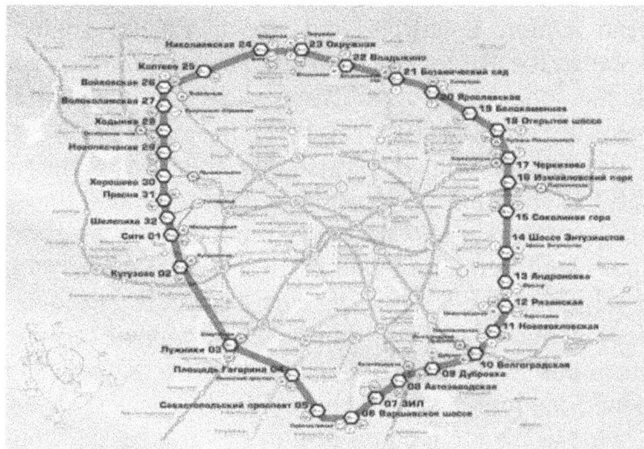

Figure 3. Railway diagram[3].

- Construction time schedule of railways, motor roads, and fiber-optic lines;
- Upgrade of existing railways and motor roads as per new requirements;
- Development of regional transport infrastructure; also in the areas of raw materials mining in the North and East Asia, as well as in regions of ongoing national projects;
- Higher cargo handling capacity and optimal specialization of sea ports to provide for larger volumes of export cargos (grain, coal, containers), transition of Russian export transportation from foreign to Russian ports, and development of high technology river ports with high speed of cargo handling;
- Setting up a network of hub distribution centers of air transportation (hereinafter – hub airports), construction of new airport cargo and passenger terminals, upgrade of runways;
- Development of Russian shipbuilding and modernization of the existing river ships;
- Increasing the capacity of existing and building additional terminals, including container terminals;

[1] http://www.pandia.ru/text/77/306/15457.php
[2] http://www.gosthelp.ru/text/SpravochnikSpravochnayaen.html
[3] http://www.allo495.ru/faq-view-186.html

- Further development of logistical and information technologies of the whole infrastructure of transit transportation to speed up guaranteed delivery of transit cargo, securing its safety and higher quality of service (Odintsov & Shumayev, 2017);
- Modernization of production basis at border passage points of transport vehicles, improving procedures of customs inspection and clearance, adapting it to the world practice;
- Implementation of modern technologies on customs clearance and management of goods transport vehicles, using electronic logistical system (Shumayev, Ilyukhina, & Galushkin, 2015) higher speed of work and efficiency of customs stations.

Proper functioning of transport routes requires various services, which can be rendered in typical multifunctional zones of road maintenance in towns and suburbs, outside of residential complexes, complexes of public catering, trade, hotels, medical and household facilities, auto service, car rental, communication services, etc. It will also be necessary to plan house construction for service personnel and corresponding social infrastructure. The above will be a component in the development of nearby territories and regions (Nardi, Silva, Ribeiro, & Oliveira, 2017).

## 4 CONCLUSIONS AND OFFERS

Transition to modern international logistical technologies means the following for any region (and the country as a whole:

1. Improving commodity flow and reduction of costs due to:
   - Completion of cycle "raw materials mining/procurement – processing/production – sales of ready products",
   - Dramatic reduction of transport, loading/unloading, reloading, expedition, customs and security costs in product price structure,
   - Reduction of redundant middlemen and cargo-free trips, storage reserves, better safety of goods, easy identification of current location and condition of goods at any moment,
   - Higher speed of cargo flow turnover (Vidal Vieira, Ramos Toso, da Silva, & Cabral Ribeiro, 2017).
2. Development of regional economy by means of:
   - Participation of large and mid-size business in development of regional economy;
   - Attracting Russian and foreign cargo flows and investment due to infrastructure readiness;
   - Higher financial turnover, taxes, and revenues in regions.
3. Production development due to:
   - Optimization of logistical operations at industrial enterprises;
   - Using process automation systems (Ni & Chen, 2017).
4. Creating new production facilities and work places in a number of sectors:
   - Processing local raw materials;
   - Construction and engineering infrastructure;
   - Development of transport;
   - Machine and instrument building
   - communication;
   - Light industry and household service;
   - Agricultural and food industry;
   - Market infrastructure (packaging, processing, wholesale–retail);
   - Business service (Akhmetshina, Vagizova, Koczar, & Kseniia, 2018).
5. Urban development, proper planning of cities, towns, and location of production forces:
   - Development of city logistics;
   - Creating modern social and business infrastructure (residential, communication, passenger transport, retail trade, recreation industry, public catering, service).
6. Improving the environment of territories by moving production outside of cities, more cargo transportation by river and railway transport, decentralization of oil and chemical terminals, use of sealed containers (Bukharova, Samusenko, & Semenova, 2017).Construction of transport routes and complex logistical hubs calls for development of industry, agriculture, construction and service sector, in particular:
   - Development of hydraulic, atomic, and fuel-based energy production; construction of wind, solar, sea wave, and other electric facilities;
   - Higher production of oil and gas; deep processing of gas and oil;
   - Export of more processed products;
   - Higher production of coking coal, minerals; development of metallurgy in Siberia, Far East, North of the European Russia;
   - Broader use of modern technologies in machine building, import substitution; development and production of modern, high speed transport facilities;
   - Rational use of wood resources in material production; recovery of wood resources;
   - Rising residential and household construction, production of construction materials;
   - Creating modern cattle breeding and poultry facilities for home consumption and export; higher agricultural yield; creating infrastructure, more and bigger storages for product processing.

It is worth noting that improved work of separate logistical links cannot be a final target of commodity flow optimization and obtaining the final result. Only united stages of the cycle "developing local resources" – "procurement" – "production" – "storage" – "transportation" – "sales" form a complete self-paid-back synergy cycle of high effect.

Social and economic effect of full logistication of the Russian economy is tantamount to such stages of the scientific and technical progress as electrification, industrialization, chemicalization, or computerization of the economy. Logistical networks can be compared to the energetic one, which conveys energy while the logistical one conveys material flows (Ratner & Nizhegorodtsev, 2017).

Setting up a highly technological Euro-Asian transport system may become a mainstream project of innovative development in industry, construction, transport, social and economic development of regions.

Here we should note the importance of a complex approach to creating integration mechanisms of Russian transport and storage facilities in the world standardized logistical system, allowing maximum access to positive foreign experience in this spheree.

It is recommended to develop a Federal target program "Logistication of the Russian Federation till 2030," providing for complex development of logistical infrastructure, covering the whole territory of the country, considering perspective development of discovered deposits, industrial and agricultural zones, transport routes and accompanying productions, service facilities, and social infrastructure. The program should include creation of the main Federal logistical center, conducting common policy of commodity flow and coordinating the work of all logistical centers of Federal and regional scale. The common logistical system of Russia must also become a center of international transportation, coordinating movement along international transport corridors with broad use of respective infrastructure. It would be correct to develop a special section on development and functioning of the North Sea route and adjacent territories. Implementation of this program will not only save Russia from crisis, but will also boost the development of energetics, industry, agriculture, constituting the core of the country's economy.

## REFERENCES


Akhmetshina, A., Vagizova, V., Koczar, J., & Kseniia, T. (2018) Infrastructure investments as a tool for implementing the strategy of sustainable development of regional economic systems. *Advances in Intelligent Systems and Computing* 657, 38–46.

Bandy, N. L., DeShields, S. C., Cunningham, T. D., & Britt, R. C. (2017) Statewide assessment of surgical outcomes and the acute care surgery model. *Journal of Surgical Research*, p. 25.

Bukharova, E. B., Samusenko, S. A., & Semenova, A. R. (2017) Integrated assessment of economic security in a resource region, *Regional Research of Russia*, 237–248.

Chang, J. S. R., Jung, D., Ross, C. L., & Kim, J. (2017) Evaluating the nonuse values of expressways. *Journal Transportmetrica A: Transport Science*, 449–466.

Dumeignil, F., Paul, J.-F., & Paul, S. (2017) Heterogeneous Catalysis with Renewed Attention: Principles, Theories, and Concepts. *Journal of Chemical Education*, 675–689.

Liu, Y., & Ren, Z. (2016) Study on site selection of cold chain logistics in northwest territories. *Journal of Investment and Management*, 17–24.

Nardi, P. C. C., Silva, R. L. M. D., Ribeiro, E. M. S., & Oliveira, S. V. W. B. D. (2017) Proposal for a methodology to monitor sustainability in the production of soft drinks in Ref PET - Elsevier Ltd, 212–218.

Ni, D. C., & Chen, P. (2017) Integration of wired and wireless lighting control system for green energy managment. *International Conference on Mechanical, System and Control Engineering, ICMSC 2017*, 206–210.

Odintsov, A. A., & Shumayev, V. A. (2017) On logistical management in economy. *Design & Technologies*, 58 (100),88–94.

Ratner, S. V., & Nizhegorodtsev, R. M. (2017) Analysis of renewable energy projects' implementation in Russia. *Thermal Engineering*, 429–436.

Sarkar, B., Ullah, M., & Kim, N. (2017) Environmental and economic assessment of closed-loop supply chain with remanufacturing and returnable transport items. *Journal Computers & Industrial Engineering*, 148–163.

Shumayev, V. A. (2014) Logistics in the theory and practice of modern economy management. Monography. Moscow: NOU VPO Vitte Moscow University.

Shumayev, V. A., Ilyukhina, S. S., & Galushkin, A. A. (2015) Innovative directions of market infrastructure and logistics development. *Euro-Asian Judicial Magazine*, 9(88),21–23.

Vecera, R., & Pribyl, O. (2017) Key denominators of success in crowdsourced logistics. *IEEE Publications*, 1–5.

Vidal Vieira, J. G., Ramos Toso, M., da Silva, J. E. A. R., & Cabral Ribeiro, P. C. (2017) An AHP-based framework for logistics operations in distribution centres. *International Journal of Production*, 246–259.

*The Social Sciences Empowered – Lumban Gaol, Hutagalung & Fong Peng (eds)*
*© 2020 Taylor & Francis Group, London, ISBN 978-0-367-00090-5*

# Feedback mechanism of digital technology using strategies of enterprises

L. Goncharenko, E. Sharko, S.A. Sybachin & M. Khachaturyan
*Plekhanov Russian University of Economics, Moscow, Russia*

Z. Prokopenko
*South Federal University, Rostov-on-Don, Russia*

ABSTRACT: The subject of this article is the process of strategy formation based on the use of digital tools. The aim of the article is to develop a mechanism for feedback of the use of digital technologies in the strategy of enterprises. The authors set out the concept and structure of the media space, highlighting the elements of the macro- and mesosphere of the enterprise. In addition, a study was conducted to identify the profile characteristics that consumers present when choosing a company for further interaction. The methodological basis taken was consumer review with Google forms. In the end, the research obtained the following results: before the purchase of products, future customers carefully read the information on the enterprise network, analyze the website, and read the reviews and blogs that influence decision-making. The obtained results can be used in the formation of concept development of the strategy.

## 1 INTRODUCTION

In 2017–2030, the Government of the Russian Federation is realizing the program of digital economy approved by the president Vladimir Putin (Order of the Government of the Russian Federation No. 1632, 2017). A main goal of the program is the creation and development of cyberspace that will facilitate the solution of problems of competitiveness and national safety of the Russian Federation. According to the priority concept during global informatization (authors comment: "informatization is policies and processes aimed at building and developing telecommunications infrastructure that combines geographically distributed information resources") management subjects are obliged to introduce modern digital technologies for realization of their strategies. Today, the borders of the physical and virtual markets are particularly deleted so marketing specialists are bringing as aggressive character to the struggle for consumers.

Digital technologies are designed as much as possible to reduce the interval between information message transmission from the enterprise to the final consumer by means of effective use of media space.

Close to the concept of "media space" are the concepts of "information space," "information and communication space," and "hyper-reality." Toffler brought the concept of infonoosphere (authors comment: "integration of information, digital, technological and intellectual spheres of human activity") – a transitional stage on the way to transformation of society in future civilizations, from the technogenesis (authors comment: "the process of transformation of the environment influenced by various types of human technical activities") before anthropogenesis of a human civilization (Toffler, 2004). Sokhatskaya marks out three options of information space: a cyberspace, an infosphere, and a noosphere; at the same time, the noosphere unites all three spaces (Sokhatskaya, 2012). We suggest considering that the concept "media space" as the closest; it is synonymous with the concept "noosphere."

The wider the market borders, the more difficult it is to follow the back reaction of consumers and to estimate the efficiency of enterprise strategy. The need for formation of a feedback mechanism for the use of digital technologies for the strategy of enterprises has also caused the actuality of this research and problems (Sharko, 2015).

To understand the weak points in the mechanism of interaction between enterprises and final consumers in the information digital environment, it is necessary to designate the conceptual concepts and basic elements of this mechanism.

The research question is to define tools and methods of interaction between clients and enterprises to ensure the implementation of strategic goals in the era of digital economy.

The research goals are:

1) to define the basic components of the feedback mechanism of digital technologies using in strategies of enterprises;
2) to organize research by interviewing potential customers of enterprises in the service sector – users of mobile devices and Internet resources – to identify effective digital technologies for enterprises;

3) to summarize the results and make recommendations for businesses that operate in the era of digital economy.

## 2 BACKGROUND

To ensure the operation of the feedback mechanism of the use of digital technologies in the strategy of enterprises, it is necessary to use the media space. There are various scientific approaches to determination of content of the concept of media space:

- sociological – set of means of social communication (Berger & Lukman, 1966; Bourdieu, 2007; Giddens, 2005);
- psychological – set of means of psychological impact on the personality (Bodriyar, 2006), (McLuhan, 2002);
- journalistic – set of mass media and mass media (Plakhty, 2017), (Matveev, 2012);
- philosophical and cultural and anthropological – media space as a multidimensional phenomenon which is considered at various levels and in different manifestations (Gritsay, 2012; Ilyashenko, 2011; Baran, 2010).

Now there is the sociological approach to definitiing the concept "media space," but, except as explicated above, philosophical, cognitive and neurophilosophical approaches to definition of this concept are more and more actively formed.

On the one hand, the media space is an object of attention of media managers, sociologists, culturologists, psychologists, lawyers etc. On the other hand, the media space itself forms such spheres as social space, culture, the legal field, the right, and so forth. Researchers define the manipulative nature of the processes proceeding in media space, the plasticity of media space (defined by various social and political structures), and its communication with social space. Creating new conditions and new reality influences not only its organizational and communication features, but also the character of public relations, which become prompter, media saturated, varied, and intensive, generating new social, informational, and psychological phenomena: information phobias, information loading and tension, aggression and information crimes. Such an approach gives us an opportunity to consider definition of the concept "media space" from positions of its multidimensionality, synthetic character, and the complexity of modern information and communication space which covers several spheres forming new conditions of public life (Gritsay, 2012):

- a technosphere that is constructed on ICT;
- an infosphere that is based on information and network highways;
- a socioinfosphere that occupies any flows of information structures that organize and operate them, create consumption, and influence the condition of social intelligence.

The structural concept of media space is interpreted by different researchers in different ways. But, in our opinion, all definitions consider its multidimensional superposition or reflection of superposition on three types of spaces: information, virtual, and physical. Nem writes about various principles of approach to a concept of media space (textocentric, structural, territorial, technological, ecological) and suggests three media space dimensions (Nem, 2013):

1) the media space –"live" transmission of information messages that represent both physical and social aspects; these are media images and media texts as a result of media "mapping" of reality;
2) the mediated space – any type of social space that assumes the use of media and/or comes under their considerable influence, that is the sphere of distribution of media technologies that change the nature and a configuration of spaces (it is possible to speak about a mediatization (authors comment: "introduction of media technologies in all spheres of human activity") of public and private space, policy, religion, work, rest, shopping, travel, and so forth);
3) space of media ("old", "new", its convergent forms) – material space of mass media networks and streams (can have both physical, and "virtual" geography).

If the media space corresponds to content, mediated by the sphere of its distribution and consumption, then the media space is channels of production and information transmission; these are media and the system of their interrelations. Borders between these three dimensions are very conditional.

Means of exchanging between enterprises and final consumers are informational digital streams.

Informational digital streams which form media space:

- universality – information doesn't have frontiers or any other borders;
- infinity – information has cumulative properties of continuous accumulation and self-restoration;
- hierarchy – information has hierarchical structure and a tendency to transition to higher levels of hierarchy and to increase in number of new communications;
- the targeting – information is always connected with some material carrier;
- orientation – from an object to the subject;
- continuity;
- uniqueness;
- variety.

Understanding the problems of the immersion of the real world to the digital virtual world demonstrates that the formation of the new basic media space is continuing – the re-inovirus. The three-dimensional world of boundless opportunities and properties continues, and there is a process of its rapid expansion on the basis of creation of new system of reproduction of signals and movements of information streams. At the same time, being the result and the stimulus of processes of globalization, the media space eliminates all traditional restrictions of physical space and cancels any geographical remoteness; further globalization opens unique information, educational, scientific, and cultural opportunities for mankind, forming the person media and a new media civilization (Goncharenko, 2017).

Transformation of media originated when industrial was succeeded by postindustrial (information) society. It became a new stage of social development, and information and computer revolution – the process of informatization of all life spheres (societies in general and the person in general) – its quintessence. The latest technologies considerably transformed material and production and social spheres of mankind. At the same time, there were also considerable changes of processing, production, and information transfer, the approach to its serving.

The computerization, internetization (implementation of Internet technologies into the company's activities), mediation, and virtualization of information as a process of optimization and new submission became driving forces of the designated transformations. For the first time, computer text was used to present information, but later it could present any type of information – text, sound, video, or graphic – with a digital form. The Internet as virtual space together with the ability of computer memory to facilitate instant reproduction, in turn, has promoted a considerable gain of knowledge, enormous expansion of information borders, and the invention of a new information picture of the world. Significantly updated media sphere in general and new media in particular became its main exponents; perhaps the factor for final transition from consumer industrial society to postindustrial, with knowledge and information in the center, and then – society informational (Wertheim, 2010).

The consumer as the most important element of the mechanism, exposed not only to information influence from the enterprises, but also undergoing psychological and emotional influence. The psychoemotional aspect is the main component the influence mechanism and feedback between clients and the enterprises. The essence of the strategy, based on digital communications, consists in formation of the consumer's desire to buy goods or service (works) from concrete enterprise, influencing his psychoemotional state by means of digital technologies in media space. It is important to note that digital methods of impact on the consumer are a little other than classical Internet marketing. As a proof, we will show its comparative characteristics in the form of Table 1.

As we see according to Table 1, the fundamental difference between Internet and digital marketing consists in target audience: Internet marketing is aimed at coverage of Internet users, and digital marketing is aimed at involvement of all possible consumers by means of digital media space.

Table 1. Comparative characteristic of Internet marketing and digital marketing.

| Features | Internet marketing | Digital marketing |
|---|---|---|
| Sphere of impact on target audience | Online sphere | Online+offline |
| Distribution channels | Internet channels | All types of digital channels (Internet, mobile applications, digital advertizing, tablets and video game consoles, digital TV, etc.) |
| Target audience | Everybody who has Internet access | Everybody who has Internet access + attraction offline; audiences on the online market |
| Way of communication with audience | E-mail newsletters, landings, websites, advertizing (search, banner, tabulated, contextual), and so forth | Digital TV, advertizing in online games and mobile applications, messengers, interactive screens, POS terminals, local networks of the large cities |

*Source:* Developed by authors on the basis of Wertheim (2010)

Early in the process, the enterprise has to conduct market researches, including some to reveal consumer needs (questioning, poll, an interview, a hot interview, the analysis of the protocol, and so forth). That is, it "addresses" the consumer and receives back information. On the basis of the obtained and processed data, the relevant marketing activities are developed. It is not necessary that they concern only policies of advancement, but can include any other elements of a marketing mix. However, instruments of marketing communications for the purpose of bringing the consumer information on new offers, sales promotion of already existing goods, formation of loyalty systems, etc., are still used more extensively.

Let us consider the structure of media space for formation of a feedback mechanism of use by digital technology enterprises (Figure 1). There are three layers: layer 1 – enterprise (micro environment), layer 2 – meso environment, layer 3 – macro environment.

Let us consider each element of structure of media space from the point of view of participation and a role in the mechanism.

Public authorities create standard and legal bases, and the normal investment environment for creation and development of the enterprises of a certain sphere. Functions: creative and control.

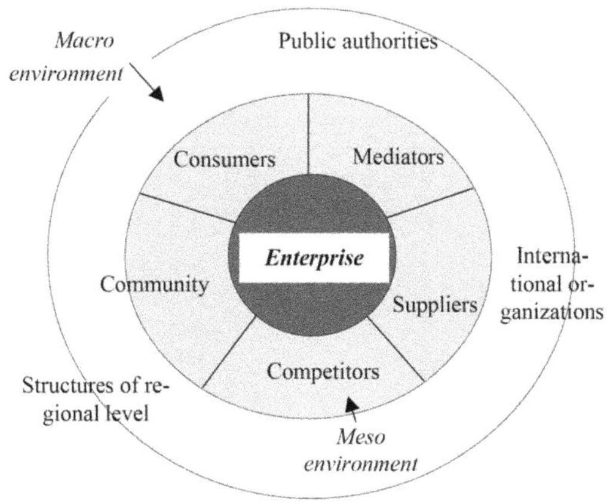


Figure 1. Structure of media space.
*Source:* Developed by authors

Structures of the regional level provide normal development and formation about the investment climate in the region (conditions in which investors want to invest in companies, and new companies – both foreign and domestic, want to carry out its activities in this region) and also respect for the principles of perfect competition in the regional markets. Functions: organizational, control, and planning.

International organizations, ifpresent at the concrete market, have to use acceptable forms and instruments of competitive activity. Functions: differential and structural (they provide different forms of interaction of participants of the market taking into account the international features).

The macro environment is formed by external factors and the company cannot influence it: managers can only develop strategies and policies to reduce the negative or enhance the positive impact of the macro environment. Within the enterprise, media space is being formed, then, to affect the meso environment and connect with it. Next we will illustrate the mechanism of interaction of elements in a meso environment (Figure 2).

## 3 METHODS AND DATASETS

A large number of marketing specialists fail in attempts to advance a brand on the Internet because they mistakenly consider stimulation of growth of sales of goods to be a main objective of social media.

Actually, the purpose of such type of media is the formation of consumer interest; creation of positive brand image; positioning of the company for a priority other than profit; and the satisfaction of consumers needs, gaining their trust and attachment. To determine the features of providing information about the company, which is displayed on the Internet and on mobile devices in modern conditions, studies were conducted among 157 consumers of various services (potential customers of enterprises in the service sector – users of mobile devices and Internet resources – were interviewed anonymously). Google form was created for this and was filled out by 157 respondents. The results have been brought in the tabular Excel form and processed.

Results of a poll of consumers concerning use of the Internet and mobile devices at the choice of the enterprise are given in Table 2.

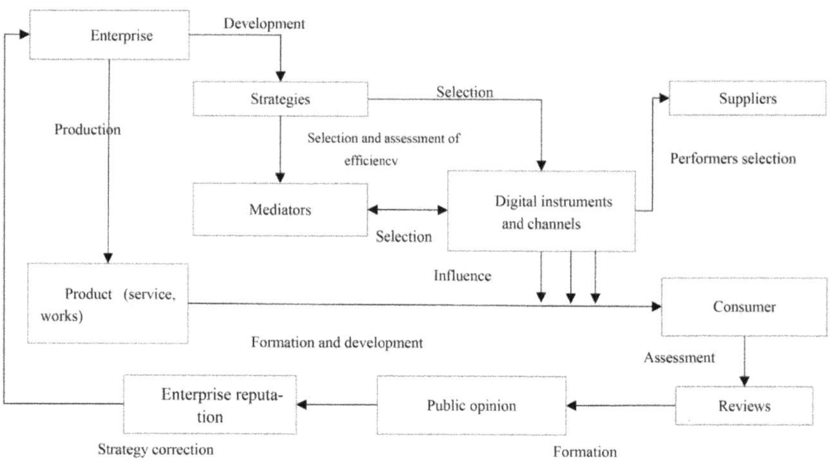


Figure 2. Mechanism of interaction of media space elements in the mesoenvironment.
*Source:* Developed by authors

Table 2. The results of poll of consumers concerning use of the Internet and mobile devices at the choice of the enterprise.

| Questions for consumers | Number of affirmative answers | Affirmative answers, % |
|---|---|---|
| How information about the enterprise is captured on the Internet: | | |
| reviews about the enterprise are read | 122 | 77.7 |
| website of the enterprise is analyzed | 35 | 22.3 |
| How information about the enterprise is perceived in blogs: | | |
| is read and taken into consideration | 109 | 69.4 |
| influences decision-making | 21 | 13.4 |
| isn't used | 27 | 17.2 |
| How information from the sites is read: | | |
| is considered consistently and in detail | 16 | 10.2 |
| only the headlines and inserts are read; if it is necessary the information is considered in detail | 141 | 89.8 |
| What influences information perception of the enterprise's site: | | |
| the availability of video content | 10 | 6.4 |
| the availability of detailed information about the enterprise and its services | 89 | 56.7 |
| the availability of user-friendly functioning | 14 | 8.9 |
| the possibility of interaction with the enterprise | 27 | 17.2 |
| the comfort of information presentation | 17 | 10.8 |
| There is an interest in mobile applications by cooperation: | | |
| there is an opportunity and wish to use | 104 | 66.2 |
| there is a wish but no opportunity | 41 | 26.1 |
| not plan to use | 12 | 7.6 |

*Source:* own elaboration by the authors

## 4 RESULTS

According to the research, the Internet gives new opportunities for consumers to obtain information about the enterprise from various sources, increases information processing speed, and promotes emergence of new requirements to information.

About 80% of respondents answered that before ordering products (services, works), they carefully study information on the enterprise network, analyze its website, and read responses and blogs which influence decision-making. This is connected with the fact that non-material services and the process of their production and receiving synchronize. It is possible to estimate services only after receiving the service or to learn about the assessment of others by means of information on the Internet. The policy of transparency increases chances of the enterprise gaining trust and cooperation with consumers and other stakeholders. Therefore, communication of personnel of the enterprises in a blogosphere has to be the most open (Nikiforova, 2013).

As researches have shown, the relationship to information content on the website and its representation has changed. Most of the respondents answered that they read information on headings and cuttings, and if necessary consider information in details. Existence of video content, convenient functionality, and opportunities of interactive communication considerably influence concentration of consumer attention. Also, interest in mobile applications that allow support communications with the enterprise is considerable, particularly to monitor freight movement and to make necessary corrections. It is connected with distribution of smartphones with new opportunities of communication.

## 5 CONCLUSION

An important answer to the question of the research is the fact that today the media space in the era of the digital economy for enterprises plays an important role in attracting customers and this fact can not be ignored in the development of the strategy for the future. But since there are many tools and methods of information transmission and display in the media space, the conducted research has shown what information and how it should be displayed in the media space in order to strengthen its competitive position in the market.

Also, as a result of the research, an important conclusion was obtained; namely, that it is necessary to regularly interact with users for feedback – to receive feedback and opinions about your image and reputation in order to recognize the risks of losing your competitive positions in time.

Changes in communication processes that contribute to enterprise development in a fast-paced environment connect with the fact that digital communications allow enterprises to be constantly present in the virtual sphere. This full-time availability to consumers and other stakeholders is crucial for creation of strong communications and relations of trust. As clients get access to content on several devices, an integrated approach that increases coherent interaction with consumers and other stakeholders becomes the urgent need.

For the constant involvement of consumers, enterprises need to create convenient functionality and modern design of the website projecting professionalism, convenience, and pleasantness at the consumer. This can happen in the use of certain colors, creation of qualitative content, and the possibility of interactive communication. In terms of marketing of trust, then the question is how, for example, with the help of color to attract consumers to the website, to create feeling of trust, to induce them to certain actions, and to an create emotional connection with a brand of the enterprise.

As a direction of further research in this area, we can highlight the improvement of forms and methods of obtaining and processing information received through the feedback mechanism with consumers by the main points: reliability, efficiency, objectivity.

During the digital era, marketing communications has been becoming more and more integrated due to lifting of previous restrictions that in consequence of fast technological development. Ways of establishing contact with consumers have moved to qualitatively new levels and give the enterprise the chance to introduce communications constantly by means of the website, mobile communication, and social networks. Digital technologies also allow carrying out analytical work, to trace consumers as they use these digital opportunities, and to integrate the obtained data with the existing sets of traditional data from CRM systems.

## ACKNOWLEDGEMENTS

This article was prepared as part of the project section of the government contract as requested by the Ministry of Education and Science of the Russian Federation on the subject formulated as "'Development of Methodological Principles and Organizational Economic Mechanism of Strategic Management of Economic Security in Russia" (Assignment No 26.3913.2017/ПЧ).

## REFERENCES

Baran, G. (2010) Latest Tools of Internet marketing. *Economy of Crimea*, 4(33), 328–331.

Berger, P., &Luckmann, T. (1966) *The Social Construction of Reality. A Treatise on sociology of Knowledge.*

Bodriyar, Zh. (2006) *The Consumer Society: Its Myths and Structures*. Translation ed. by E. Samarskaya. Moscow.

Bourdieu, P. (2007) *Sociology of Social Space*. Translation ed. by N. Shmatko. Moscow: Institute of Experimental Sociology.

Giddens, E. (2005) *Sociology*. 2nd ed. Moscow.

Goncharenko, L., Sybachin, S., & Ionkin, S. (2017) The state of the research complex of Russia as a fundamental basis for the transition to an innovative type of development of the national economy. *National safety/nota bene*, 2, 109–119.

Gritsay, C. (2012) Definition of "media" from the standpoint of interdisciplinary approach. *Bulletin. Kharkov*, 36, 235–243.

Ilyashenko, S. (2011) Modern trends in the use of Internet technologies in marketing. *Marketing and Innovation Management*, 2(4), 64–74.

Matveev M. (2012) Viral marketing in the international market of banking services. *Journal of European Economy*, 11(3) 360–368.

McLuhan, M. (2002) *The Medium is the Massage: An Inventory of Effects*. New York: Ginko Press.

Nem, E. (2013) Media: Main directions of research. *HSE: Business Society. Authority*. No 14. https://www.hse.ru/mag/27364712/2013-14/83292427.html.

Nikiforova, S., & Sovershaeva, S. (2013) The Effectiveness of marketing communications in the digital environment. *Problems of Modern Economy*, 2(46). http://cyberleninka.ru/article/n/effektivnost-marketingovyh-kommunikatsiy-v-didzhital-srede.

Order of the Government of the Russian Federation. "About the approval of the program "Digital economy of the Russian Federation" no. 1632, 2017. http://government.ru/rugovclassifier/614/events.

Plahtiy, I. S. (2017) Development of media in modern society. *Young Scientist*, 17, 204–207.

Sharko, E. R. (2014) Approaches and methods for selecting BSC indicators for the enterprise, taking into account the specifics of the business. *International Scientific Publication Modern Fundamental and Applied Research*, 2(13), 147–152.

Sokhatskaya, A. (2012) Monetization of social media in the global information space. *Journal of European Economy*, 11(1), 104–114.

Toffler, A. (2004) *The Third Wave*. Moscow.

Wertheim, K. (2010) Digital Marketing: *How to Increase Sales Using Social Networks, Blogs, Wikis, Mobile Phones and Other Modern Technologies*. Moscow.

Wipperfurth, A. (2007) *Brand Engagement. How to Make the Buyer Work for the Company*. St. Petersburg.

*The Social Sciences Empowered – Lumban Gaol, Hutagalung & Fong Peng (eds)*
 *ISBN 978-0-367-00090-5*

# A conceptual model of problem-based learning in software modeling course in information technology higher education

Bahar
*Universitas Negeri Jakarta, Jakarta, Indonesia*
*STMIK Banjarbaru, Banjarbaru, Indonesia*

B. Wibawa & R. Situmorang
*Universitas Negeri Jakarta, Jakarta, Indonesia*

ABSTRACT: The Direct-Instruction model which is still widely used in the learning of software modeling in higher education places more emphasis on classroom interactions initiated by the teacher. Generally, it only involves a small portion of student-to-student interaction. The direct learning systems that depend on the reflection ability of instructors only provide few opportunities for students to be actively involved in the learning process. Thus, it is different from software modeling which emphasizes Student-Centered Learning. Consequently, learning becomes ineffective and students cannot reach the minimum competency standard stated in the learning design.

This paper proposes a Problem-Based Learning (PBL) model that is integrated in software modeling learning at three levels: Curriculum level, emphasizing the use of problems as the starting point of student learning; group level, emphasizing collaboration systems (group discussion-based-learning); and student level, stressing the Student-Directed Learning (SDL) system.

Based on the results of the literature review, it can be concluded that the characteristics of the Software Engineering field whose material is continually changing due to changes in end-user needs and changes in technological requirements (hardware and other software), gave rise to new literature. This aligns with the characteristics of the Problem-Based Learning model which emphasizes active and meaningful learning for students. Thus, it can facilitate students to become lifelong learners following the development of Software Engineering field. Therefore, the PBL model can be further examined in terms of its application for learning modeling software.The PBL model can also be used for learning modeling software.

*Keywords*: Conceptual Model, Problem-Based Learning, Software Modeling

## 1 INTRODUCTION

Software Engineering (SE) education (curriculum, outcomes, and delivery) has received considerable attention from the Institute of Electrical and Electronics Engineers (IEEE) and Association for Computing Machinery (ACM) societies. SE is defined as "the application of a systematic, disciplined, quantifiable approach to the development, operation, and maintenance of software." Guiding Principles of IEEE and ACM clearly state that SE education needs to combine computer science foundations with engineering, organizational, teamwork, communication and project management issues. Guidelines for curriculum delivery from the same reference point to the need for a practical plan and team-oriented exercises with a significant capstone project (IEEE Computer Society, 2004).

Software Engineering teaches software development skills for industrial and business purposes, such as for automation in manufacturing activities, sensor technology, interconnection (Prasetyo, 2017), and significant volume data analysis (Pujianto, 2018), which underlies the emergence of a 4.0 industrial revolution era.

Software Engineering has characteristics that are different from other fields of science. Software Engineering is a science whose technology is continuously changing, due to the following factors. First, software is related to the advances in technology in relation to other hardware and software that demand compatibility, as well as changes in the company's business strategy and the behavior of end-user demand support changes to the need for software. It requires high adaptation skills in the development process (Juman, 2018). Second, the development of Information and Communication Technology operating by software has caused some difficulties during the software development process. The large volume of code (program) and the complexity of the system architecture cause small software development systems to be individually abandoned, and transferred to technology development collectively and collaboratively. Third, software design and programming technology also develop dynamically. Various development models and software programming languages have encouraged the emergence of a variety of new libraries in the software development environment, thus demanding lifelong learning adaptation (Kazimov, 2017).

Software Engineering applies a systematic and disciplined approach to the development, operation, and maintenance. Thus, software developers need technical and social skills in their work. Traditional teaching methods that focus on lectures and tutorials for Software Engineering students are not enough to develop the skills to solve real-world problems. In the traditional learning approach, most students must complete their assignments, and this is contrary to professional practice in the collaborative environment of software development teams (Krusche, 2017; Oliveira, 2013). Su, Jodis, and Zhang (in Fertalj, 2013) suggest that the importance of providing students with real problems and a real teamwork environment must be a concern in software learning in college.

It is essential for software developers to have the ability to understand user needs, cooperate in teams, and participate in the overall software engineering development process. Related to this, students feel that software engineering is a complex science because, in addition to mastering technical skills, it also requires social aspects. Žagar (2008) suggests that software engineers are required to have soft skills, including the ability to present knowledge, learn from independent sources, and listen to what others have to say.

The characteristics possessed by software as described previously, require an approach in the learning process of Software Engineering which emphasizes active learning, leads to the concept of lifelong learning, emphasizes the resolution of problems in the real world and unstructured problems, and emphasizes collaboration skills. It is contrary to the learning system that is widely used in learning Software Engineering in Higher Education, namely the direct learning model (direct instruction).

Direct learning model (direct instruction) is a learning model that uses teacher demonstrations and explanations combined with training and student feedback to help students obtain real knowledge and skills needed for further learning (Wahono, 2012). Direct-Learning Model emphasizes class interaction which is primarily initiated by the teacher and generally does not involve student interaction (Ewing, 2011). The direct learning model also highlights the achievement of basic/declarative and procedural pedagogical goals (Ekasari, 2016). The direct learning system that depends on the reflection ability of instructors only provides little opportunity for students to be actively involved in the learning process, causing learning to be ineffective.

We have conducted preliminary research at several Information Technology campuses in Indonesia using questionnaires distributed to teachers and students. We also make observations in classes that hold lectures on Software Modeling. The results of the preliminary study indicate that the direct learning method in the modeling software course results in a passive learning process, the low critical nature of students, low understanding of students in certain parts of the competence taught, and dull atmosphere of learning in the classroom. Due to such factors, students cannot reach the minimum competency standards planned in the learning design.

Problem-Based Learning (PBL) is one of the learning models that can be applied to engineer disciplines. PBL is an educational strategy that encourages students to get to know how to learn and cooperate in groups to find solutions of problems in real situations. Problem simulation is used to activate students' curiosity before studying a subject. PBL makes students think critically and analytically to get and use the source of learning appropriately. PBL correlates with cognitive functions that contain various kinds of thinking activities in the learning stages, including utilization of existing knowledge (prior knowledge), reorganizing new knowledge in cognitive structures, processes of analysis and synthesis, structuring and developing ideas, and problem-solving (Dickens, 2009).

This paper proposes a conceptual PBL model implemented in Software Engineering life cycle phase (the analysis phase and design phase), which Pressman called "the System Modeling phase" (Pressman, 2002).

## 2 PROBLEM-BASED LEARNING

Several learning models have been classified into discovery learning systems, such as Collaborative-Learning, Case-based Learning, Inquiry-based Learning, Cooperative-Learning, and Problem-Based Learning (Schneider, 2014).

Discovery-Learning models have pedagogical objectives that emphasize deep learning, meta-cognitive skills (developing problem-solving skills and creativity), and active involvement of students. According to van Joolingen in Schneider, discovery learning is a type of education where students build their knowledge by experimenting on specific domains and deducing rules from the results of their experiments. The basic idea of this kind of learning is because students can design their operations in a particular area and infer the rules from the domain itself. Because of this productive activity, it is assumed that they will understand the field at a higher level than when the information needed is only presented by the teacher or the expository learning environment (Schneider, 2014). Borthwick and Jones in Daniel (2014) suggest that in discovery learning, students learn to recognize problems, characterize what solutions will be seen, find relevant information, develop solution strategies, and implement selected policies. The idea is that students are more likely to remember the concepts they found themselves.

According to Borthwick and Jones (in Daniel, 2014), Discovery-Based Learning has several advantages. Discovery learning supports the active involvement of students in the learning process, fosters students' curiosity, and enables the development of lifelong learning skills because most adults learn in work and real-life situations. Discovery-Learning which personalizes the learning experience is very motivating because it provides an opportunity to experiment with finding something for its own sake and builds prior knowledge/experience. Discovery learning develops a sense of independence and student autonomy, fosters a sense of responsibility for students' mistakes, and records learning experiences. Discovery learning also develops problem-solving skills, creative skills, and improves cognitive meta-skills (including some cognitive strategies at a higher level) that are useful in lifelong learning.

One learning model included in the category of discovery learning models is Problem-Based-Learning model (PBL). Christopher (2018) argues that the philosophy behind Problem-Based Learning is that knowledge and skills are obtained by following sequences of contextual problems along with learning material and support from instructors. The point lies in collaboration and personal reflection as one of its primary goals to encourage independent learners. As a form of active learning, Problem-Based Learning promotes the construction of knowledge and integrates classroom learning with real-life dynamics, where students learn how to develop flexible knowledge, and practical problem-solving skills, gain intrinsic motivation, exchange ideas and collaborate. Through collaboration, students can identify what they already know, what they need to know, and the ways and sources of information they need, to succeed in solving problems. Teachers facilitate students by supporting, guiding and monitoring the development of their students, building their confidence, encouraging them to participate and expand their understanding actively.

PBL gives students the opportunity to learn to solve problems that are relevant to the context, which is appropriate to skills in the workplace, develop initiatives, develop the ability to find and use appropriate resources for problem-solving, teach effective self-motivating and self-learning skills, and proactive thinking to continue learning. PBL monitors continuously and assesses the adequacy of their knowledge, their problem-solving skills, and trains to practice critical thinking. PBL gives students the opportunity to collaborate efficiently in a group, and make students skilled in communication.

Many PBL models have been implemented, such as in Nilson (2010), Daniel (20140), and Richard (2010). Apart from different PBL implementations, several key elements are not unusual in their application according to Barrows in Loyens et al., which can be categorized into three levels: (1) curriculum level; (2) group level, and (3) individual student level (Loyens, 2011). The main features of PBL at the curriculum level are the use of problems as a starting point for student learning, small group collaboration as an accessible curriculum format, learning objectives that guide students to independent learning, and learning resources used during independent learning. At the group level, there are two key elements, namely the teacher (tutor) and students, and the learning problems that are generated. The teacher facilitates and stimulates group discussion group discussion, ensures that content issues considered in depth, and evaluates the contributions of group members to solve problems. After discussing the initial question, students produce learning problems that guide their independent learning activities. Learning problems must cover the main material underlying the problem and what is related to the learning objectives. There are three characteristics of learning problems; they contain keywords that are useful for literature search, concise, and clear. PBL features at the individual level of students are independent learning (Student-Directed Learning/SDL). The sense of responsibility is a critical element of SDL. SDL can be a student characteristic and design feature of a learning environment. So, the situation must provide a space for student autonomy. In the learning process, students provided with more learning resources, so that there is more freedom of choice. It can increase higher achievement values. There is also a tendency for more time to study independently. SDL is also determined by curriculum elements, such as discussions in tutorial groups, content tested and learning objectives.

## 3 RELATED WORK

There have been many studies on engineering and learning software courses. Some reports are presented in this section.

Kamthan proposes a systematic approach towards integrating the Social Web environment in software engineering education, as a medium of interaction between teachers and students, and between students both inside and outside the classroom, to foster a social environment in software engineering learning. The web-based-learning environment has more advantages because it can offer learning tools that can be personalized inside and outside the classroom (Kamthan, 2009). Unfortunately, the study does not present a step-by-step problem-solving process in collaborative learning concepts.

Murphy, Phung and Kaiser reported the results of their investigation on The COMS W4156 Advanced Software Engineering course at Columbia University. The the study only focused on programming topics (programming work in pairs) with distance learning methods. The results of the investigation also do not clearly explain the PBL method used, and only present a number of challenges faced when teaching extreme Programming (XP) with a distance learning system using PBL for students who are not physically in one area, such as students' aversion to aspects of XP and difficulties in scheduling (Murphi, Phung & Kaiser, 2008)

Collazos at.al. introduced A Collaborative and Distributed Learning Activity Applied to Software Engineering Courses. Students collaborated with other team members from different campuses, facilitated by groups of facilitators coming from various colleges in a distance learning system using information technology. There are five levels of learning introduced in this system, namely: (1) Preparation, (2) Lecture, (3) Local Practice, (4) Distributed Practice and (5) Evaluation. At the preparation level, the instructors will collaborate to determine the parameters of the activity. At the lecture level, the instructor plays a role of providing knowledge about a particular topic to all students participating in the event (similar to traditional classes), where the instructor teaches the problem to a group of local participant students and remotely face to face mediated computer video conferences. Local practice was carried out after a face-to-face lecture session, involving a team of 3-4 students from the same institution to complete the task collaboratively, helping students to assimilate the knowledge conveyed by experts during the lecture. The distributed practice must be started by the session after the local tradition, involving 1 or 2 weeks of the students work, including students from different institutions using technology. In the final stage, the learning process carried out by students was evaluating collaboratively with the active participation of all instructors from different campuses (Collazos at.al, 2010). This model applies to learn systems independently in collaboration and does not emphasize individual independent learning.

Richardson and Delaney reported the results of their investigation regarding the use of Problem Based Learning in Software Engineering Classroom. The study focused on the stages of system analysis and design in the cycle of Software Engineering. The application of PBL in the learning cycle begins with presenting authentic problems at the beginning of the learning sequence with the aim of developing a set of skills to not only solve problems in the future but also to bring these skills to the workplace. Furthermore, interaction is more focused on collaborative learning systems through small group discussions in the classroom to share knowledge from their work experience in the school. The instructor, in this case, does not act as a facilitator as in the normal PBL environment. So, student independence is reduced. There is little individual learning done by students in this model. Learning is only done in the classroom. In addition to the socialization skills obtained from the collaborative learning system, individual learning skills to find learning resources outside the school are also needed so that students are accustomed to being learners at all time. This investigation also does not present the impact of PBL on improving students' competence in learning (Richardson & Delaney, 2009).

Shim introduced collaborative-learning in Software Engineering by implementing PBL strategies to help students understand the importance of social aspects and a systematic framework for improving teamwork skills. The concept offered is to integrate 12 stages of PBL into the four main phases of the Pressman model (Communication, Planning, Modeling & Construction, and Deployment). However, the 12 PBL phases are not implemented in any particular stage in the Pressman model. The initial three steps of PBL (Introducing PBL, introducing role instructors/learners, and identifying the problem) were integrated into the initial phase of RPL (communication phase). The next three PBL phases (establishing activity plans, learning goals settings, assigning tasks) were combined into the second phase of RPL (Planning). The next PBL Phase (scanning and collecting information, analyzing the collection information, and deriving a solution) was integrated into the third stage of RPL (modeling & construction). The final three phases of PBL (presenting a solution, evaluating the process, promoting adaptation to similar environments) were integrated into the fourth step of deployment (Shim, 2009).

## 4 THE PBL CONCEPT PROPOSED IN LEARNING SOFTWARE MODELING

Activities of Software System Analysis and Modeling in Software Engineering disciplines consist of System Analysis and System Design phases (Jogianto, 2011):

The System Analysis Phase includes 7 activities, namely: (1) Studying the existing system; (2) Assessing the problems (weaknesses) that exist in the system; (3) Analyzing issues in the order; (4) Formulating alternative system requirements; (5) Proposing alternative solutions to existing problems; (6) Analysis Modeling; and (7) Documentation (reporting) of analysis results. The 1st to 5th activities in the System Analysis phase are field investigation activities, which require interaction with users (software users) to understand their real problems and needs. This activity is a very complex. Users often do not understand the software specifications needed, so system analysts must refer to the theoretical concepts of general requirements in conducting a study of User needs. Thus, analysts are required to provide an extensive insight into the system developed. The development of information technology (software and hardware) and changes in user requirements are very fast, which also become a challenge in this phase. Compatibility between software built with software or other device and synchronization of needs and proposals must always be maintained so that the developed software still has high usability, and this requires analysts to learn all the time.

Modeling (design) phase of the system, consisting of 3 activities, namely: (1) Conceptual diagram of software; (2) design of software procedural logic; and (3) software physical design. Each activity phase ends with activities to document the results of actions. The documentation of each event is inseparable from the definitions attached to the term "Software," namely computer programs along with accompanying documents (planning documents, analysis and design documents, and operation and maintenance documents).

Learning modeling software courses are done by combining learning theory concepts and practice concepts of software modeling development projects. In general, the stages in learning modeling software courses consist of: (1) Explaining learning objectives; (2) General description of the software concept and its development method; (3) Explanation of the concepts of problem analysis and system requirements. Theoretically, students were introduced to unstructured problems that underlie the development of software based on user needs. Various alternative solutions to the issues underlying the development of software were also discussed at this stage. The discussion focused on in-depth case studies on the technique of defining problems, problem-solving techniques and formulating solutions/proposed needs. For field investigations relating to the real conditions of the theoretical concepts discussed, study groups were formed for each particular type of problem explained. At this stage, students were also directed to learn independently about a case/theme being discussed, and facilitated with links to learning resources; (4) Field Investigation. At this stage, the study groups had formulated a specific case for further practice of investigating problems, needs in real terms in the field, and the study results reflected in small groups; (5) Software Modeling. At this stage, the students were taught how to make conceptual-models, logical-models, and physical-models by referring to the results of system requirements analysis using system modeling tools. Then, the students practiced the system model building; (6) Prepare reports on System Analysis and Modeling Projects; (7) Project presentations.

In stages (1) and (2) adequate learning references are needed from various sources to deepen students' insight before conducting investigations in the field (real world). Discussions between groups were also performed by involving the instructor as the director to increase the understanding level of the particular problem. Stage (3) focuses on students to understand the condition of the real field (real world) through interactions with various parties outside the class. Step (4) requires technical skills in mastering modeling theory and software modeling tools. At this stage, collaborative learning is needed between students and instructors directly (face to face), given the nature of this material is technical material that cannot be easily understood if it is only independent learning. Stage (5), (6) and (7) are the stages of reflection on the results of the study as outlined in the project report presented through presentations and group discussions. The instructor acts as a director while assessing the activeness and critical level of students in participating in a series of lecture activities (the final assessment outside the assessments takes the form of formative tests and quizzes at the end of each learning session in specific competency fields). To accommodate the learning characteristics of the Software Modeling disciplines, the PBL model aligned with the learning structure or phase in learning the scientific subject of software modeling.

As Loyens stated (2011), the PBL model includes 3 main aspects, namely: (1) at the Curriculum-level, the use of problems as a starting point for student learning; (2) at the Group-level, there is a collaboration (group discussion); and (3) at the Student-level, independent learning (Student-Directed Learning/SDL). Integration PBL levels with the software modeling phase in software modeling course presented in Figure 1.

In Figure 1, there are 3 phases in the software modeling proposed learning concept model, which are the results of alignment between the main PBL components (level) and the software modeling phase:

1) Stage 1 (initial phase)
   The initial period is a learning phase with a face-to-face system in class, with presentation and brainsstorming methods. In the software modeling course, this initial phase was used for introducing lectures that explain the basic concepts of analysis and modeling of software and the objectives and objectives of analysis and modeling. Direct face-to-face models in class (direct learning) were used to convey conceptual and structured matters. In PBL, this phase is at the curriculum level, namely the use or explanation of the problem as the starting point of student learning.

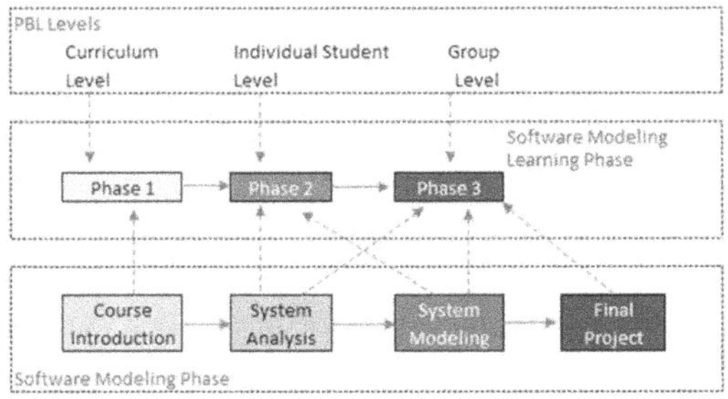


Figure 1. The concept of PBL levels integration with the software modeling phase in software modeling course.

2) Stage 2 (mid phase)
   Phase 2 is the phase of independent learning by individual learning participants. The second phase in PBL is called the period at the Student level, which is independent learning (Student-Directed Learning/SDL). In learning Modeling Software, this activity is identical to some of the events of problem analysis and system requirements and software system modeling activities, which emphasizes the activeness of students in independently reviewing things that are not structural and real in the field related to the concept of software requirements model well developed.
3) Stage 3 (final stage)
   Stage 3 is a collaborative learning phase involving students in small groups, both in class and outside the classroom. The third phase in PBL is called the period at the Group level, characterized by collaboration (group discussion), while in learning modeling software, this activity is identical to several events in system analysis and system modeling which emphasize the partnership of students (small groups) in formulating and synthesizing the results of studies previously carried out independently. This 3rd phase in learning software modeling is also identical to the stages of software modeling preparation that will be accounted for in groups in the general discussion forum in the class, as a medium to reflect on all conceptual issues that have been studied both individually and in small groups.

## 5 DISCUSSION

The software must be able to solve problems faced by users and must be able to accommodate changes in user needs, changes in technology requirements for hardware and other software. Software modeling is no longer implemented only in the classroom with conventional learning techniques (which focuses on the teacher). Learning modeling software must emphasize students' active involvement inside and outside the school, learning that leads to the concept of independent learning, lifelong learning, learning that emphasizes solving real-world problems and unstructured problems, and emphasizes the skills of collaboration under the control and direction of the instructor.

The Problem-Based Learning (PBL) model is one of the teaching methodologies that involves students in complex and challenging problems, working collaboratively to find solutions. PBL connects students' knowledge with real-world problems. Motivation to solve problems is the basis for forming students' motivation to learn. PBL creates a learning environment where instructors train students thinking and guide students to critically ask questions, and facilitate students' level of understanding to be more profound (Torp & Sange in Fatima & Abdullah). The characteristics of PBL are taken into account in this paper to propose a PBL conceptual model in learning modeling software.

Phase 1 (in Figure 1) is a face-to-face activity directly in class to form unstructured problems, to challenge and arouse student motivation in developing solutions related to issues in the stages of analyzing the system requirements of the Software. The steps of analyzing user needs are very crucial stages in the success of Software Engineering. Sommerville (2001) argues that good software is software that fits the user's needs. As explained by Savery (2006), the problem simulations used in Problem-Based Learning must be ill-structured and allow free inquiry. Therefore, in this study, we adapted this concept to the initial phase for the PBL model that we proposed. Stage 1 can also be used by teachers to emphasize learning rules to stay within the planned PBL corridor because some studies found that students sometimes take shortcuts in the PBL process (West, 2013). Phase 1 ensures that PBL runs under the control and direction of the instructor.

Phase 2 and phase 3 (in Figure 1) the PBL-based conceptual modeling learning model proposed in this paper are the phases of student self-learning, both individually and collaboratively. The learning phase independently trains students to interact with the real world, which is very difficult to be presented at face-to -face meetings in class. With the support of information systems technology devices, students can easily obtain and build their knowledge through various learning resources available outside the classroom. Students can also collaborate with other students in reflecting on the experience acquired by each.

Phase 2 and Phase 3 (in Figure 1) intended for the stages of user requirements analysis and modeling software in the cycle of software engineering. The activity of analyzing user needs which is the basis for building a software model is a very complicated activity. The requirements of end-user software that is continually changing, as well as the development of hardware technology related to software developed, has triggered the emergence of new literature on software development. System analysts are required to have extensive insight into the software system produced. System analysts also need to always update their knowledge related to software development theory.

Phase 3 (in Figure 1) PBL-based modeling software learning is also intended for collaborative learning stages through the software modeling final project. Collaborative skills are needed in developing large-scale software. The large volume of code (program) and the complexity that emerges in software system architecture cause small-scale software development systems to be individually abandoned and transferred to collective and collaborative software development technologies, known as Pair Programming (Kazimov, 2017).

In line with the statement of Fatima and Abdullah (2013) that in today's dynamically changing Information Technology environment, it is no longer sufficient to only teach students the concepts of System Analysis and Design in classroom meetings, because the content and context will change drastically before students reach their first job after graduation. Students need to be lifelong learners in this software development era. They need cognitive skills, critical thinking, and problem-solving practice skills. A very effective way to achieve this is to make students work on a problem that requires them to practice these skills, utilize knowledge and obtain new information as needed. The PBL model proposed in phase 2 and phase 3 learning modeling software is designed to facilitate students to become independent learners and learners of all time.

To overcome the changing trends and needs of the software market, in addition to knowledge about the subject and content of Software Engineering material, practical knowledge and experience in software development are more critical. In this context, universities which are suppliers of skilled labor in the field of Software Engineering must consider specific learning models in the curriculum that will make students become lifelong learners.

Problem-Based Learning is a learning model that is expected to be an alternative solution that can be applied in the development of learning design modeling software at higher dducation. However, some researchers argue that problem-based learning is only suitable as a teaching strategy at the end of the learning process, whereas for the beginning and middle part, it does not function, such as Kirschner in Schneider (2014) who introduces problem-based learning directed at the final stages of instructional design.

Open debate about the implementation of the PBL model in learning also deals with: (1) how many guidelines are needed; (2) the potential to confuse the learner if there is no initial framework available; (3) weak students have a tendency to lag behind, while teachers fail to detect situations that require more remediation or encouragement. Some studies also found that the PBL model was most successful when students had the prerequisite knowledge and had undergone some structured experience in Schneider (2014).

Related to Students' skill in collaborating in teams, because students have different backgrounds and abilities, the teacher's efforts to form a balanced group can face obstacles. Students may eventually join team members that they do not want. So, intensive collaboration between team members sometimes does not occur. However, teachers usually choose leaders for each team to promote strong teamwork. This action slightly reduces conflicts among team members (Zu, 2007).

Begel (2008) asserts that the use of Pair Programming (a technique in which two programmers work together on one computer), can streamline costs, scheduling problems, and personality conflicts. However, Pair Programming has fewer excess bugs and overall results in higher program code quality.

Long discussions that lead to different views regarding the application of the PBL concept in learning provide opportunities for further research to find PBL models that are genuinely effective in learning Software Engineering.

## 6 CONCLUSION

The field of Software Engineering always raises a variety of new libraries because the material is always changing following changes in end-user needs and changes in the requirements of other hardware and software technology. That is in line with the PBL Characteristics which emphasize active and meaningful learning for students, facilitating students to become lifelong learners following the development of the field of Software Engineering.

The stages of learning Software Engineering (Software Analysis & Modeling) require first explanation and mentoring of students by Teachers in the classroom, independent assignments for students, and collaborative teamwork in completing the Final Software Project in line with the three levels of activity in the model PBL. The three levels: the level of the curriculum initiated by the teacher, the level of the individual who emphasizes the independence of students individually, and the level of the group that highlight the progressive attitude of students in collaborating with fellow students.

The PBL model can be further tested to be applied in the learning design plan of software modeling. However, some previous findings explained that the PBL model is only suitable as a teaching strategy at the end of the learning process, while for the beginning and middle part, it does not work. So, it is necessary to emphasize careful assessment of the strategies for implementing PBL at the beginning and middle of the learning process.

## REFERENCES


Begel, A., & Nagappan, N. 2008. "Pair Programming": What's in it for Me? *Proceedings of the Second ACM-IEEE international symposium on Empirical software engineering and measurement*, 2008: 120–128.

Collazos, C.A., at al. 2010. CODILLA: A Collaborative and Distributed Learning Activity Applied to Software Engineering Courses in Latin American Universities, *Proceedings of 6th International Conference on Collaborative Computing: Networking, Applications and Work sharing (CollaborateCom, 2010)*, Chicago, IL, USA.

Cristopher, P. 2018. Instructional Design Models and Theories: Problem-Based-Learning. https://elearningindustry.com/problem-based-learning. 2 April 2018.

Daniel K.S. 2014. *Educational Design Models*, (Edutec WiKi): 94.

Dickens, J. & Arlett, C. 2009. *A Hand Book for Teaching and Learning in Higher Education*. New York: Routledge.

Ekasari, R.R., Gunawan & Hairunnisyah, S. 2016. The Effect of Direct Learning Model Assisted by Laboratory Media on the Creativity of Physics of High School Students. *Journal Pendidikan Fisika dan Teknologi*. II (3): 106–110.
Ewing, B. 2011. Direct Instruction in Mathematics: Issues for Schools with High Indigenous Enrolments: A Literature Review. *Australian Journal of Teacher Education*. 36(5): 64–91.
Fatima, S. & Abdullah, S. 2013. Improving Teaching Methodology in System Analysis and Design Using PBL for ABET. *International Journal of Modern Education and Computer Science*. 7: 60–68.
Fertalj, K., Milasinovic, B., & Kosovic, I.N. 2013. Problems and Experiences Width Students Projects Based on Real-Word Problems. *Journal Technics Technologies Education Management*. 8(1): 176–186.
IEE Computer Society. 2004. "Software Engineering 2004": Curriculum Guidelines for Undergraduate Degree Programs in Software Engineering, The Joint Task Force on Computing Curricula, IEEE Computer Society, Association for Computing Machinery, August 2004 http://sites.computer.org/ccse/SE2004Volume.pdf
Jogianto, H.M. 2011. *Information System Analysis and Design*. Yogyakarta: Andi Offset.
Juman, K.K. 2018. Development of Multimedia-Based Software Engineering Learning Models in Higher Education. *Journal Forum Ilmiah*. 15(1): 84–94.
Kamthan, P. 2009. A Methodology for Integrating the Social Web Environment in Software Engineering Education, International Journal of Information and Communication Technology Education (IJICTE), 5(2): 21–35.
Kazimov, T. H. & Bayramova, T.A. 2017. Problems of Teaching Software Engineering in Azerbaijan, *Problems of Information Society*, (1): 92–96.
Krusche, S., Frankenberg, N.V., & Afifi, S. 2017. Experiences of a Software Engineering Course based on Interactive Learning, Semantic Schooler, https://www.semanticscholar.org: 32–40.
Linda, B. N. 2010. Teaching at Its Best – *A Research-Based Resource for College Instructor*, Third Edition, San Francisco: Jossey Bass: 187–188.
Loyens, S.M. 2011. Problem-based Learning, in K. R. Harris, S. Graham & T. Urdan (Eds.), *APA Educational Psychology Handbook*: Vol. 2. Washington: American Psychological Association: 6.
Murphy, C., Phung, D. & Kaiser, G. 2008. A distance learning approach to teaching extreme programming, *ITiCSE Proceedings of the 13th annual conference on Innovation and technology in computer science education*, Madrid, Spain, 2008: 199–203.
Oliveira, A.M.C.A., Santos, S.C.D., & Garcia, V.C. 2013. PBL in Teaching Computing. *Proceeding of Frontiers in Education Conference*. October 2013: 1–6.
Prasetyo, H. & Sutopo, W. 2017. Scientific Development of Industrial Engineering Towards an Industrial Age 4.0. *Proceeding of National Conference IDEC 2017*. Mei 2017: 488–496.
Pressman, R.S. 2002. Software Engineering. McGraw-Hill Book. Yogyakarta: Andi Offset.
Pujianto, A., Mulyati, A., & Novaria, R. 2018. Utilization of Big Data and Privacy Protection of Consumers in the Digital Economy Era, *Journal BIJAK*, 15(2): 127–137.
Richard, I.A. & Kilcher, A. 2010. *Teaching for Student Learning*, New York: Routledge: 330–334.
Richardson, I. & Delaney, Y. 2009. Problem Based Learning in the Software Engineering Classroom, *IEEE Proceedings of* 22nd Conference on Software Engineering Education and Training, Hyderabad, Andhra Pradesh, India, 2009.
Shim, C.Y., Choi, M. & Kim, J.Y. 2009. Promoting Collaborative Learning in Software Engineering by Adapting the PBL Strategy, *World Academy of Science, Engineering, and Technology*, (29): 1161–1164.
Su, H., Jodis, S., & Zhang, H. 2007. Providing an integrated software development environment for undergraduate software engineering courses, *Journal of Computing Sciences in Colleges*, 23(2): 143–149.
Summerville, I. 2001. *Software Engineering*, 6th Edition, United Kingdom: Pearson: 8.
Wahono, S., 2012, *Learning Strategies and Models*, Jakarta: Permata Puri Media, pp.363.
West, R.E., Williams, G.S. & Williams, D.D. 2013. Improving Problem-based Learning in Creative Communities Through Effective Group Evaluation, *The Interdisciplinary Journal of Problem-based-Learning*, 7(2): 102–127.
Žagar, M., Bosniü, I., & Orliü, M. 2008. Enhancing software engineering education: a creative approach, *Proceedings of the 2008 international workshop on Software Engineering in east and south Europe*, 2008: 51–58.

*The Social Sciences Empowered – Lumban Gaol, Hutagalung & Fong Peng (eds)*
*© 2020 Taylor & Francis Group, London, ISBN 978-0-367-00090-5*

# Relationship between parenting styles and social skills among preschoolers

Jamaliah Kassim* & Fonny Dameaty Hutagalung*
*Department of Educational Psychology and Counselling, Faculty of Education, University of Malaya, Malaysia*

Chin Hai Leng
*Department of Curriculum & Instructional Technology, Faculty of Education, University of Malaya, Malaysia*

Abd Razak Zakaria
*Department of Educational Foundation & Humanities, Faculty of Education, University of Malaya, Malaysia*

ABSTRACT: Research about preschoolers' social skills is an early effort to detect deficiencies before any intervention can be implemented. Parents affect the development of children, including social skills, through parenting styles. However, there was little information regarding preschoolers' level of social skills and its association with parenting styles in Malaysia. Thus, this study aims to examine the level of preschoolers' social skills and how it relates to parenting styles. The respondents consisted of 339 preschoolers in Hulu Langat, Selangor, and their parents. The stratified random sampling method was used and the questionnaires are distributed to parents through the children. Descriptive and correlational analyses were conducted. The findings show that the level of social skills of preschool children in Malaysia is average. The authoritative style is most frequently adopted compared to authoritarian and permissive styles. There is a significant positive correlation between paternal and maternal authoritative style with social skills. Only maternal authoritarian style has a significant negative relationship with preschoolers' social skills. These findings highlight the importance of enhancing preschoolers' social skills by intervention in the problematic dimension of social skills as a first step.

*Keywords*: parenting styles, social skills, preschool children

## 1 INTRODUCTION

The development of good social skills is the foundation of children's success at home, schools, neighborhoods, their careers and adult lives (Merrell, 2003; Guralnick, 2010). Children with good social skills are adaptable to the environment and receive peer acceptance. However, children who lack social skills often face peer rejection and have behavioral problems like aggressive behavior, hyperactivity, anxiety, and over dependence on others (Hay, Hudson, & Liang, 2010; Poulou, 2014). Furthermore, social skills deficits during childhood will lead to a range of high-risk behavioral problems at later ages, such as crime, school dropout, job dismissal, and unemployment (Hawkins, Kosterman, Catalano, Hill, & Abbott, 2008; Niepel, 2010).

Childhood is a great time to develop a variety of sociocultural skills including social skills. At home, the learning of social skills happens naturally (Adams, Womack, Shatzer, & Caldarella, 2010) and involves a variety of situations (Schulze, 1989). Children's social skills are built through their interactions with parents, siblings, and communities around them. The involvement of parents in nourishing children's social skills is very important as parents have more time to spend with children and have a diverse environment to apply social skills naturally (at home and community), as well as parents can expose the children to social interaction in various groups other than school (Schulze, 1989).

According to the report from Department of Social Welfare Malaysia (JKKM) in 2016, in Malaysia, there were 4,886 children involved in crimes, and 1,582 juveniles and 456 children that were beyond control. These statistics of social problems of children and adolescents in Malaysia may give a sign of social skills deficits since many studies found the association between the the deficiency of social skills and behavioral problems (Hosokawa & Katsura, 2017). It is more effective if the problems are detected at an early age and an intervention implemented from early childhood (January, Casey, & Paulson, 2011).

*Corresponding authors: jamaliah9815@gmail.com & fonny@um.edu.my

## 2 CONCEPT OF SOCIAL SKILLS AND PARENTING STYLES

Social skills are certain behaviors that lead to social reactions desired by individuals who initiate the interaction (Merrell, 2003). Social skills include interpersonal behavior, self-reliance, academic-related skills, assertive behaviors, peer-acceptance, and communication skills (Gresham & Elliott, 1987). In this study, social skills refer to the behavior of preschool children encompassing three aspects of social skills, namely social cooperation, social interaction, and social independence (Merrell, 2003). Social cooperation refers to cooperative and self-restraints behaviors and the ability to follow adults' instruction; social interaction describes social initiation behaviors including the ability to obtain and sustain friendship and acceptance; social cooperation reflects behaviors in gaining independence within the peer group, such as confidence, assertiveness, and the ability to separate from caregivers.

Parenting style is the parent's attitude towards the child through communication with them and the emotional climate that results when the behaviors of the parents are expressed (Darling & Steinberg, 1993). Baumrind (1971) had introduced three types of parenting styles; namely authoritative, authoritarian, and permissive styles. These three parenting styles differed based on two main dimensions: level of demandingness (control, supervision, and maturity) and responsiveness (warmth, acceptance, and involvement) (Maccoby & Martin, 1983). In this study, parenting style refers to the three parenting styles – namely authoritative, authoritarian, and permissive – practiced by parents.

## 3 STATEMENT OF THE PROBLEMS

A lack of social skills since childhood will affect the later age groups. Children with social skills deficits risk a lack of non-cognitive skills when they become adults (Jones, Greenberg, & Crowley, 2015). Adults with a lack of social skills have difficulty finding jobs or carrying out their duties. This can be seen in Malaysian studies where employers say that graduates lack social skills thus it affects self and company performance (Yahaya, Rasul, & Yasin, 2016). Steps to improve the social skills of these graduates should be implemented immediately and as early as possible. Interventions from early childhood are a great step because early interventions will save time and are more effective (January et al., 2011).

Information regarding the level of social skills of children in Malaysia should be reviewed before any intervention is implemented, however, such information is very limited. Lack of information regarding the children's social skills means early interventions to improve social skills can not be effectively implemented. Therefore, this study will examine the level of social skills at the early stage of childhood in Malaysia.

Studies have found that Malaysian parents adopt more authoritarian styles than parents in western countries (Winskel, Salehuddin, & Stanbury, 2013). The adoption of authoritarian parenting styles among Asian societies, including Malaysia, is related to the collective culture which values social support and believes in the concept of filial piety. This collective culture appreciates the culture of mutual assistance, adherence, and dependency amongst others while the concept of filial piety emphasizes respect and concern for the older family institution (Ho, 1994). However, lately, exposure to international culture has influenced parent–child interaction and the Western culture of individualism is being slowly accepted (Keshavarz & Baharudin, 2009). Looking at movement in the practice of parenting style, researchers want to identify the kind of parenting style practiced by parents of preschoolers in Malaysia.

Many studies have found that authoritative parenting styles are associated with good social skills in children (Hosokawa & Katsura, 2017) such as assertiveness, self-reliance and self-control (Baumrind & Black, 1967), prosocial behavior (Patterson, Debaryshe, & Ramsey, 1990), and communication skills (Connell & Prinz, 2002). The permissive and authoritarian parenting styles are associated with low social skills (Abu Taleb, 2013) and negative behavior (Rinaldi & Howe, 2012; Roopnarine, Krishnakumar, Narine, Logie, & Lape, 2014). However, some studies in different cultures and ethnic groups showed inconsistent findings. Bartholomeu, Montiel, Fiamenghi, & Machado's (2016) study on the Brazilian population showed that negative parenting styles expected the assertiveness and confidence of children. A study conducted by Rao & Pearson (2003) found that the relationship of authoritative styles with the child's peer relationships was not significant. Moreover, a study in the Middle-East found no significant relationship between authoritarian style and negative child behavior (Kol, 2016).

In addition, some studies have found different effects between paternal and maternal parenting styles on the development of children. In the study of Roopnarine, Krishnakumar, Metindogan, & Evans (2006) on Caribbean immigrants, the paternal authoritative style has greater influence than the maternal authoritative style on children's social behaviors such as social skills and self-confidence. Talib, Mohamad, & Mamat's (2011) studies found that paternal authoritarian style had a greater effect on children's behavior problems than maternal authoritarian style. Given that there is still a lack of study on parenting style and social skills of preschool children Malaysia, researchers are keen to know the relationship between paternal and maternal parenting styles and social skills of preschoolers.

This study will look for answers to some of the issues raised by studying the level of social skills of preschool children and the type of parenting style practiced. Knowledge of the relationship between parenting style and social skills can provide basic information and recommendations to the parents, teachers, and the Ministry of Education Malaysia to improve the social skills of preschool children in preparation to providing individuals with competency in 21st century skills. Hence, this study will answer some of the following questions:

1. What is the level of social skills of children?
2. What type of parenting style does a child's parents practice?
3. Is there a significant relationship between parenting style and the level of social skills of children?

## 4 THEORY OF THE STUDY

The Social Learning Theory by Bandura (1977) states that an individual learns a behavior based on imitation of an individual's or significant model's behavior. Based on this theory, as a significant individual in a child's life, parents become models of behavior and child referents. Children not only learn the behavior of parents while parents interact with other people but also learn the behaviors of parents in interacting with them, which are called parenting styles. Children learn the behavior by looking at behavior of the parents in controlling and managing children and providing responses, support, and intimacy to children. Behaviors of children are also built from the response of the parenting style they received; for example, children become obedient because of the fear of harsh punishment imposed by parents (Zhou et al., 2002), or children show empathetic behavior because they feel safe when their parents are always supportive and responsive to their feelings (Zhou et al., 2002).

Based on this theory, imitation is a way of learning abstract information as well as learning concrete skills (Bandura, 1977). Social skills can be learned as well as other skills. The learning process of this skill will go through the various interactions between individuals and the environment, either directly or indirectly, involving observation and imitation of the behavior of other people or significant models. Parents are their children's primary references and there are more parent–child interactions, thus children learn a lot of social skills from parents, indirectly or directly, through observation and imitation.

## 5 METHODOLOGY OF STUDY

### 5.1 *Research instrument*

A set of questionnaires was distributed to parents through their preschool children. The questionnaires consists of Part A, Part B, Part C, and Part D. Part A contains demographic information of children and their parents. Part B contains a social skills instrument named Preschool and Kindergarten Behavior Scales: - Second Edition (PKBS-2) from Merrell (2002). PKBS-2 contains two parts: which are social skills and behavior problems part. In this study, researchers use only the social skills questionnaire. This questionnaire has been translated and repeatedly translated by two translator experts in the study conducted by Zaharah (2015). The questionnaire consisted of 34 items comprising three dimensions, namely social cooperation, social interaction and social independence. All items use a four-point Likert Scale: i.e. (0) Never, (1) Rarely, (2) Occasionally, and (3) Always. Part C and D contains parenting style questionnaires translated and adapted by Nellie (2012) from the Parenting Styles Dimension Questionnaire (PSDQ) (Robinson, Mandleco, Olsen, & Hart, 2001). The questionnaire has been gone through the translation process using the back-translation method as suggested by Brislin (1986). The process of translation and review of this questionnaire has been done by professors who are proficient in both languages and have a background in social science education (Nellie, 2012). The PSDQ questionnaire consists of three dimensions (authoritative, authoritarian, and permissive style) and contains questions using the a five-point Likert scale. The score for each dimension of parenting style are obtained separately.

A pilot study was conducted with 31 questionnaires distributed to parents. The Cronbach's alpha value obtained is 0.89 for social skills, 0.92 for authoritative style, 0.78 for authoritarian style, and 0.60 for permissive style. Hence, all subscale instruments have fulfilled the reliability test (Mohd Majid, 2000) and can be used in the study.

### 5.2 *Respondents of the study*

The stratified random sampling method was used in this study. In the first stage, the location of the study has been divided based on location (urban and rural). Next, preschool selection is made randomly from the group. A total of 339 preschool children in Hulu Langat District were selected.

Respondents of the study consisted of preschoolers aged five and six and their respective parents. A total of 231 children (68.1%) from urban preschoolers and 108 (31.9%) from rural preschools. The total number of boys is 191 (56.3%) while the girls are 148 (43.7%). Based on the number of siblings, 108 children (31.9%) have fewer than three siblings, 225 children (66.4%) have three to six siblings and only 6 (1.8%) have more than six siblings.

Based on mothers' demography, 145 (42.8%) are under 35 years old, 181 (53.4%) are 35–45 years old and 13 (3.8%) are above 45 years old. Most of the fathers were 35–45 years old (214 people, 63.1%), 90 (26.5%) were under 35 years old, and 35 (10.3%) are over 45 years old. Generally, the families involved in this study were of low socioeconomic class. According to the level of education, 64.5% of mothers and 71.9% of fathers had SPM education and below. Nearly 52.8% of mothers did not work and stated no income. Among the fathers, 4.1% earned monthly income less than RM1,000, 70.8% earned between RM1000 to RM3000, and 25.1% earned more than RM3000.

## 6 FINDINGS

### 6.1 *Research question 1: What is the level of social skills of a child?*

Descriptive analysis is used to determine the level of social skills of the children, interpreted based on the functional level proposed by Merrell (1994) as in Table 1:

Based on Table 2, the findings show that the level of social skills of preschool children is average. The three dimension of social skills (social cooperation, social interaction, and social independence) are also at average level.

### 6.2 *Research question 2: What kind of parenting style is practiced by the parents of the preschool children?*

Descriptive analysis is also used to determine the type of parenting style practiced by the parents. The findings are shown in the following Table 3:

Based on the maternal parenting style, it is found that the practice of authoritative style is most frequent, followed by permissive style and authoritarian style. Similar findings are also found in the paternal parenting style: the authoritative styles is most frequently practiced by the father, followed by permissive style and authoritarian style. In addition, the findings of the study also showed that the maternal authoritative styles (M = 4.03, SP = 0.52) was higher than the paternal authoritative styles (M = 3.82, SP = 0.63). The findings also showed that the authoritarian styles was the least practiced by both mothers (M = 2.27, SP = 0.55) and fathers (M = 2.19, SP = 0.56).

Table 1. Functional level of social skills scale.

| | Range score | | | |
|---|---|---|---|---|
| Level | SS | SC | SInt | SInd |
| 1. High functioning | 95–102 | 35–36 | 31–33 | 32–33 |
| 2. Average | 76–94 | 25–34 | 23–30 | 21–31 |
| 3. Moderate deficit | 59–75 | 20–24 | 15–22 | 20–25 |
| 4. Significant deficit | 0–58 | 0–19 | 0–14 | 0–19 |

SS = Social skills, SC = Social Cooperation, SInt = Social Interaction, SInd = Social independence

Table 2. Level of social skills children.

| Social skills | Min | SP |
|---|---|---|
| Social Cooperation | 28.31 | 4.88 |
| Social Interaction | 24.91 | 4.77 |
| Social Independence | 27.49 | 3.58 |
| Total Social Skills | 80.71 | 11.63 |

Table 3. Type of parenting parenting style.

| | Parenting Style | Mean | SP |
|---|---|---|---|
| Mother | Authoritative | 4.03 | 0.52 |
| | Authoritarian | 2.27 | 0.55 |
| | Permissive | 2.73 | 0.56 |
| Father | Authoritative | 3.82 | 0.63 |
| | Authoritarian | 2.19 | 0.56 |
| | Permissive | 2.70 | 0.56 |

### 6.3 *Research question 3: Is there a significant relationship between parenting style and the level of social skills of children?*

Correlation between parenting styles and social skills, were examined. Based on Table 4, the findings show that there is a significant positive relationship between the maternal authoritative style ($r = 0.47$; $p < 0.01$) and the paternal authoritative style ($r = 0.35$; $p < 0.01$) with children's social skills. Only the maternal authoritarian style shows a significant negative relationship ($r = 0.11$; $p < 0.05$) with the social skills of children.

## 7 DISCUSSION & INTERPRETATION

The findings show that preschool children have average levels of social skills, including all three dimensions (social cooperation, social interaction, and social independence). This finding also found that the level of children's social interaction was lowest compared to other dimensions and approaching moderate deficit. Major, Seabra-Santos, & Albuquerque (2017) also found that Portuguese children's social skills were at average level and social interaction at the lowest. The low level of social interaction shows that preschool children face problems in interacting with peers and adults. This finding shows the need for social skills, especially the social interaction, of preschool children to be improved.

This study also found that the adoption of authoritative style was the most frequent for both mother and father. The authoritarian is least practiced by both parents. Carapito, Ribeiro, Pereira, & Roberto (2018) also found that the authoritative style was most practiced while the authoritarian style was least practiced. The study of Iranian parents found that the authoritative styles was most practiced but the permissive style was least practiced (Alizadeh, Abu Talib, Abdullah, & Mansor, 2011). Overall, the findings were consistent with other studies that stated that the authoritative style was the most adopted by parents.

Table 4. Correlation analysis between parenting styles and social skills.

| Constructs | | Authoritative | Authoritarian | Permissive |
|---|---|---|---|---|
| Social Skills (Mother) | correlation (r) | 0.49** | –0.13* | 0.08 |
| | Sig | 0.00 | 0.01 | 0.15 |
| Social Skills (Father) | correlation (r) | 0.35** | –0.08 | –0.04 |
| | Sig | 0.00 | 0.14 | 0.48 |

** $p < 0.01$, * $p < 0.05$

The correlation between authoritative style of mother and father and social skills was positive and significant at ($r = 0.49$; $p < 0.01$) and ($r = 0.35$; $p < 0.01$) respectively while only the authoritarian style of the mother was negatively correlated and significant with children's social skills. This means that children's social skills are effected by both paternal and maternal authoritative style and the maternal authoritarian style. This finding is in line with most studies that state the authoritative styles is positively and authoritarian style is negatively associated with the social skills of children (Baumrind & Black, 1967; Connell & Prinz, 2002; Hosokawa & Katsura, 2017; Patterson et al., 1990). In addition, this finding found that the maternal authoritative style has a stronger relationship with children's social skills than paternal authoritative style, and the authoritarian style of the mother alone was negatively associated with the social skills of the child. The findings of this study differ from Roopnarine et al., (2006), where paternal parenting style affects children's social behavior compared to maternal parenting style. This is likely because mothers in this study are mostly housewives so they have more time interacting with their children thus their parenting style might affect the development of the children.

## 8 CONCLUSION

The research questions had been answered. This study found that Malaysian preschoolers had average social skills. The level of social skills of these children also has a relationship with paternal and maternal authoritative and the maternal authoritarian style. The permissive style has no significant relationship with the social skills of children. The study also found that the of authoritative style is most frequently adopted by Malaysian parents followed by permissive and authoritarian styles. From the study, the parents from collectivist cultures did not necessarily predominantly adopt authoritarian styles. Future studies through qualitative research such as in-depth interviews and focus groups may give more insights into the parenting styles adopted and the social skills of preschoolers.

## REFERENCES


Abu Taleb, T. F. (2013). Parenting styles and children's social skills as perceived by Jordanian mothers of preschool children. *Early Child Development and Care, 183*(11), 1646–1660. https://doi.org/10.1080/03004430.2012.744988

Adams, M. B., Womack, S. A., Shatzer, R. H., & Caldarella, P. (2010). Parent involvement in school-wide social skills instruction: Perceptions of a Home Note Program. *Education, 130*(3), 513–528.

Alizadeh, S., Abu Talib, M., Abdullah, R., & Mansor, M. (2011). Relationship between parenting style and children's behavior problems. *Asian Social Science, 7*(12), 195–200. https://doi.org/10.5539/ass.v7n12p195

Bandura, A. (1977). *Social Learning Theory.* Englewood Cliffs, New Jersey: Prentice Hall.

Bartholomeu, D., Montiel, J. M., Fiamenghi, G. A., & Machado, A. A. (2016). Predictive power of parenting styles on childrens social skills: A Brazilian sample. *SAGE Open*, 1–7. https://doi.org/10.1177/2158244016638393

Baumrind, D. (1971). Current patterns of parental authority. *Developmental Psychology Monograph, 4*(1), 1–103.

Baumrind, D., & Black, A. E. (1967). Socialization practices associated with dimensions of competence in preschool boys and girls. *Child Development, 38*(2), 291–327.

Brislin, R. W. (1986). The wording and translation of research instruments. *Field Methods in Cross-Cultural Research*, 137–164. https://doi.org/10.1037/0022-3514.90.4.644

Carapito, E., Ribeiro, M. T., Pereira, A. I., & Roberto, M. S. (2018). Parenting stress and preschoolers' socio-emotional adjustment: the mediating role of parenting styles in parent–child dyads. *Journal of Family Studies*, 1–17. https://doi.org/10.1080/13229400.2018.1442737

Connell, C. M., & Prinz, R. J. (2002). The impact of childcare and parent–child interactions on school readiness and social skills development for low-income African American children. *Journal of School Psychology, 40* (2), 177–193.

Darling, N., & Steinberg, L. (1993). Parenting style as context : An integrative model. *Psychological Bulletin, 113* (3), 487–496.

Gresham, F. M., & Elliott, S. N. (1987). The relationship between adaptive behavior and social skills: Issues in definition and assessment. *The Journal of Special Education, 21*(1), 167–181. https://doi.org/10.1177/002246698702100115

Guralnick, M. J. (2010). Early intervention approaches to enhance the peer-related social competence of young children with developmental delays: A historical perspective. *Infants and Young Children,*

*23*(2), 73–83. https://doi.org/10.1097/IYC.0b013e3181d22e14.Early
Hawkins, J. D., Kosterman, R., Catalano, R. F., Hill, K. G., & Abbott, R. D. (2008). Effects of social development intervention in childhood 15 years later. *Arch Pediatr Adolesc Med*, *162*(12), 1133–1141.
Hay, D. F., Hudson, K., & Liang, W. (2010). Links between preschool children's prosocial skills and aggressive conduct problems: The contribution of ADHD symptoms. *Early Childhood Research Quarterly*, *25*(4), 493–501. https://doi.org/10.1016/j.ecresq.2010.01.003
Ho, D. (1994). Filial piety, authoritarian moralism and cognitive conservatism in Chinese societies. *Genetic, Social, and General Psychology Monographs*, *120*(3), 349–365.
Hosokawa, R., & Katsura, T. (2017). Marital relationship, parenting practices, and social skills development in preschool children. *Child and Adolescent Psychiatry and Mental Health*, *11*(1), 4–11. https://doi.org/10.1186/s13034-016-0139-y
January, A. M., Casey, R. J., & Paulson, D. (2011). A meta-analysis of classroom-wide interventions to build social skills: Do they work? *School Psychology Review*. https://doi.org/0279-6015
Jones, D. E., Greenberg, M., & Crowley, M. (2015). Early social-emotional functioning and public health: The relationship between kindergarten social competence and future wellness. *American Journal of Public Health*, *105*(11), 2283–2290. https://doi.org/10.2105/AJPH.2015.302630
Kagitcibasi, C. (1996). *Family and Human Development across Cultures*. Mahwah, NJ: Lawrence Erlbaum.
Keshavarz, S., & Baharudin, R. (2009). Parenting style in a collectivist culture of parenting style in a collectivist culture of Malaysia. *European Journal of Social Sciences*, *10*(1), 66–73.
Kol, S. (2016). The effects of the parenting styles on social skills of children aged 5-6. *Malaysian Online Journal of Educational Sciences*, *4*(2), 49–58.
Maccoby, E. E., & Martin, J. A. (1983). Socialization in the context of the family: Parent-child interaction. In P. H. Mussen (Ed.), *Handbook of Child Psychology* (Vol. 4, p. 1983). New York: Wiley.
Major, S., Seabra-Santos, M. J., & Albuquerque, C. P. (2017). Validating the Preschool and Kindergarten Behavior Scales-2: Preschoolers with autism spectrum disorders. *Research in Developmental Disabilities*, *65* (June2016), 86–96. https://doi.org/10.1016/j.ridd.2017.04.008
Merrell, K. W. (1994). *Prechool and Kindergarten Behavior Scales*. Brandon, Vermont: Clinical Psychology Publishing Company. https://doi.org/10.1017/CBO9781107415324.004
Merrell, K. W. (2002). *Preschool and Kindergarten Behavior Scales: Second Edition*. Austin: TX:PRO-ED.
Merrell, K. W. (2003). *Behavioral, Social, and Emotional Assessment of Children and Adolescents*. New Jersey: Lawrence Erlbaum Associates, Inc.
Nellie. (2012). *Hubungan di antara Cara Gaya Keibubapaan dan Temperamen dengan Masalah Tingkah Laku Kanak-kanak*. Universiti Kebangsaan Malaysia.
Niepel, V. (2010). *The Importance of Cognitive and Social Skills for the Duration of Unemployment*. Stockholm.
Patterson, G. R., Debaryshe, B., & Ramsey, E. (1990). A developmental perspective on antisocial behavior. *American Psychologist*, *44*, 329–335. https://doi.org/10.1037/0003-066X.44.2.329
Poulou, M. S. (2014). The effects on students' emotional and behavioural difficulties of teacher–student interactions, students' social skills and classroom context. *British Educational Research Journal*, 1–19. https://doi.org/10.1002/berj.3131
Rao, N., & Pearson, E. (2003). Socialization goals, parenting practices, and peer competence in Chinese and English preschoolers. *Early Child Development and Care*, *173*(1), 131–146. https://doi.org/10.1002/icd.341
Rinaldi, C. M., & Howe, N. (2012). Mothers' and fathers' parenting styles and associations with toddlers' externalizing, internalizing, and adaptive behaviors. *Early Childhood Research Quarterly*, *27*(2), 266–273. https://doi.org/10.1016/j.ecresq.2011.08.001
Robinson, C. C., Mandleco, B., Olsen, S., & Hart, C. H. (2001). The Parenting Styles and Dimensions Questionnaire (PSDQ). In & G. W. H. B. F. Perlmutter, J. Touliatos (Ed.), *Handbook of Family Measurement Techniques: Vol. 3. Instruments & Index* (pp. 319–321). Thousand Oaks: Sage.
Roopnarine, J. L., Krishnakumar, A., Metindogan, A., & Evans, M. (2006). Links between parenting styles, parent-child academic interaction, parent-school interaction, and early academic skills and social behaviors in young children of English-speaking Caribbean immigrants. *Early Childhood Research Quarterly*, *21* (2), 238–252. https://doi.org/10.1016/j.ecresq.2006.04.007
Roopnarine, J. L., Krishnakumar, A., Narine, L., Logie, C., & Lape, M. E. (2014). Relationships between parenting practices and preschoolers' social skills in African, Indo, and mixed-ethnic families in Trinidad and Tobago: The mediating role of ethnic socialization. *Journal of Cross-Cultural Psychology*, *45*(3), 362–380. https://doi.org/10.1177/0022022113509884
Schulze, K. A. (1989). Coincidental teaching: Parents promoting social skills at home. *Teaching Exceptional Children*, *21*(2), 24–27. http://eric.ed.gov/ERICWebPortal/recordDetail?accno=EJ382263
Talib, J., Mohamad, Z., & Mamat, M. (2011). *Effects of Parenting Style on Children Development*, *1*(2), 14–35.
Winskel, H., Salehuddin, K., & Stanbury, J. (2013). Developmental milestone expectations, parenting styles and self-construal in Malaysian and Australian caregivers. *Kajian Malaysia*, *31*(1), 19–35.
Yahaya, N., Rasul, M. S., & Yasin, R. M. (2016). Social skills and social values in Malaysian dual training system apprenticeship. In *International Conference on Innovation in Engineering and Vocational Education (ICIEVE 2015)* (pp. 170–175). Atlantis Press.
Zaharah Osman. (2015). Kesan Implementasi Modul Kompetensi Sosioemosi terhadap Kemahiran Sosial Kanak-Kanak Prasekolah. *Journal of Social Science*, *1*, 75–82.
Zhou, Q., Eisenberg, N., Losoya, S. H., Fabes, R. A., Reiser, M., Guthrie, I. K., ... Shepard, S. A. (2002). The relations of parental warmth and positive expressiveness to children's empathy-related responding and social functioning: A longitudinal study. *Child Development*, *73*(3), 893–915.

*The Social Sciences Empowered – Lumban Gaol, Hutagalung & Fong Peng (eds)*
 *ISBN 978-0-367-00090-5*

# Transition into childcare: From secure base to separation individuation: A case study in a Malaysian childcare centre

Karen Chiew Jhong Tan* & Fonny Demeaty Hutagalung
*Department of Psychology & Counselling, Faculty of Education, University of Malaya, Malaysia*

ABSTRACT: The usage of institutional childcare is a necessity these days to enable modern mothers in society to get back to work. Hence the transition of children from home care to childcare centers has become an inevitable process. The process of transition is often overlooked but it is one of the most crucial process for a child, especially if the child is younger than 24 months. Managing an appropriate transition process will assist in helping a child in terms of social emotional development, attachment and a future development of a healthy self-individuation of the child. John Bowlby and Mary Ainsworth's work on attachment and Secure Base, alongside Margaret Mahler's theory on separation-individuation is applied within the framework of this study. This study is aimed to observe the current transition process of a child from in-home care into an institutional childcare, with hopes that the findings will later lead to the formation of a transition handbook to aid stakeholders in the industry, particularly parents and child caregivers to manage the transition process in a way that will benefit the child's development and wellbeing. A qualitative approach utilizing the method of observation, interview and journaling is undertaken for this research, based heavily upon the theory and concept of secure base and separation individuation.

*Keywords*: Childcare, Transition, Secure Base, Separation Individuation, Attachment

## 1 INTRODUCTION

A statistics report issued by the National Family and planning board in 1999 showed that less than 20% of Malaysian parents took care of children on their own. By the time the child is 3 months of age, alternative and multiple care is usually being sought after (Chiam, 2008). In today's modern world, women are looking out to prioritize their career. A recent survey conducted in 2014 by the Malaysia Statistic department and LPPKN indicated that more than 50% of women between the ages of 15 and 44 years old are in the workforce. Recent studies have also indicated that 75% of mothers return to work 2 months after post-delivery and outsource care of their child to individuals like maids, hired nannies and relatives (Sulaiman, Liamputtong, & Amir, 2017). Laws in Malaysia regarding maternity leave provide for only 2 months mandatory maternity leave. All these facts indicate that women, mothers have to not only place their child in alternative care, but they are also forced by circumstances to separate from their child at an early age.

Being separated from parent or a main caregiver is a daunting issue for every child. While attachment and eventually separation is inevitable to every human being, early childhood development theorists have posited theories and research that children should not be separated from their mothers or main caregivers at an early age, or at least not before the child turns 2 (Ainsworth, 1979; John Bowlby, 1983; Pandey & Patnaik, 2014; Van Der Horst, Leroy, & Van Der Veer, 2008; Waters, Merrick, Treboux, Crowell, & Albersheim, 2000) Child-mother attachment or child primary caregiver attachment is important because this initial attachment forms the base of the child's social emotional growth.

For a child and the parent, the need of using a childcare facility would bring both parties to experience a form of separation anxiety. The child's first couple of weeks in the childcare center can be proved to be rather stressful not only for the child but also the parents (Borelli, Rasmussen, John, West, & Piacentini, 2015). However, despite knowing that the transition will be stressful and quite possibly a traumatic event for a small child, parents place much emphasis on selecting the right childcare or nursery institutions based on facility, schedules and programmes, instead of looking into the child's readiness to separate. The child's readiness to separate and detach from the primary caregiver is important to the development of the child's social

*Corresponding authors: ktcj85@siswa365.um.edu.my

emotional wellbeing and should be done right from the start (Howard, Martin, Berlin, & Brooks-Gunn, 2011; Kobak, 1999). Past research has indicated that premature separation are usually the cause of attachment disorders (Cassidy, 2008; Waters et al., 2000). Hence, how can daily separation and transition for a child take place whilst minimizing the trauma of separation and transition? The objective of this study is to investigate the current separation and transition process of a child into a Malaysian childcare based on the attachment theory and the theory of separation-individuation.

## 2 CONCEPT OF TRANSITION

Transition is a process whereby one encounters change from one state or a condition to another. Transition in early years is a child's experience as he or she makes the move from home into an institutional settings, i.e. preschool or a school. This experience will involve leaving behind familiar routines, people, roles and relationship to start a "new life" in a place that is likely to include differences in routines, people, roles and relationships (Hussain, Hanin. et.al, 2018). For a child in early years, the transition they experience is often a dramatic event that has elements of challenges and learning opportunities which are crucial to the child's social emotional and intellectual development. In most research in regard to transition, transition is viewed as a one-off experience. (e.g. first day in primary one). However, transition in the early years is a process involving multiple continuities and discontinuities of experience (Vogler, P., Crivello, G., and Woodhead, 2008).

In the process of transition, it is not only the role of the child making the adjustment but also the parent, mainly the primary caregiver, and in the Asian context, adjustment may also have to be made by the multiple caregivers that were involved in caring for the child in the household prior to sending the child into a formal childcare setting.

## 3 STATEMENT OF THE PROBLEMS

- Past studies of transitions have mainly focused on children transiting out of preschool into primary school. However, the fact remains that the biggest and the most daunting transition a child will ever experience is the transition from home into an out of home care i.e. a formal childcare instituition.

Transition in early years has been quite a hot research topic in the last 5 years. It has been proven as a fact that positive experience in early year's transition encourages a better overall development in a child, hence contributing to a better school experience in their later years of childhood (Choy & Karuppiah, 2016; Early Child Care Research Network, 2003; Hussain et al., 2018; Peters, n.d., 2000; Tay-Lim & Lim, 2015). However, majority of these early year's transition studies have mostly focused on preschoolers transiting into primary schools. Only a handful of these studies looked into transitions from homecare to childcare (Jovanovic, 2011) and even fewer studies were ever made in an Asian context. Transition into childcare is deemed more as an education associated transition. However, this particular "peripheral" transition is a key social transition that shapes the child's experience, pathway and wellbeing (Dunlop & Fabian, 2007). Study has shown that since the early 1990's, there has been a steady increase in females entering the labour force (KPWKM, 2016). With nearly 50 percent of women entering the labour force, Malaysia has seen a steady growth in childcare centers in the last 2 decades. According to (Chiam, 2008), only 20 percent of mothers are able to care for their child exclusively. The remaining 80 percent are placed into out of home care. In addition, data has also indicated that children are now entering childcare at a younger age, as early as 3 months old.

This raises the question about the prolonged effect on the child when he or she encounters premature separation. Studies in the early 1980's and 1990's tried to address this particular issue. However, many of these studies focused on the care provided in the centers rather than the separation or the transition process (Belsky, 1990; Howes, 1990; Roggman, Langlois, Hubbs-Tait, Rieser & Rieser-Danner, 1994; "The Effects of Infant Child Care on Infant-Mother Attachment Security: Results of the NICHD Study of Early Child Care," 1997). Belsky's study on children in childcare indicated that most children in childcare institution develop insecure attachment. He came to this conclusion by looking at groups of children in childcare vs maternal care at home. It was found that most studies failed to look into the background of the child or the process of the secure base pattern of the child prior to entering the institution. Another fact noted in his study was that it was mostly conducted in the 1980's and the 1990's where childcare centers in America are generally known to lack in terms of quality care, e.g. low caregiving ratio, lack of trained caregivers etc. Belsky's findings were eventually conflicted by later research which states that childcare may not necessarily contribute to child attachment disorder, but other factors may include multiple caregiving, lack of security at home or inconsistent family dynamic (Meins et al., 2001; Sullivan, 2012; Zachrisson et al., 2013). Looking closer at these factor, I note that these elements in his research showed a lack of secure base investigation of a child at home which may contribute to later attachment disorders. For instance, some of Belsky's sample come from household where the parents are low income earners often spending very little time with their children. These factors were not considered when he concluded that children who attend childcare mostly develop an insecure attachment.

Through Belsky's research, many developmental psychologists, parent advocates chose to advise parents to keep their children in home care and provide exclusive maternal care till the age of 2 years. This was supported by the attachment theorist original theory, which states that a child with early maternal deprivation will in turn develop social emotional problems (American Academy Of Pediatrics, 2009; Belsky, 1988; Cárcamo et al., 2016; Datler et al., 2012; Fein, 1995; Findings for Children up to Age 4½ Years, 2006). No doubt, early separations has been linked to the formation of insecure avoidant attachments and subsequent mental health problems in life (Howard et al., 2011). Children who form insecure avoidant attachments to their primary caregivers are often expected to be at heightened risk in forming social relations with others, and show lesser compliance and cooperation with adults (John Bowlby, 1973; Godwin et al., 1993; Roggman et al., 1994; Waters et al., 2000)

However, looking at the dynamics of today modern society, this solution of keeping the child home till the age of 2 is no longer practical. The needs of childcare and out of home care is inevitable. If we consider the 2.6 million children under 4 years of age and take a very conservative approach, with 54% of women currently employed according to the Malaysian statistics, this would mean about 1.3 million children are either in alternative care or out of home care. Managing the transition process or bringing light to parents and main caregiver on the correct transition process to minimize trauma within the child becomes a necessity. Experts have raised opinions that separation, if handled well, can be a positive learning experience for a child's resilience, independence, sense of agency, and sense of belonging to their group as well as a family (Child-care and family predictors of preschool attachment and stability from infancy, 2001; Stonehouse, 2009).

Distress at separation in childcare or out of home care is unpleasant not only for the child, but also the parents and the professional and possibly other children. If this event is not dealt with in a constructive manner, the child will likely feel insecure, unhappy and in worst case scenario, social emotional detriment and develop negative attachment behaviors and distrust to his or her primary caregiver. The initial separation when children enter into a childcare, as well as the daily separations that occur on a daily basis to the point the child settles in, are important parts to be understood to enable stakeholders like teachers, child caregivers and parents to ease the separation process and to reduce the possibility of a traumatized separation process.

- Though past studies on separation and transition of young infants have been conducted, these studies looked into traumatic separations rather than routine separation. The fact also remains that a young child transiting into and out of home care in today's modern society is an inevitable event.

Studies which focus on separation and transition which were conducted in the past placed more emphasis on tragic separation (Heinicke, 1956; Pinckney & Robertson, 1969; Tizard & Hodges, 1978) or short term separation (Ainsworth, 1979; Tizard & Hodges, 1978) or tragic separation i.e. permanent departure of primary caregiver (Ahnert et al., n.d.; Pinckney & Robertson, 1969; Tizard & Hodges, 1978) rather than routine separation. There has been lack of research done in the context of routine separation in the last 2 decades and very few of these researches were done in an Asian context (Godwin et al., 1993; Jovanovic, 2011; King & Perrin, 1987; Klein, Kraft, & Shohet, 2010). The latest research into separation and transition was conducted by Jessie Jovanovic in 2011 with a small sample of 11 children in an Australian childcare. Jovanovic looked into pattern of separation and behavior of the child during transition at the childcare center. Her observation concluded that parents play a large role into helping their child to transit and manage routine separation. In her observation, she also noted that the child seemingly has control over the speed of transition. However, Jovanovic also acknowledged that her observation was limited in the center and further investigation should look into the child parent dyad home dynamics.

Hence, observation based research on young children transitioning into childcare generally were very limited. (Jovanovic, 2011; Jovanovic et al., 2016; King & Perrin, 1987; Klein et al., 2010) and the focus were mainly on the separation behaviour demonstrated by the child and parent at the center, rather than looking into the whole transition process. Past research conducted has also always been in the western context where children usually are reared in a nuclear family environment. In contrast, a child in Malaysian household may have multiple and inconsistent caregiving at home prior to transiting into childcare. All these cumulative facts indicated a need to look into the patterns of separation of young children transiting into childcare in the Malaysian context.

The main objective of this study is as follows:

i. To investigate the separation and transition process of a Malaysian child from homecare into childcare. Within the main objective, these are the other two sub-objectives of this study:

a. To observe how the child's established secure base pattern affects his or her transition and separation into a childcare
b. To observe how the process of separation-individuation takes place during the process of separation and transition from homecare into a childcare.

## 4 THEORY OF THE STUDY

As the transition process for the youngsters in the early year's sector highly relates to events of routine separation and adjustment to new caregivers and surroundings, this study has adapted 2 main theories which are attachment theory and separation-individuation theory.

The idea of how early separation affects a child's social and mental development into his adulthood was a theorized by Bowlby in his research of attachment, loss and separation (J. Bowlby, 1969; John Bowlby, 1958).

Eventual works from Bowlby's theory led to many researches that indicated premature separation from mother or primary caregiver, and inconsistent care giving will lead to development of social maladies within an individual. A handful of research has indicated that institutionalized children separated from their mothers are unable to recover emotionally, hence not only affecting their emotional well-being but also their language and cognitive performance (John Bowlby, 1944; Zeanah, Smyke, Koga, Carlson, & Bucharest Early Intervention Project Core Group, 2005).

Longitudinal researches were also made on children being sent away to childcare centers and nurseries. Results, although not conclusive, indicated that children being sent away for care in nurseries and childcare center at a younger age developed more aggressive social behavior in preschool as compared to peers who were cared for at home (Belsky, 1988; Roggman et al., 1994).

One of the biggest breakthroughs that came out of the attachment research was the concept of "secure base". The secure base concept explains how the mother or the primary caregiver, with its sensitivity and attachment to the child, will develop a secure base for the child, enabling the child to explore his or her surroundings. The idea of secure base is similar to Erik Eriksson psychosocial stages model (Crittenden, 2017) as one of the values emphasized in creating the secure base for the child is the value of trust to its mother or primary caregiver and eventually to the child's surrounding (Robert, 1994; Salter Ainsworth, n.d.; Salter Ainsworth & Bell, 1970).

> *"Warm, sensitive care, Ainsworth insisted, does not create dependency; it liberates, and enables autonomy. "It's a good thing to give a baby and a young child physical contact," she says, "especially when they want it and seek it. It doesn't spoil them. It doesn't make them clingy. It doesn't make them addicted to being held." (Ainsworth, 1979)*

Trust is an important value cultivated through the attachment process. With sufficient amount of trust, it seems that a child will be more willing to explore its surroundings, in other words, this translates to a better separation process. Ainsworth in the same study indicated that a child displaying secure attachment may not necessarily display clingy attachment behaviour; instead the child will display confidence in exploring his or her surrounding knowing his or her mother or caregiver is there. Here we see the child has developed a sense of trust (Cassidy, 2008).

The idea of cultivating trust in the model of secure base of the attachment theory is also similarly agreed by Margaret Mahler, who was the framer of separation-Individuation theory. Erikson's concept of basic trust (Erikson, 1968) is also related to Mahlerian theory, as it is to attachment theory. Basic trust developed from the primary object relationship as a result of the experience of the caretaker person as a coherent being who recognizes and reciprocates the child's physical and emotional needs, and whose face is recognized. Basic trust requires maternal sensitivity and is clearly related to both secure attachment and the later achievement of self and object constancy (Blum, 2004).

Development of object constancy, better known as object permanence through Jean Piaget's theory of Cognitive maturation (Piaget, 1964), is crucial prior to putting a child into nursery. Object permanence is an ability of a child knowing that an object exists even though it is no longer in sight. Secure attachment, and development of a reliable secure base, enables the growth of the cognitive maturation function of a child, thus enabling the child to grasp the concept of object permanence at an earlier stage of infancy.

The inclusion of Mahler's theory was subject to the fact that the attachment theory does not account for the essential development of separateness, and separation individuation is important for the promotion of autonomy, independence and identity (Blum, 2004; Chodorow et al., 2003). The separation of a child from the mother from home care to childcare does not only constitute a physical separation, but also a psychological separation, knowing that although mother is not there, the mother, the individual, still exists for the child.

### 4.1 *Theoretical framework*

This study attributes its theoretical framework mainly to Ainsworth's work on the secure base concept from the attachment theory and Margaret Mahler's theory on Separation-Individuation. The elements of the secure base framework are derived from Gillian and Beek's (Schofield & Beek, 2005) conceptual

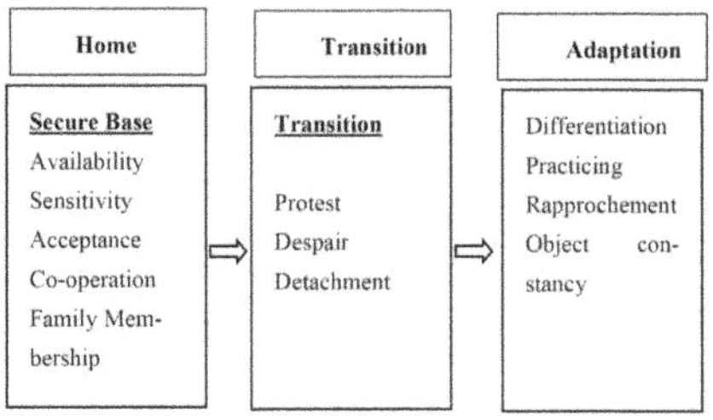


Figure 1.

framework, which is highly based on Ainsworth's work on secure base. The theoretical framework will also serve as a conceptual framework for this study.

Figure 1 indicates the conceptual framework utilized throughout the whole study. The observation of this study is divided into 3 parts, which are home, transition and adaptation. The elements of the secure base framework (adapted from Schofield & Beek, 2005) are used to observe the established secure base of the child at home. Elements observed in the transition process is acquired from John Bowlby's attachment theory and his writing on attachment and separation. Observation in part 3 of the transition, which is during the period of adaptation, utilizes elements from Margaret Mahler's theory of separation and individuation.

## 5 RESEARCH DESIGN

This study will be based on the qualitative philosophy and it applies the strategy of an explorative and causal case study. A case study is generally conducted to deepen the understanding of phenomena. An explorative case study studies an event which usually lacks detailed preliminary research. It is defined by Merriam as an empirical inquiry of a phenomenon within its real life context. It is an intensive, holistic description and analysis of a single instance phenomenon or a social unit.

The study of the case is viewed in a bounded system, allowing the researcher to fence in the unit of the study. Case studies in the exploratory method generally are distinguished by the absence of proposition and hypothesis. (Mills et al., 2010) In the case of this study, research and studies on transition of children is rather limited, especially in this region, coupled with the fact that not much research has been conducted to observe the nature of a young child transiting from home care to out of home care. The exploratory method in my view will serve as a good base for future empirical research, if needed, on the similar subject matter.

Causal case study strategy is also applied in this research design. According to the encyclopedia of case studies, causal case studies tell a story of a sequence of events and processes (Mills et al., 2010) Often in a scientific research, causal statements are often made e.g., "The infection is caused by the bacteria". In the fields of social science, however, explanations involved a purposeful act by a human being, known as volitional actors. In this research, separation is triggered by the decision of the parent to transit the child from a home care into out of home care. Past research has also indicated that there is high possibility that parental relationship and action could possibly affect the outcome of how a child separates and settles into his or her out of home care.

In terms of data collection, 3 main instruments were utilized for this research, which are interview protocol, observation protocol and document review

The interview and observation protocol utilized in this research has been validated and verified by individuals who are focused on subject matter in the field of psychology, early childhood education and family counselling. The secure base interview utilized in this study is extracted from the research of Dr. Mary Beek and Gillian Schofield (Schofield & Beek, 2005). The secure base interview outline is provided for the use of general practitioners and researchers of the industry.

### 5.1 *Selection of participants*

The intent of this research is not to be inferred to the general population, rather to look in depth into the case of this study, which is the transition of a child into a childcare in Malaysia. Sampling in this study is based on the purposive sampling model. In purposive sampling method, individuals studied are intentionally selected to ensure rich information is provided into the study being conducted. In this study, the child parent dyad are selected based on a set background and theoretical criterion to fulfil the intent of this research.

The form of purposeful sampling that is administered is known as typical and theoretical sampling. Typical sampling form is administered to ensure that the samples selected fulfill the typical background of a Malaysian child in a typical urban Malaysian family dynamic. The theoretical sampling form is administered to ensure the samples that are selected will be able to assist this research to unfold and discover the concepts of the theory being applied (Creswell, 2012). Since this study has its foundation heavily set on the theories of attachment and separation individuation, attaining samples that will assist me to understand the practicality of this theory is very important

There are 4 parent child participants in this study, each with a rather similar background.

Table 1 below lists a summary of case selection. The reason children in this study selected are below the ages of 24 months is because prior research has indicated events of separation are most traumatizing for children below the age of 24 months (John Bowlby, 1958). The fact also remains that plenty of children entering out of home care for the first time

Table 1. Case selection criteria.

| Participants | Selection Criteria |
|---|---|
| Child & Parent Dyad | I. Child must be below the age of 2 years old or 24 months<br>II. Cared for by multiple caregivers prior to attending childcare<br>III. Child is attending childcare for the first time |

Table 2. Participant table summary.

| Participant Pseudonym | Age (in Months) | Primary caregiver | Past caregivers/ |
|---|---|---|---|
| Alex | 18 | Mother | Grandmother |
| Bryan | 21 | Mother | Grandmother & Aunt |
| Chloe | 24 | Mother | Nanny & Maid |
| Desiree | 24 | Mother | Maid & Grandmother |

are between the ages of 18 and 24 months (Chiam, 2008). Criteria 2 is a prerequisite in this study as this study would like to focus on a typical Malaysian child that is normally cared for by multiple caregivers prior to attending childcare. Multiple caregivers would mean that the child is not fully cared for by a primary caregiver but rather cared for by different people within the household who step in as primary caregivers at different parts of the day. Criteria 3 is considered as transition for the first time from home care into a childcare and is deemed by the researcher to be a more traumatizing event than transiting for the 2nd or 3rd time. Table 2 below lists the summary of the participant background selected for this case.

## 5.2 *Research procedures*

### 5.1.1 *Parent child observation*

The intent and purpose of home observation was to corroborate the secure base interview claimed by the parents in terms of their relationship with their child. Also, since the researcher views that the home is the most natural environment, the child and parent will be at their most natural behavior.

However, I have noted that due to my presence and the presence of a video camera in a home setting, the parents tend to not perform their daily routine; rather, the parents seemed more cautious with their behavior.

Seeing this as a problem in research biasness, I attained permission from the principal of the childcare center to observe parent and child relationship at the center. The childcare center has a policy whereby the child and parent can attend play session in the center garden prior to starting childcare. Video recorder is positioned discretely and the researcher plays a role as a participant observer in the group, helping other children while observing the child parent dyad. Journaling and reflection is done by the researcher immediately after the session has ended and once again after viewing of the video observation.

Observation protocol is undertaken while viewing the recorded video clip.

### 5.1.2 *Observation in childcare*

Childcare observation is conducted 3 times a week over 3 months, with each session lasting one hour. A camera is installed in the classroom, and I assume the role of a non-participant observer. Occasionally, if help is required, I will assist in the classroom. Observation protocol is observed during this session and reflective journal entry of the session are updated at the end of the day that the observation session has taken place.

### 5.1.3 *Interview with parents*

Interview with parent or parents take place in 3 stages. The initial stage will be an interview focused on the background of the child and parent. Parents who are selected according to the criteria needed for the study will go through with the secure base interview of this study. The secure base interview has been reworded without alterating its original meaning to ensure that Malaysian parents understand the questions asked during the interview process. Another parent interview is conducted after 3 months the child has attended the childcare. The interview method with the parents is utilized to corroborate the observation method in this study. It is also part of the triangulation process utilized in this study.

Interview transcripts are then reproduced and given to interviewee for assent and for the purpose of member checks.

### 5.1.4 *Interview with child care giver*

Interview is conducted with the childcare givers at the center working with the subjects of the research after 3 months of observation. Similar to the interview protocol of the parents, the intent and purpose of this interview is to corroborate and triangulate the observation method that was undertaken in this study.

### 5.1.5 *Journal diary entry by parent*

Each parent who are subject in this study are presented with a journal diary to record their thoughts on their journey as their child transits from home care to out of home childcare. I have encouraged that parents note down their thoughts on a daily basis. However, I found that in the pilot study, the parent is not able to note down their thoughts on a daily basis due to time and personal constraint. Hence, instead of daily entry, I have asked the parents to submit a weekly entry to me instead. These writings are used to corroborate and triangulate with the other methods of data collection I have administered throughout this research. The data collected through the diary journal entry are also transcribed and eventually coded to be utilized as part of this study.

5.1.6 *Data analysis*

According to Creswell, majority of qualitative research depends on the coding technique for data analysis (Creswell, 2012). Data coding is process where one examines images or text database line by line and assigns these interpretations into the relevant codes next to the text segments.

Data collected will be transcribed by me. The Atlas Ti software will be utilized to analyze the data collected. Data Coding will be the primary technique of managing the data for this research. Axial coding or deductive coding method are pre-determined by looking at the unit of analysis determined in this study. Inductive coding are eventually added on when there are new data categories discovered during data processing.

Inductive codes from the data are formed based on the operational definition and unit of analysis used in this study. Themes are subsequently formed, grouping all the codes that have a similar theme.

## 6 CONCLUSION

The study is currently in its intermittent stage. The data gathering process, which consists of observing toddlers and children younger than 24 months, is a time consuming process, and is often subjected to the temperament and welfare of the child. The targeted observation period is at 3 months.

This particular study is intended to be the basis of future study and research in assisting young children transiting from in home care to out of home care. The role of a small qualitative study as such is of importance and is with hoped that this study is able to provide better insight for a mass quantitative study to measure and evaluate the most effective form of transition from homecare to childcare which can be practiced generally in childcare centers across the country.

The need and the use of childcare facility is an inevitable process in every developing country. Hence providing a framework for successful and healthy transition that can benefit the welfare of the child and at the same time emphasizing and advocating the role of community is crucial. This study also intends to assist stakeholders to create a safe, supportive and inclusive environment to foster social emotional development during the whole transition process.

All this in hopes that parents, childcare centers and the Ministry of Women Family and Community of Malaysia may promote awareness and eventually produce a guidebook outlining roles of the individuals in the society into helping a child to have a healthy transition into an out of home care, i.e. institutional childcare centers, consequently minimizing the social emotional trauma and bringing forth a better, more well- rounded developed child to serve the future nation.

## REFERENCES


Ahnert, L., Gunnar, M. R., Lamb, M. E., & Barthel, M. (n.d.). Transition to Child Care: Associations With Infant – Mother Attachment, Infant Negative Emotion, and Cortisol Elevations. Retrieved from https://pdfs.semanticscholar.org/a581/6742ad06e4a6fcb04d1e76f6cfc8a4df5d60.pdf

Ainsworth, M. D. S. (1979). Attachment as Related to Mother-Infant Interaction. In J. S. Rosenblatt, R. A. Hinde, C. Beer, & M.-C. Busnel (Eds.), *Advances in the Study of Behavior* (Vol. 9, pp. 1–51). Academic Press. https://doi.org/https://doi.org/10.1016/S0065-3454(08)60032-7

American Academy Of Pediatrics. (2009). Caring for Your Baby and Young Child, Revised Edition.

Belsky, J. (1988). The "Effects" of infant day care reconsidered. *Early Childhood Research Quarterly*, *3*(3), 235–272. https://doi.org/10.1016/0885-2006(88)90003-8

Belsky, J. (1990). Parental and Nonparental Child Care and Children's Socioemotional Development: A Decade in Review. *Journal of Marriage and Family*, *52*(4), 885–903. https://doi.org/10.2307/353308

Blum, H. P. (2004). Separation-Individuation Theory and Attachment Theory. *Journal of the American Psychoanalytic Association*, *52*(2), 535–553. https://doi.org/10.1177/00030651040520020501

Borelli, J., Rasmussen, H., John, H., West, J., & Piacentini, J. (2015). Parental Reactivity and the Link Between Parent and Child Anxiety Symptoms. *Journal of Child & Family Studies*, *24*(10), 3130–3144. https://doi.org/10.1007/s10826-015-0117-7

Bowlby, J. (1944). Forty-Four Juvenile Thieves: Their Characters and Home Life. *International Journal of Psychoanalysis*, *25*, 19–52. https://doi.org/10.3917/psye.491.0007

Bowlby, J. (1958). The Nature of the Child's Tie to His Mother. *International Journal of Psycho-Analysis*, *39*(1), 350–373. Retrieved from http://www.psychology.sunysb.edu/attachment/online/nature of the childs tie bowlby.pdf

Bowlby, J. (1969). *Attachment: Attachment and Loss. The British Journal of Psychiatry* (Vol. 1). https://doi.org/10.1177/000306518403200125

Bowlby, J. (1973). *Attachment and Loss VOLUME II SEPARATION ANXIETY AND ANGER* (3rd) ed.. United States of America: Basic Books. Retrieved from https://www.abebe.org.br/files/John-Bowlby-Separation-Anxiety-And-Anger-Attachment-and-Loss-Vol-2-1976.pdf

Bowlby, J. (1983). *Attachment and loss*. New York: Basic Books.

Cárcamo, R. A., Vermeer, H. J., van der Veer, R., & van IJzendoorn, M. H. (2016). Early Full-Time Day Care, Mother–Child Attachment, and Quality of the Home Environment in Chile: Preliminary Findings. *Early Education and Development*, *27*(4), 457–477. https://doi.org/10.1080/10409289.2016.1091971

Cassidy, J. (2008). Handbook of Attachment. *Handbook of Attachment: Theory, Research, and Clinical Appication*, 3–22. https://doi.org/10.1016/B978-012445860-4/50008-3

Chiam, H. (2008). Child Care in Malaysia: Then and Now. *International Journal of Child Care and Education Policy, 2*(2), 31.Retrieved from http://www.diglib.um.edu.my/interaktif/default.asp?url=http://search.ebscohost.com/login.aspx?direct=true&db=edo&AN=ejs35002912&site=eds-live

Child-care and family predictors of preschool attachment and stability from infancy. (2001). *Developmental Psychology, 37*(6), 847–62.Retrieved fromhttp://www.ncbi.nlm.nih.gov/pubmed/11699758

Chodorow, N., Noshpitz, J., Bergman, A., Harris, A., Kerr, J., Nachman, P., Steinberg, Z. (2003). JOHN BOWLBY AND MARGARET S. MAHLER: THEIR LIVES AND THEORIES. Retrieved from http://citeseerx.ist.psu.edu/viewdoc/download?doi=10.1.1.1009.7864&rep=rep1&type=pdf

Choy, M. Y., & Karuppiah, N. (2016). Preparing Kindergarten Two children for Primary One in Singapore: perceptions and practices of parents, kindergarten teachers and primary schoolteachers. *Early Child Development and Care, 186*(3), 453–465. https://doi.org/10.1080/03004430.2015.1040989

Creswell, J. W. (2012). *Educational Research: Planning, Conducting and Evaluating Quantitative and Qualitative Research. Pearson Education, Inc.*

Crittenden, P. M. (2017). Gifts from Mary Ainsworth and John Bowlby. *Clinical Child Psychology and Psychiatry, 22*(3), 436–442. https://doi.org/10.1177/1359104517716214

Datler, W., Ereky-Stevens, K., Hover-Reisner, N., & Malmberg, L.-E. (2012). Toddlers' transition to out-of-home day care: Settling into a new care environment. *Infant Behavior and Development, 35*, 439–451. https://doi.org/10.1016/j.infbeh.2012.02.007

Dunlop, A.-W., & Fabian, H. (2007). *Outcomes of good practice in transition processes for children entering primary school : working paper 42*. Retrieved from https://strathprints.strath.ac.uk

Early Child Care Research Network, N. I. of C. H. (2003). Does Amount of Time Spent in Child Care Predict Socioemotional Adjustment During the Transition to Kindergarten? *Child Development, 74*(4), 976–1005. https://doi.org/10.1111/1467-8624.00582

Erikson, E. (1968). *Youth: Identity and crisis. New York, NY: WW.* https://doi.org/10.1002/yd.29

Fein, G. G. (1995). Infants in group care: Patterns of despair and detachment. *Early Childhood Research Quarterly, 10*(3), 261–275. https://doi.org/https://doi.org/10.1016/0885-2006(95)90007-1

*Findings for Children up to Age 4½ Years*. (2006). Washington. Retrieved from https://www1.nichd.nih.gov/publications/pubs/Documents/SECCYD_06.pdf

Godwin, L. J., Groves, M. M., & Horm-Wingerd, D. M. (1993). "Don't leave me": Separation distress in infants, toddlers, and parents. *Day Care & Early Education, 20*(3), 13–17. https://doi.org/10.1007/BF01620625

Heinicke, C. M. (1956). Some Effects of Separating Two-year-old Children from their Parents: A Comparative Study. In *Brief separations* (1st ed., pp. 105–176). London: New York International University Press. Retrieved from http://journals.sagepub.com/doi/pdf/10.1177/001872675600900201

Howard, K., Martin, A., Berlin, L. J., & Brooks-Gunn, J. (2011). Early mother–child separation, parenting, and child well-being in Early Head Start families. *Attachment & Human Development, 13*(1), 5–26. https://doi.org/10.1080/14616734.2010.488119

Howes, C. (1990). Can the age of entry into child care and the quality of child care predict adjustment in kindergarten? *Developmental Psychology, 26*(2), 292–303. https://doi.org/10.1037/0012-1649.26.2.292

Hussain, H., Jesuvadian, M. ., Lim, J. ., & Karuppiah, N. (2018). *Supporting Children In Their Transition to Primary 1. A guide for Teachers and Parents in Singapore.* Singapore: National Institute of Education Singapore.

Jovanovic, J. (2011). Saying Goodbye: An Investigation into Parent-Infant Separation Behaviours on Arrival in Childcare. *Child Care in Practice, 17*(3), 247–269. https://doi.org/10.1080/13575279.2011.571237

Jovanovic, J., Brebner, C., Lawless, A., & Young, J. (2016). Childcare educators' understandings of early communication and attachment. *Australasian Journal of Early Childhood, 41*(4), 95–105.Retrieved from http://www.diglib.um.edu.my/interaktif/default.asp?url=http://search.ebscohost.com/login.aspx?direct=true&db=ehh&AN=121089746&site=eds-live

King, M. A., & Perrin, M. S. (1987). Arrival time behavior of two-year-olds in child care. *Child and Youth Care Quarterly, 16*(4), 235–240. https://doi.org/10.1007/BF01085826

Klein, P. S., Kraft, R. R., & Shohet, C. (2010). Behaviour patterns in daily mother–child separations: possible opportunities for stress reduction. *Early Child Development and Care, 180*(3), 387–396. https://doi.org/10.1080/03004430801943290

Kobak, R. (1999). The emotional dynamics of disruptions in attachment relationships: Implications for theory, research, and clinical intervention. In *Handbook of attachment: Theory, research, and clinical applications*. (pp. 21–43). New York, NY, US: Guilford Press.

KPWKM. (2016). *Statistcs on women family and community*. Putrajaya. Retrieved from https://www.kpwkm.gov.my/kpwkm/uploads/files/Penerbitan/Buku Perangkaan/Perangkaan 2016.pdf

Meins, E., Fernyhough, C., Fradley, E., & Tuckey, M. (2001). Rethinking Maternal Sensitivity: Mothers' Comments on Infants' Mental Processes Predict Security of Attachment at 12 Months. *Journal of Child Psychology and Psychiatry, 42*(5), S0021963001007302. https://doi.org/10.1017/S0021963001007302

Mills, A., Durepos, G., & Wiebe, E. (2010). *Encyclopedia of Case Study Research*. 2455 Teller Road, Thousand Oaks California 91320 United States: SAGE Publications, Inc. https://doi.org/10.4135/9781412957397

Pandey, S. C., & Patnaik, S. (2014). Establishing reliability and validity in qualitative inquiry: a critical examination. *Journal of Development and Management Studies, 12*(1), 5743–5753.Retrieved from http://www.researchgate.net/profile/Satyendra_Pandey2/publication/266676584_ESTABLISHING_RELIABILITY_AND_VALIDITY_IN_QUALITATIVE_INQUIRY_A_CRITICAL_EXAMINATION/links/543779b40cf2dc341db4d7fb.pdf

Peters, S. (n.d.). *Literature Review: Transition from Early Childhood Education to School Report to the Ministry of Education*. Retrieved from www.waikato.ac.nz

Peters, S. (2000, September). Multiple Perspectives on Continuity in Early Learning and the Transition to School. Retrieved from https://eric.ed.gov/?id=ED447916

Piaget, J. (1964). Part I: Cognitive development in children: Piaget development and learning. *Journal of Research in Science Teaching, 2*(3), 176–186. https://doi.org/10.1002/tea.3660020306

Pinckney, P.-C., & Robertson, J. (1969). Section of Pcediatrics Going to Hospital with Mother A Documentary Film. *Proceedings of the Royal Society of Mediwtne*. Retrieved from http://journals.sagepub.com/doi/pdf/10.1177/003591575905200518

Robert, K. (1994). *Becoming attached: unfolding the mystery of the infant mothet bond and it's impact on later life*. New York: Warner. Retrieved from http://www.psychology.sunysb.edu/attachment/online/karen.pdf

Roggman, L. A., Langlois, J. H., Hubbs-Tait, L., Rieser, L. A., & Rieser-Danner, L. A. (1994). Infant Day-Care, Attachment, and the "File Drawer Problem" *Child Development*, *65*(5), 1429–1443. Retrieved from http://www.jstor.org/stable/1131509

Salter Ainsworth, M. D. (n.d.). Attachments Beyond Infancy. Retrieved from https://pdfs.semanticscholar.org/5431/41e657bda74736ff87ac10d70643cd639892.pdf

Salter Ainsworth, M. D., & Bell, S. M. (1970). Attachment, Exploration, and Separation: Illustrated by the Behavior of One-Year-Olds in a Strange Situation ILLUSTRATED BY THE BEHAVIOR OF ONE-YEAR-OLDS IN A STRANGE SITUATION. *Source: Child Development*, *41*(1), 49–67.Retrieved from http://www.jstor.org/stable/1127388

Schofield, G., & Beek, M. (2005). Providing a secure base: Parenting children in long-term foster family care. *Attachment & Human Development*, *7*(1), 3–26. https://doi.org/10.1080/14616730500049019

Stonehouse, A. (2009, December). Positive goodbyes: Helping children and families to manage separations. *Putting Children First, the Magazine of National Childcare Accreditation Council*, 10–12. Retrieved from http://ncac.acecqa.gov.au/educator-resources/pcf-articles/Positive_goodbyes_Dec09.pdf

Sulaiman, Z., Liamputtong, P., & Amir, L. H. (2017). Timing of return to work and women's breastfeeding practices in urban Malaysia: A qualitative study. *Health and Social Care in the Community*, (May 2017), 48–55. https://doi.org/10.1111/hsc.12460

Sullivan, R. M. (2012). The Neurobiology of Attachment to Nurturing and Abusive Caregivers. *The Hastings Law Journal*, *63*(6), 1553–1570.Retrieved from http://www.ncbi.nlm.nih.gov/pmc/articles/PMC3774302/

Tay-Lim, J., & Lim, L. (2015). *Early years transitions : building bridges for children/by Joanna Tay-Lim and Levan Lim. – National Library*. Singapore: Association for Early Childhood Educators. Retrieved from http://www.nlb.gov.sg/biblio/202803121

The Effects of Infant Child Care on Infant-Mother Attachment Security: Results of the NICHD Study of Early Child Care. (1997). *Child Development*, *68*(5), 860–879.Retrieved from http://www.jstor.org/stable/1132038

Tizard, B., & Hodges, J. (1978). The effect of early institutional rearing on the development of eight year old children. *Journal of Child Psychology and Psychiatry, and Allied Disciplines*, *19*(2), 99–118.Retrieved from http://www.ncbi.nlm.nih.gov/pubmed/670339

Van Der Horst, F. C. P., Leroy, H. A., & Van Der Veer, R. (2008). "when strangers meet": John Bowlby and Harry Harlow on attachment behavior. *Integrative Psychological and Behavioral Science*, *42*(4), 370–388. https://doi.org/10.1007/s12124-008-9079-2

Vogler, P., Crivello, G., and Woodhead, M. (2008). *Early childhood transitions research: A review of concepts, theory, and practice*.Retrieved from www.greenink.co.uk

Waters, E., Merrick, S., Treboux, D., Crowell, J., & Albersheim, L. (2000). Attachment Security in Infancy and Early Adulthood: A Twenty-Year Longitudinal Study. *Child Development*, *71*(3), 684–689.Retrieved from https://pdfs.semanticscholar.org/94bc/044720ad1edda02ac3f1d9fe130d52205682.pdf

Zachrisson, H. D., Dearing, E., Lekhal, R., & Toppelberg, C. O. (2013). Little Evidence That Time in Child Care Causes Externalizing Problems During Early Childhood in Norway. *Child Development*, *84*(4), 1152–1170. https://doi.org/10.1111/cdev.12040

Zeanah, C. H., Smyke, A. T., Koga, S. F., Carlson, E., & Bucharest Early Intervention Project Core Group. (2005). Attachment in Institutionalized and Community Children in Romania. *Child Development*, *76*(5), 1015–1028. https://doi.org/10.1111/j.1467-8624.2005.00894.x

*The Social Sciences Empowered – Lumban Gaol, Hutagalung & Fong Peng (eds)*
*© 2020 Taylor & Francis Group, London, ISBN 978-0-367-00090-5*

# The relationship between orientation, attitudes, intensity of motivation and Chinese competency

Tan Chor Ter*, Fonny Dameaty Hutagalung*, Chew Fong Peng* & Zanariah Hamid*
*Department of Language and Literacy Education, Faculty of Education, University of Malaya, Malaysia*

ABSTRACT: Chinese competency is an important language skill for every preservice Chinese teacher. Multilingual usage and a multicultural society are factors that affect the integrative and instrumental orientation, attitude towards a learning situation, intensity of motivation and Chinese competency of Chinese preservice teachers. The analysis involved 205 questionnaires from preservice teachers. The structural model showed that instrumental orientation positively predicted Chinese competency. However, the intensity of motivation was not a mediator in the relationship between integrative and instrumental orientations, attitude towards a learning situation and Chinese competency.

*Keywords*: orientation, attitudes, motivation, Chinese competency, multilingual

## 1 INTRODUCTION

The quality of teachers and school leaders is part of the 2012-2025 Malaysia Education Blueprint. The Ministry of Education emphasizes that highly qualified teachers and school leaders should effectively promote the achievement of each student's learning outcome. Therefore, the preservice teacher selection requirement for 2010 was 119,782 candidates with only 1% of them achieving outstanding academic results in the Malaysian Examination Certification, whereas in 2012, 9% of the 96,974 candidates achieved outstanding academic results. The Ministry of Education raised the academic requirements for undergraduate degree programs for teacher education in order to enhance the quality of teachers and provide effective education for all children (Ministry of Education, 2013).

## 2 STATEMENT OF THE PROBLEM

Generally, language environment and social-culture are inputs that affect learners' orientation, attitude and motivation during the language learning process as well as the results of language learning (Gardner, 2001). Therefore, role of affective variables, especially integrative and instrumental orientations, attitude towards a learning situation and motivation intensity is crucial. Song and Pornsima (2017) and Lai (2012) highlighted that the cultural model, language environment, politics and economic power situation influenced the learner's orientation and language attitude.

The Chinese language environment and cultural problems in the Malaysian Chinese community.

According to Wang (2012), Malaysian Chinese education has successfully spread the effective use of Chinese. The study highlighted the change in language attitude in the new Chinese generation, whereby most of them used Chinese as a language during learning and generated an interest in using Chinese. However, Koh (2008) stated that the ethnic Chinese still maintained the language dialect of their ancestors and inherited the sub-ethnic Chinese culture. Parents used language dialect when communicating with children. Neo (2012) stated that the majority of Chinese families practice their dialect in daily communication. The Chinese family uses language dialect to choose the family name; for example, the Chinese family surname "Chen" (陈) in Chinese is changed to "Tan" in the Min dialect. Fan (2012) stated that the Chinese language used by the Malaysian Chinese is influenced by sub-ethnic dialect and culture. The language dialect environment directly influences Chinese proficiency and language attitude.

According to Ng, Wong, Guek, Lim & Tan (2017) and Lam, Yau, Yoe, Chew, Lee and Lee (2006), Chinese reading skills is a major problem encountered by Chinese language teachers. Both studies above found that Chinese preservice teachers during practicum training could not exhibit their mastery in Hanyu Pinyin (汉语拼音) phonetics. The preservice teachers could not distinguish the exact phonetics of

*Corresponding authors: tchorter@gmail.com, fonny@um.edu.my, fpchew@um.edu.my, zanariahhamid@gmail.com

the word and made repeated errors when reading the same word. In such cases, weaknesses in phonetics and intonation affects preservice teachers when pronouncing the word correctly and misconducting Chinese language lessons. Similarly, it affects teaching and learning activities, creates language flaws and makes pupils mispronounce words during lessons. Therefore, the problem of phonetics and intonations in Chinese language mastery provides a negative impression on school children and disrupts communication and discussions between teachers and students (Mora & Darcy, 2017).

The reading skill issue was due to the problem of mastering vocabulary. The lack of vocabulary has led to preservice teachers being unable to communicate with students fluently and actively as well as having to use the Malay language to explain something (Ng et al., 2017). According to Wen (2016), the Chinese Hanyu Da Cidian (汉语大词典) dictionary, has 3,700,000 Chinese words, Zhongwen Da Cidian dictionary (中文 大 词典) published in Taiwan, has 4,400,000 Chinese words and the Dahan He Cidian dictionary (大汉和 辞典) that is issued in Japan, has 5,500,000 words. By comparing these dictionaries, it would be easy to master Chinese words due to the huge reservoir of Chinese words in language learning. Moreover, the Chinese words included the number of similar character words, synonym words as well as phonetic words that could confuse learners.

Based on the problem of learning the Chinese language, Gardner and Lambert (1972) and Gardner (1985) stated that orientation, language attitude and motivation are non-cognitive factors that lead to language achievement. Gardner (2001, 2006) also stated that integration orientation has a positive correlation that enhances a positive attitude towards language as well as self-motivation to achieve learning outcomes. However, the social education model by Gardner (2001) has not been tested on Chinese competency in preservice teachers.

## 3 CONCEPTUAL FRAMEWORK

The conceptual framework was based on social psychology (Gardner and Lambert, 1972) and social learning models (Gardner, 1985). Generally, integration orientation refers to individuals who are interested in communicating and interacting with other language speakers, having a positive attitude towards other languages and cultures as well as a desire to learn other languages (Gardner & Lambert, 1972). According to Gardner (2001), integrativeness refers to an individual who has an open mind, is interested in target language speakers and is willing to engage in target language communities; whereas, instrumental orientation refers to the interest in learning a target language as being pragmatic. In other words, instrumental orientation emphasizes on the benefits of learning a language, such as future job opportunities, educational opportunities and external incentives (Gardner, 1985; Gardner & Lambert, 1972).

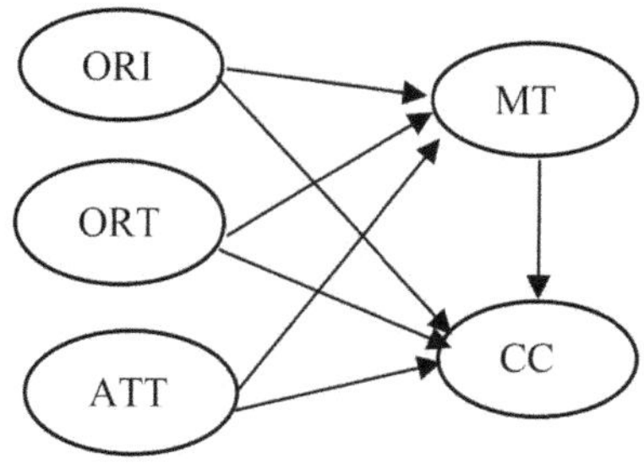

Figure 1. Hypothesized relationship of factors predicting Chinese competency.

Attitude towards a learning situation reflects the student's self-assessment of the learning environment in the learning process. Attitude towards Chinese lecturers, the Chinese language and teacher education courses are the main focus of an attitude related to the preservice teacher learning process. A positive attitude towards learning situations has a direct relationship with learning orientation. Motivation intensity is the desire or ability to enhance self-confidence in order to achieve learning goals. Gardner (2001) suggested that the relationship between intra-orientation and attitude towards positive learning situations influences the intensity motivation to achieve language achievement.

Based on the above discussion, a prior model (Figure 1) was established to carry out the study and determine the relationship between integration (ORI), instrumental orientation (ORT) and attitude towards a learning situation (ATT), which were the independent variables, while Chinese competency (CCH) domination was the dependent variable and intensity motivation (MTV) as a mediator in this study.

## 4 RESEARCH OBJECTIVE

The purpose of this study was to predict the integrative orientation, instrument orientation, attitude towards a learning situation that affect Chinese Competency. This study also aimed to determine the effect of intensity motivation between integrative orientation, instrument orientation and attitudes towards a learning situation, such as Chinese Competency.

## 5 RESEARCH HYPOTHESES

1. Integrative orientation is positively related to Chinese competency.
2. Instrumental orientation is positively related to Chinese competency.

3. Attitudes toward the learning situation is positively related to Chinese competency.
4. Intensity motivation is positively related to Chinese competency.
5. Intensity motivation mediates the relationship between integrative orientation, instrumental orientation and attitudes toward the learning situation and Chinese competency.

## 6 INSTRUMENTS

There were 29 items in the questionnaire and the details are discussed below.

The integrative and instrumental scale. This questionnaire was adapted from a previous study entitled 'Construction and validation of a questionnaire on language learning motivation' (Larisa Niktina, Zuraidah Mohd Don, & Loh, 2016), consisting of two subscales: (a) a 5-item Integrative orientation scale (ORI), and (b) a 5-item Instrumental orientation scale (ORT).

The Motivation Intensity Scale. This 7-item Motivation Intensity questionnaire (MTV) was adapted from a previous study entitled 'Motivation, Motivation Intensity, use of Chinese and Self- Rated Chinese' Competence (Liu, 2017).

The Attitude Towards a Learning Situation scale. This 12-item Attitude Towards a Learning Situation (ATT) questionnaire was adapted from a previous study entitled 'Culture Studies and Motivation in Foreign and Second Language Learning in Taiwan' (Meng-Ching Ho, 1998).

The scale for the ORI, ORT, MTV and ATT items was a 5-point Likert scale, ranging from 'Strongly Disagree' (1) to 'Strong Agree' (5), with values of 1 to 5 assigned to the alternatives respectively.

The Chinese competency scale (CCH). The 3 items required the Chinese preservice teacher to self-rate the competence level in communication skills, reading skills and writing skills on a scale of 1-4, ranging from "Poor" to "Excellent" to each alternative respectively.

## 7 PROCEDURE

The questionnaire was translated into Chinese, rechecked, and then together with a consent form, distributed to undergraduate Chinese preservice teachers studying in two Malaysia Teacher Education Institute (N=205) campuses. The Chinese preservice teachers filled in the questionnaires in about 25 minutes during lessons in class. A total of 205 questionnaires were collected for further statistical analyses.

## 8 DATA ANALYSES

All survey data were analysed using Smart PLS 3.0, to evaluate the relationship between constructs, and estimate both the measurement and structural parameters in the prior model.

## 9 MEASUREMENT MODEL

The measurement model for constructs with reflective measures was assessed by examining the individual item reliability, internal consistency, and discriminant validity.

Tables 1, presented the outer loading for all item in the reflective measurement model is great than .5 on the corresponding factors. The present result shows that 2 items in ORT (ORT1 and ORT 3) and ATT (ATT1 and ATT 2) were eliminated due to a better fit according to the rule of thumbs (Hair et al., 2014).

As for the internal consistency for this measurement model, the composite reliability value was greater than .7, which was consistent with this study

Table 1. Convergent validity.

| Variable | Items | λ | α | CR | AVE |
|---|---|---|---|---|---|
| CCH | Com | 0.773 | 0.792 | 0.872 | 0.696 |
| | Rdg | 0.901 | | | |
| | Wrt | 0.822 | | | |
| MTV | MTV1 | 0.757 | 0.876 | 0.904 | 0.573 |
| | MTV2 | 0.672 | | | |
| | MTV3 | 0.767 | | | |
| | MTV4 | 0.740 | | | |
| | MTV5 | 0.796 | | | |
| | MTV6 | 0.800 | | | |
| | MTV7 | 0.760 | | | |
| ORI | ORI1 | 0.676 | 0.787 | 0.852 | 0.537 |
| | ORI2 | 0.692 | | | |
| | ORI3 | 0.765 | | | |
| | ORI4 | 0.776 | | | |
| | ORI5 | 0.747 | | | |
| ORT | ORT2 | 0.787 | 0.612 | 0.793 | 0.562 |
| | ORT4 | 0.687 | | | |
| | ORT5 | 0.771 | | | |
| ATT | SBC3 | 0.628 | 0.925 | 0.936 | 0.596 |
| | SBC4 | 0.699 | | | |
| | SPC1 | 0.834 | | | |
| | SPC2 | 0.824 | | | |
| | SPC3 | 0.848 | | | |
| | SPC4 | 0.812 | | | |
| | SPY1 | 0.770 | | | |
| | SPY2 | 0.770 | | | |
| | SPY3 | 0.772 | | | |
| | SPY4 | 0.735 | | | |

Note: λ=Outer loading; acceptable values for λ are above .50 (Hair et al., 2014).
α= Cronbachs Alpha.
CR="composite reliability"; > .60 (Hair et al., 2014).
AVE="Average variance extracted"; > .50 (Hair et al., 2014).

Table 2. Fornell-larcker criterion.

| | ATT | CCH | MTV | ORI | ORT |
|---|---|---|---|---|---|
| ATT | **0.772** | | | | |
| CCH | 0.105 | **0.834** | | | |
| MTV | 0.653 | 0.105 | **0.757** | | |
| ORI | 0.383 | 0.058 | 0.478 | **0.732** | |
| ORT | 0.443 | 0.193 | 0.521 | 0.382 | **0.750** |

and good for the composite reliability of the model. The AVE is an appropriate measure of a construct convergent validity. In this case, all constructs had exceeded the minimum threshold of .5, which shows that over 50% of the variance in the measurement items were accounted for by the latent variables (Fornell & Lacker, 1981).

Discriminant validity refers to cross loading (Table 1) between constructs, indicated by all indicators being more than .1 (Chin, 1998) and the loadings of indicators on the assigned latent variable higher than the loading on other latent variables.

The results of Fornell-Larckers criteria (Table 2) shows the correlations between constructs. The diagonal value of the square roots of AVE was greater than the correlation between the pair constructs. Thus, the discriminant validity of the indicators used in the model was supported.

## 10 STRUCTURAL MODEL

The structural model describes the relationship between the model constructs when assessing the structural model to test the level of variance in the dependent variable that was caused by the independent variables.

Table 3. Path coefficients and hypothesis test.

| | β | ST DEV | T values | P Values | $R^2$ |
|---|---|---|---|---|---|
| CCH | | | | | 0.018 |
| ORI -> CCH | –0.044 | 0.071 | 0.627 | 0.265 | |
| ORT -> CCH | 0.183 | 0.084 | 2.180 | 0.015 | |
| ATT -> CCH | 0.020 | 0.084 | 0.242 | 0.405 | |

Note: * significant at 0.05 level ($p<0.05$)

Based on the results of the path coefficient assessment without mediators, the integrative orientation (β=-0.044, t=0.627, p>0.05) was negatively related to Competence in Chinese. The instrumental orientation (β=0.183, t=2.180, p>0.05) and Attitude Towards a Learning Situation (β=0.020, t=0.242, p<0.05) was positively related to Chinese competency. Thus, H2 was supported whereas H1, and H3 were not supported; thus, rejected. Next, The R2 value was 0.018, suggesting that 1.8% of the variance can be explained by integrative orientation, instrumental orientation and attitude towards a learning situation. With a Q2 value larger than 0 (Q2=0.026), the result confirms that the model has respective relevance (Hair et al., 2014).

Table 4. Result of mediator test.

| | *β* | ST DEV | T values | P Values | $R^2$ |
|---|---|---|---|---|---|
| MTV | | | | | 0.492 |
| CCH | | | | | 0.013 |
| ORI-> MTV | 0.215 | 0.065 | 3.279 | 0.001 | |
| ORT->MTV | 0.243 | 0.059 | 4.132 | 0.000 | |
| ATT->MTV | 0.440 | 0.051 | 8.562 | 0.000 | |
| MTV->CCH | 0.018 | 0.106 | 0.170 | 0.433 | |
| ORI->CCH | –0.048 | 0.076 | 0.634 | 0.263 | |
| ORT->CCH | 0.179 | 0.084 | 2.119 | 0.017 | |
| ATT->CCH | 0.012 | 0.100 | 0.125 | 0.450 | |
| ORI-> MTV-> CCH | 0.004 | 0.024 | 0.160 | 0.436 | |
| ORT-> MTV-> CCH | 0.004 | 0.027 | 0.160 | 0.436 | |
| ATT-> MTV-> CCH | 0.008 | 0.048 | 0.166 | 0.434 | |

The R2 value was 0.492, which suggests that 49.2% of variance in intensity motivation could be explained by integrative orientation, instrumental orientation and attitude towards a learning situation. A close look shows that integrative orientation (β = 0.215, p<0.05), instrumental orientation (β=0.243, p<0.05) and attitude towards a learning situation (β=0.440, p<0.05) were positively related to intensity motivation.

Next, the R2 value of 0.013 (1.3%) for the variance in Chinese competency was explained by the extent of intensity motivation (β = 0.018, p>0.05) compared to Chinese competency. However, the relationship between intensity motivation and Chinese competency was not significant (β=0.018, p>0.05). When controlling the effect of intensity motivation, the direct effects of integrative orientation (β=-0.048, p>0.05), instrumental orientation (β=0.179, p>0.05) and attitude towards a learning situation (β=0.012, p>0.05) changed to being non-significant, as well as the indirect effect showed that it was not statically significant in the relationship between integrative orientation (β=0.004, p>0.05), instrumental orientation (β=0.004, p>0.05) and an attitude towards a learning situation (β=0.008, p>0.05). Thus, H5 was not supported. The effect size of intensity motivation on Chinese competency was 0.000, leading to the conclusion that intensity motivation had no effect on Chinese competency.

Lastly, R2 Chinese competency values (R2=0.013) were larger than 0 (Hair et al., 2017; Fornell & Cha, 1994), the effect size (f2) to R2 was ORI=0.002 (small), ORT=0.023 (moderate) and ATT=0.001 (small) (Cohen, 1981). The result indicates that instrumental orientation has a close relationship in producing the R2 for Chinese competency. The predictive relevance of the model was examined using the blindfolding procedure. Chinese competency (Q2=0.018) and motivation intensity (Q2=0.264) were both more than 0, indicating that the model had sufficient predictive relevance (Hair et al., 2017; Stone, 1974; Geisser, 1974).

## 11 DISCUSSION

The analysis has shown that instrumental orientation towards learning motivation was the main factor affecting Chinese competency among Chinese language preservice teachers. This finding in line with Gursoy (2013), who found that preservice EFL (English as a Foreign Language) teachers' instrumental orientation values were related to a language teacher's education. This study shows that the respondents were generally instrumentally oriented to learn Chinese. The findings showed that preservice teachers learned Chinese more for personal interests and needs, such as to further their education, to excel in a major study or research and to increase their opportunities in the future job market. The efforts they expended on learning Chinese was subject to the needs and interests of their major study. Nevertheless, Gardner and Lambert (1972) mentioned that integrative orientation is essential for understanding different cultures and lifestyles that would improve linguistic awareness.

The study also indicated that attitude towards a learning situation has a small positive effect ($<0.2$ %, Cohen, 1988) on Chinese competency but the data analysis showed that it was not significantly ($p>.005$) related to Chinese competency because preservice teachers lacked the autonomy to engage in a learning situation. The preservice teachers must follow regulations in order to complete the course and pass the examinations. Therefore, the learning situation has made them less active in their thinking and proactive in their study.

The integrative orientation element showed a negative effect and had no significant correlation to Chinese competency. This revealed that integrative orientation was not associated with the use of Chinese language learning. It showed that preservice teachers are native speakers, are used to speaking Chinese and adopt the Chinese culture as a lifestyle. Previous researchers (Sugita McEown, Sawaki, & Harada, T., 2017; Taie & Afshari, 2015; Dörnyei, 2003; Gardner, 1985, 2001) have stated that the integrative dimension reflects a favourable attitude towards the foreign language community, where learners are psychologically comfortable with the native speaker.

The findings also revealed some unique observations in this area of study. It revealed that motivation intensity was not relative to the preservice teachers' Chinese competency. One possibility was that Chinese preservice teachers live in a language environment and are able to frequently use Chinese in various situations. This could be the reason why they do not need any extra effort to learn Chinese. According to Liu (2017), those with a low command of Chinese competency might need to study Chinese. In such cases, probably the lessons do not need to target the preservice teachers' needs or interest and hence, the preservice teachers do not have to depend on their motivation to achieve their goals.

Finally, the least instrumental orientation was found to be positively significant and had a low relationship with Chinese competency. This means that these inconsistencies could be due to some different factors, such as the nationality context of the situation, level of proficiency, motivation and strategies. This is an indication to researchers as to which factors need to be deeply examined and also which other constructs can be developed for examining factors that motivate learning Chinese.

## 12 CONCLUSION

Several suggestions were derived from the research.

First, the present study showed that there was a weak positive attitude towards the learning situation regarding Chinese competency. Hence, further studies are necessary for identifying factors, such as the curriculum, instruction, and examination system that affects Chinese preservice teachers' positive attitude in a learning situation.

Second, a negative integrative orientation should be considered as an indication of professional readiness because an internal desire to learn Chinese culture and language is related to a teacher's teaching efficacy. Thus, reasons for a negative integrative orientation should be identified because it might affect a preservice teacher's classroom performance.

## REFERENCES


Chin, W.W. 1998. Commentary: Issues and Opinion on Structural Equation Modeling. *MIS Quarterly* 22(1):1.

Cohen, J. 1988. Statistical Power Analysis for the Behavioral Sciences, Lawrence Erlbaum Associates, Hillsdale, NJ.

Damar, E.A. 2018. Identifying motivational factors of preservice EFL teachers. *Academic Journal of Interdisciplinary Studies* 7(1): 147–151.

Deci, E. L., & Ryan, R.M. 2000. The "What "and "Why "of goal pursuits: Human needs and the self-determination of behavior. *Psychological Inquiry* 11(4): 227–268.

Dörnyei, Z. 2003. Attitudes, orientations, and motivations in language learning: advances in theory, research, and applications. *Language Learning* 53: 3–32.

Fan, P.S. 2012. Components and Characteristics of Dialect Vocabulary in the Malaysian Chinese Language (马来西亚华语的方言词语成分与特色). *Institute of China Studies*. Kuala Lumpur: University of Malaya Press.

Fornell, C. & Cha, J. 1994. Partial least square. In R. P. Bagozzi (eds.), *Advanced methods of marketing research*: 52–78. Cambridge, MA: Blackwell Publishers.

Furnell, C. & Larcker, D. F. 1981. Evaluting structural equation models with unobservable variables and measurement error. *Jurnal of Marketing Research* 18(1): 39–50.

Gardner, R.C. 1985. *Social psychology and second language learning: The role of attitudes and motivation*. London: Edward Arnold.

Gardner, R.C. 2001. Language learning motivation: The student, the teacher, and the researcher. *Foreign Language Education* 6: 1–18.

Gardner, R.C. 2006. Motivation and Attitudes in Second Language Learning. *Encyclopedia Of Language & Linguistics*: 348–355. doi:10.1016/B0-08-044854-2/00625-8

Gardner, R.C. & Lambert, W.E. 1972. *Attitude and motivation in second language learning*. Rowley, MA: Newbury House.

Geisser, S. 1974. A predictive approach to the random effect model. *Biometrika* 6(1): 101–107.

Hair, H.J. Hult, G.T. Ringle, C.M. & Sarstedt, M. 2014. *A primer on partial least squares structure equation modelling (PLS-SEM)*. Thousand Oaks, CA: Sage.

Hair Jr.J.F. Babin, B.J. & Krey, N. 2017. Convariance-based Structural Equation Modeling in the Jurnal of Advertising: Review and Recommendations. *Journal of Advertising*: 1–15.

Ho, M.C. 1998. Culture studies and motivation in foreign and second language learning in Taiwan. *Language, Culture & Curriculum* 11(2): 165–182.

Koh, S.T. 2008. Attitudes towards the Mandarin usage among Chinese in Malaysia: A case study. Dissertation (M.M.L.S). Universiti Malaya, 2008.

MOE. 2013. *Malaysian Education Blueprint 2012-2025*. Putrajaya: Kementerian Pendidikan Malaysia.

Lai, M.L. 2013. Impacts of medium of instruction on language attitudes: A case of Hong Kong. *Asia-Pacific Edu Res* 22: 61–68. http://:doi10.1007/s40299-012-0025-1.

Lam, Y.C. Yau, Y.S. Yeo, C.E. Chew, C.K. Lee, S.P. & Lee, V.L. 2006. The preservice teachers' efficiency in teaching and learning Chinese reading. *Jurnal Penyelidikan IPBL* 7: 48–54.

Larisa Niktina, Zuraidah, Mhd.D. & Lau, S.C. 2016. Construction and validation of a questionnaire on language learning Motivation. *Зборник Института за педагошка истраживања Година* 48(2): 284–300. http://:doi 10.2298/ZIPI1602284N

Liu, M.H. 2017. Motivation, motivation intensity, use of Chinese and self-rated Chinese competence. *College Student Journal* 51(1): 63–76.

Mora, J.C. & Darcy, I. 2017. The relationship between cognitive Control and pronunciation in a second Language. In Mora, J.C. & Darcy, I. (eds), *Second Language Pronunciation Assessment: Interdisciplinary Perspectives*: 95–120. Bristol: Multilingual Matters.

Neo, K. S. (2011). *Chinese language and Chinese culture*. Selangor: Pusat Penerbitan Universiti (UPENA), UiTM.

Ng, P.H. Wong, S.B. Tan, K.C. Guek, Y.H. & Lim, W. H. 2017. The case study on the level of oral proficiency of the teacher of the Post Graduate Diploma Program who graduated from the Chinese Language as a second language from Beijing. *Jurnal Pendidikan Bahasa*: 79–86.

Song, S. & Pornsima, D. 2017. Motivation for learning among Thai students studying Korean as a foreign language in Thailand. *Scholar: Human sciences* 8(2): 154–164.

Stone, M. 1974. Cross-validation and multinomial prediction. *Biometrika* 61(3): 509–515.

Sugita McEown, M. Sawaki, Y. & Harada, T. 2017. Foreign Language Learning Motivation in the Japanese Context: Social and Political Influences on Self. *Modern Language Journal* 101(3): 533–547.

Taie, M. & Afshari, A. 2015. A critical review on the socio-educational model of SLA. *Theory and Practice in Language Studies* 5(3): 605–612.

Wang, H. & Huang, C. 2016. Those who can, teach: the academic quality of preservice students in teacher education programmes in Taiwan. *Asia-Pacific Journal Of Teacher Education* 44(1): 66–79.

Wang, X.M. 2012. *Mandarin spread in Malaysia*. Kuala Lumpur: University of Malaya Press.

Wen, H. 2016. The Theoretical Methods and Principles of Teaching Chinese as a Foreign Language (对外汉语词汇教学的理论方法和原则), *Morden Chinese* 4: 41–43.

*The Social Sciences Empowered – Lumban Gaol, Hutagalung & Fong Peng (eds)*
*© 2020 Taylor & Francis Group, London, ISBN 978-0-367-00090-5*

# The relationship between well-being and EFL classroom management of Chinese university EFL teachers: The moderating role of teaching experience

Wu Yong
*Zhejiang Yuexiu University of Foreign Languages & Department of Language and Literacy Education, Faculty of Education, University of Malaya, Malaysia*

Mohd Rashid Bin Saad, Fonny Dameaty Hutagalung & Chan Choon Tak
*Faculty of Education, University of Malaya, Malaysia*

ABSTRACT: The purpose of this research is mainly to explore the influence of Chinese university EFL teachers' well-being on their classroom management, and moderation effects of their teaching experience on the influential relationship between well-being and EFL classroom management. Some 486 Chinese university EFL teachers from 19 universities in Zhejiang province were investigated, and the extracted data underwent analysis of structural equation modeling with use of AMOS software. The data analysis indicated that Chinese university EFL teachers' well-being has a significantly positive influence on their EFL classroom management, which means Chinese university EFL teachers possess more well-being (PERMA) and their EFL classroom management will become better correspondingly. Additionally, the teaching experience among EFL teachers has a significant moderating effect in the causal correlation of well-being to EFL classroom management, suggesting that experienced EFL teachers have more well-being and better classroom management. Finally, some suggestions are given on how to improve well-being and EFL classroom management of Chinese university EFL teachers in Zhejiang province.

## 1 INTRODUCTION

As the largest group of foreign language teachers, EFL (English as a Foreign Language) teachers always attract much attention from researchers. However, compared to the research situation of EFL social value, the development of EFL teachers' well-being is relatively ignored and most research works focus on the linguistic aspects of EFL teachers (Gabryś-Barker & Gałajda, 2016). In fact, a number of studies internationally have found EFL teachers are at a relatively high risk of common work-related stress compared to other jobs (Stansfeld et al., 2011). In China, the incidence of EFL teachers' well-being has been reported as an unresolved problem. For instance, the Chinese famous website Sina conducted a survey on Chinese university EFL teachers' well-being that showed nearly 90% of Chinese university EFL teachers reflected stress and ill-being. These problems have come to the attention of the Chinese government and promulgated the "outline of the National Medium and Long-term Education Reform and Development Plan (2010–020)," which put great emphasis on increasing the level of Chinese teachers' well-being; however, there has been relatively little research focused on the relationship between Chinese university EFL teachers' well-being and classroom management. Additionally, some previous studies indicated experienced teachers have high-level well-being and classroom management, while other studies reflected no difference between experienced teachers and novice teachers on their well-being and classroom management. Therefore, it is hard to identify whether teaching experience effects the influential relationship between EFL teachers' well-being and classroom management because the results are inconsistent.

Hence, the present study focuses on the relationship between Chinese university EFL teachers' well-being and EFL classroom management, and the demographics as a moderator effecting EFL teachers' well-being and their classroom management. The findings of this study will contribute to the general body of knowledge on the relationship between well-being and EFL classroom management, and present a new view to explore the complex construct in foreign language teaching, as well as providing additional understanding of the process of EFL classroom phenomenon in foreign language teaching for teachers.

## 2 STATEMENT PROBLEMS

The issue of teacher well-being has received much attention over the past century (Borg &

Riding, 1991). In 1998, Seligman was elected as a president of the American Psychology Association and began to push the movement of positive psychology so that positive psychology started to focus on well-being. More recently, Seligman (2011) defined his concept of well-being with his five pillars and stated the purpose of well-being theory is to increase flourishing by increasing positive emotion, engagement, relationships, meaning, and accomplishment (summarized by the acronym PERMA). Some scholars have proven valid, reliable, and practicable PERMA models among athletes, education leaders, and service workers. However, PERMA theory and model have been little researched in terms of Chinese university EFL teachers.

On the other hand, although classroom management seems to be one of the most common concerns of teachers, the study of foreign language education classroom management seems to be obviously lacking in the literature (Wright, 2005). There is also scant research on Chinese EFL classroom management especially in the Chinese higher education context.

Although there is some literature that has investigated well-being among teachers more generally, and provided evidence of its importance for their job stress, burnout, job performance, and job satisfactions (Collie, Shapka, & Perry, 2012), literature on the relationship between teachers' well-being (PERMA) and their classroom management is scant because PERMA is still its infancy.

In empirical studies, many scholars consider that demographic differences affect outcome variables (Tsui & Gutek, 1999). In current research, there are little research using teaching experiences as a moderator effecting well-being and EFL classroom management. More than a quarter of Australian novice teachers lost control of classroom management due to emotional exhaustion after starting their career and hope to leave the profession within the first five years of teaching (Marshall, 2013). Thus, beginning teachers need more guidance of well-being so that they can contribute to their classroom management (McCallum & Price, 2010). Other scholars explained that no matter whether novice or experienced teachers, classroom management has consistently been a concern of teachers (Kafman & Moss, 2010). Ritter and Hancock (2007) reported that level of teaching experience is not necessarily related to classroom management. There is no clear and consistent conclusion on whether teaching experience is the prosed moderator (teaching experience) on the relationship between well-being (PERMA) and EFL classroom management. It is necessary to explore teaching experience as the moderator on the relationship between Chinese university EFL teachers' well-being and classroom management in Zhejiang province.

## 3 METHODOLOGY

### 3.1 *Participants*

This research is a nonexperimental quantitative research using survey technique for data collection. Nonexperimental research is described as a systematic inquiry in which the research does not have direct control of the independent variable because the change in the independent variables has already occurred (Hoy, 2010). According to Zhejiang Education Development Statistics report (2016), there are 2,931 university EFL teachers in 37 universities. By means of random sampling, 486 university EFL teachers in Zhejiang province were extracted as research respondents. Ages range from below 30 years (n = 121, 24.9%), 31–40 years (n = 209, 43%), 41–50 years (n = 121, 24.9%), and over 50 years (n = 35, 7.2%). Among them, 131 are male (27 %) and 355 are female (73 %). The majority of the participants have more than five years teaching experience (52.7 %, n= 256), and 47.3% (n=230) of them have less than five years teaching experience.

### 3.2 *Instruments*

The instruments used in this study consists of three sections that corresponded to the three variables with 42 items in this study. Section 1 is Chinese university EFL teachers' demographic details with 5 items; while Sections 2 measured the Chinese university EFL teachers' well-being which used PERMA profiler with 15 items (Butler & Kern, 2016). And EFL classroom management questionnaire (Akbari & Bolouri, 2015) is used to measure in Section 3. The SEM procedure with Analysis of Moment Structure (AMOS) version 22.0 was carried out to answer research questions.

To make sure this study is reliable, reliability tests using Cronbach alpha test and composite reliability (C.R) were conducted. As Table 1 shows, well-being (PERMA) and EFL classroom management constructs are reliable with Cronbach alpha value above 0.70. The construct with lowest reading of Cronbach alpha value is 0.722 with reading of achievement dimension and the construct with highest value of Cronbach alpha value is the construct of meaning of 0.922. In terms of composite reliability, all readings are above 0.70 and they are deemed to be reliable (Hair et al, 2010). Therefore, all of the constructs measurements are proven to be reliable.

In this model, the lowest C.R of 0.769 is for relationship and highest C.R is 0.924 for the meaning dimension. The value of average variance extracted above 0.50 shows that the construct has high content

Table 1. The validity and veliability of the measurement model.

| Constructure | CR | AVE | Cronbach alpha |
|---|---|---|---|
| Postive emotion | 0.867 | 0.686 | 0.863 |
| Engagement | 0.807 | 0.678 | 0.804 |
| Relationship | 0.769 | 0.624 | 0.768 |
| Meaning | 0.924 | 0.801 | 0.922 |
| Achievement | 0.778 | 0.636 | 0.722 |
| Langugage management | 0.839 | 0.567 | 0.837 |
| Behavior mangement | 0.846 | 0.525 | 0.843 |
| Instructional management | 0.922 | 0.518 | 0.920 |

validity. For this model, the lowest average variance extracted is on the construct of instructional management with the value of 0.518 and the highest average variance extracted value is 0.801 for the meaning dimension. Therefore, the content validity of this model is proven (Hair et al., 2010). As mentioned by Hair et al (2010), when the constructs of SEM model have average variance extracted (AVE) and composite reliability (C.R) value above their cut off point (0.5 and 0.7 respectively), the convergent validity of the construct is proven. As in the case of this study, the AVE and C.R of the model are above the cutoff point of 0.5 and 0.70, therefore, the constructs of this model have met the requirement of convergent validity.

## 3.3 *Validity and reliability of research instruments*

Table 2. The discriminate validity index summary for measurement models.

| P | IM | LM | BM | M | A | R | E |
|---|---|---|---|---|---|---|---|
| 0.828 | | | | | | | |
| 0.394 | 0.720 | | | | | | |
| 0.406 | 0.718 | 0.753 | | | | | |
| 0.281 | 0.342 | 0.275 | 0.725 | | | | |
| 0.605 | 0.481 | 0.472 | 0.320 | 0.895 | | | |
| 0.465 | 0.456 | 0.526 | 0.258 | 0.763 | 0.798 | | |
| 0.634 | 0.422 | 0.462 | 0.282 | 0.763 | 0.627 | 0.790 | |
| 0.651 | 0.448 | 0.451 | 0.297 | 0.668 | 0.595 | 0.753 | 0.823 |

Notes: Bolded numbers are square root of AVE
P: positive emotion; IM: Instructional management; LM: Language management; BM: behavior management; M: meaning; A: achievement; R: relationship

# 4 FINDINGS

*Research question 1: Is there a significant relationship between Chinese university EFL teachers' well-being and EFL classroom management?*

In structural equation modeling analysis, model-fitting tests determine whether the model is acceptable and applicable. Due to the factor loading of item E1, R3, A1, LM1 and IM9 are too low to be deleted. The output of the modification indices of AMOS suggest that model modifications need to be carried out. In this model, there are some significant correlation between e25 and e26 (Modification Indices = 40.331, Par Charge = 0.177), e34 and e40 (Modification Indices = 23.139, Par Charge = 0.096), e19 and e20 (Modification Indices = 22.163, Par Charge = 0.353), e29 and e27 (Modification Indices = 13.098, Par Charge = 0.126).

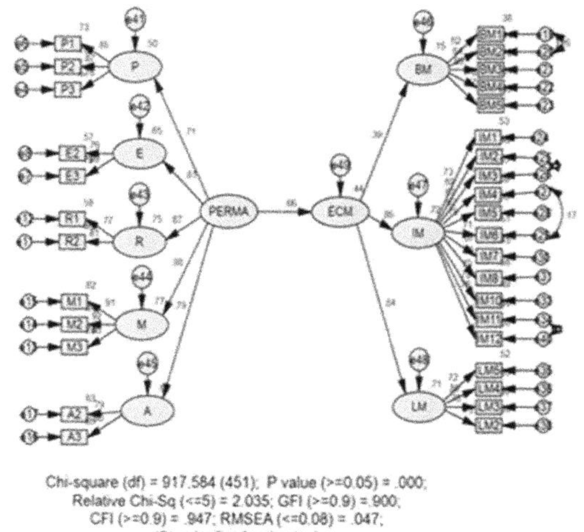


Table 3 shows the values of regression path in the hypothesized model. The relationship between Chinese university EFL teachers' well-being and EFL classroom management are positive significant ($\beta = 0.664, p < 0.001$). A one-unit input of well-being and would cause a 0.664 unit increase in EFL classroom management. In addition, the current study reflects that a total of 44% of variance of EFL classroom management is predicated by Chinese university EFL teachers' well-being. Another 56% of variance is unpredicted ($R^2 = 0.44$).

*Research question 2: Is there a causal relationship between Chinese university EFL teachers' well-being and EFL classroom management moderated by teaching experience?*

According to Zainudin (2015), the effects of the independent variable on its corresponding dependent variable must exist and be significant before

Table 3. The relationship between well-being and EFL classroom management.

| | | | $\beta$ | Estimate | S.E. | C.R | P |
|---|---|---|---|---|---|---|---|
| ECM | ← | PERMA | .664 | .266 | .046 | 5.76 | *** |

Note: *** is significant at the 0.001 level

analyzing the moderator. The previous study showed that there is a significant and positive relationship between Chinese university EFL teachers' well-being and EFL classroom management. In this study, teaching experience is a moderator in the relationship between Chinese university EFL well-being and EFL classroom management. If the value for each testing differs by more than 3.84, then the moderation occurs in the path (Zainudin, 2015). The differences in chi-square values between constrained model and unconstrained model for path were obtained. The results are presented in the Table 4.

Referring to Table 5, the chi-square value for the constrained model for the experienced group is 1,661.385, while for the unconstrained model is 1,599.684. The moderation test for the experienced group was conducted by obtaining the difference between both values. The results are presented in Table 5.

Based on Table 6, the difference in Chi-square value is 61.701 (1661.385–1599.684), while the difference in Degree of Freedom is 933 – 902 = 31. In order to make the test significant, the difference in chi-square value must be higher than the chi-square value of 1

Table 4. The chi-square value and DF for the constrained and unconstrained model for experienced teachers group.

Chi-Square Value and DF

| | Model | NPAR | CMIN | DF | P | CMIN/DF |
|---|---|---|---|---|---|---|
| Constrained | Default Model | 123 | 1,661.385 | 933 | 0.000 | 1.781 |
| | Saturated Model | 1,056 | .000 | 0 | | |
| | Independence Model | 64 | 10,154.558 | 992 | 0.000 | 10.236 |
| Unconstrained | Default Model | 154 | 1,599. 684 | 902 | 0.000 | 1.773 |
| | Saturated Model | 1,056 | .000 | 0 | | |
| | Independence Model | 64 | 10,154.558 | 992 | 0.000 | 10.236 |

Table 5. The moderation test for experienced teacher group data.

| | Constrained Model | Unconstrained Model | Chi-Square Difference | Result on Moderation |
|---|---|---|---|---|
| Chi-Square | 1,661.385 | 1,599.684 | 61.701 | Significant |
| DF | 933 | 902 | 31 | |

The hypothesis statements:
H01: Teaching experience is not a significant moderator for the relationship between Chinese university EFL teachers' well-being and EFL classroom management. (Rejected)
H1: Teaching experience is a significant moderator for the relationship between Chinese unversity EFL teachers' well-being and EFL classroom management. (Fail to reject)

Table 6 . The chi-square value and DF for the constrained and unconstrained model for novice teachers group.

Chi-Square Value and DF

| | Model | NPAR | CMIN | DF | P | CMIN/DF |
|---|---|---|---|---|---|---|
| Constrained | Default Model | 123 | 1,661.385 | 933 | 0.000 | 1.781 |
| | Saturated Model | 1,056 | 0.000 | 0 | | |
| | Independence Model | 64 | 10,154.558 | 992 | 0.000 | 10.236 |
| Unconstrained | Default Model | 154 | 1,599.684 | 902 | 0.000 | 1.773 |
| | Saturated Model | 1,056 | .000 | 0 | | |
| | Independence Model | 64 | 10,154.558 | 992 | 0.000 | 10.236 |

Table 7. The moderation test for inexperienced teachers group data.

| | Constrained Model | Unconstrained Model | Chi-Square Difference | Result on Moderation |
|---|---|---|---|---|
| Chi-Square | 1,661.385 | 1,599.684 | 61.701 | Significant |
| DF | 933 | 902 | 31 | |

The hypothesis statements:
H01: Teaching experience is not a significant moderator for the relationship between Chinese university EFL teachers' well-being and EFL classroom management. (Rejected)
H1: Teaching experience is a significant moderator for the relationship between Chinese unversity EFL teachers' well-being and EFL classroom management. (Fail to reject)

Table 8. The effect of EFL teachers' well-being on EFL classroom management is for experienced and novice teachers group.

| | Group | Standardized Beta Estimate | P | Result |
|---|---|---|---|---|
| ECM←PC | Experienced | 0.726 | 0.000 | is significantly different from zero at the 0.10 level |
| | Novice | 0.581 | 0.000 | (two-tailed) |

degree of freedom, and the value is 3.84 (Zainudin, 2015). The results show that the moderation test for experienced teachers group is significant. Therefore, the null hypothesis is rejected. The novice teacher group was then carried out. The results are presented in Table 6.

Referring to Table 7, the chi-square value for the constrained model for the experienced group is 1,661.385 while for the unconstrained model is 1,599.684. The moderation test for the novice group was conducted by obtaining the difference between both values. The results are displayed in Table 7.

Based on Table 7, the results show that the moderation test for the novice teachers group is significant. Therefore, the null hypothesis is rejected. In order to identify which group is more pronounced, the standardized parameter estimates and its significant for both groups were obtained.

Based on Table 8, teaching experience as a mdoerator is significant on the relationship between EFL teachers' well-being and EFL classroom management. The standardized parameter estimate for "experienced" is 0.726 while the same estimate for "novice" is 0.581. Thus, it can be concluded that the effect of EFL teachers' well-being on EFL classroom management is more pronounced in the experienced teachers group compared to the novice group.

## 5 DISCUSSION AND SUGGESTIONS

This empirical study reveals the general situation of Chinese university EFL teachers' well-being and their EFL classroom management in Zhejiang province. The results of this study indicated there is a positive and significant direct relationship between Chinese unversity EFL teachers' well-being and EFL classroom management (see Figure 1 and Table 3). The findings of this study are consistent with several previous studies (Mercer, Oberdofer & Saleem, 2016) that teachers possessing high well-being have a positive effect on their classroom management. According to the structural modeling analysis (Figure 1), it is indicted that when the Chinese university EFL teachers have more well-being, and they will elevate their classroom management. The teachers possessed positive emotions which can help them to establish their "personal resources" to better equip them for their teaching performance. And engagement would promote teachers to work actively and form good relationships between students so that they find the relationships meaningful and have initial motivation to be good teachers. Above all, teachers' well-being is not a castle in the air: it exists and can be improved. Therefore, it needs educators to be more concerned about EFL teacehers' well-being and adopt a series of useful strategies to reduce ill-being to improve their classroom management in order to have a positive influence on language learners.

On the other hand, it is also found that teaching experience has a moderator effect on the relationship between well-being and EFL classroom management (see Table 8). The extent that the experienced Chinese university EFL teachers' well-being raises their EFL classroom mangement is superior to that of novice teachers. More experienced teachers might have more well-being to show higher EFL classroom management to ensure their occupation existential advantages. This finding implies that in Chinese culture, experienced teachers got more respect and recognition from colleagues, students, and parents. Those experienced teachers had a deeper understanding of students, colleagues, and parents. Therefore, they can set the achievement goal to inspire students' language learning and have more positive emotions to deal with all sorts of problems in classroom. To take it from another perspective, since experienced teachers have more problem-solving ability than novice teachers, they are more engaged in classroom activities, and encourage student involvement in the class activities, especially developing communication interactions. Hence, providing well-being education is necessary for novice teachers in the early states of education training. Universities would review novice teachers' performance in the classroom management and assist in alleviating identified concerns. Experienced teachers would share more teaching experience and provide multiple kinds of assistance so that novice teachers have confidence to manage and instruct effectively in EFL classroom management.

## 6 CONCLUSION

The results of this study have indicated a significant and positive relationship between Chinese university EFL teachers' well-being and their classroom management. The final model (see Figure 1 and Table 3) can serve as an alternative to interpreting and supporting these findings. Although Seligman brought well-being (PERMA) into the education field, language educators should be more aware of the associations between improving language teachers' well-being and their teaching performance (MacIntyre & Vincze, 2016). Therefore, further studies into EFL classroom management can include the well-being (PERMA) aspect and undertake further research in this field. What's more, teaching experience is a significant moderator on the relationship between Chinese university EFL teachers' well-being and EFL classroom management (See Table 8), therefore, further studies can pay more attention to the novice teachers' well-being and their EFL classroom mangement.

## ACKNOWLEDGMENT

This research was support by Zhejiang Education Science Planning Research Project "Research on Optimization of Undergraduate Students Training Mode of Foreign Language Education in Local unversities under the Post-method era (2017SCG387).

## REFERENCES

Akbari, R., & Bolouri, M. 2015. Development and validation of an ELT-based classroom management questionnaire. *Journal of Linguistic and Language Research* 4.196–201.

Borg, M. G., & Riding, R. J. 1991. Occupational stress and satisfaction in teaching. *British Educational Research Journal*, 17, 263–281.

Butler, J., & Kern, M. L. 2016. The PERMA-profiler: A brief multidimensional measure of flourishing. *International Journal of Well-Being*, 6(3), 1–48.

Collie, R.J., Shapka, J.D., & Perry, N.E. 2012. School climate and social–emotional learning: Predicting teacher stress, job satisfaction, and teaching efficacy. *Journal of Educational Psychology*, 104, 1189–1204.

Gabrys-Barker, D. 2016. Caring and sharing in the foreign language class: On a positive classroom climate. In D. Gabrys-Barker, & D. Galajda (Eds.), *Positive Psychology Perspectives on Foreign Language Learning and Teaching* (pp. 133–153). Switzerland: Springer.

Hair, J. F., Black, W. C., Babin, B. J., & Anderson, R. E. 2010. *Multivariate Data Analysis* (7th ed.). Upper Saddle River, NJ: Prentice Hall.

Hoy, W. K. (2010). *Quantitative Research in Education: A Primer*. Los Angeles: Sage.

Kaufman, D., & Moss, D. M. 2010. A new look at preservice teachers' conceptions of classroom management and organization: Uncovering complexity and dissonance. *The Teacher Educator*, 45, 118–136.

MacIntyre, P & Vincze. 2016. Positive and negative emotions underlie motivation for L2 learning, doi: https://doi.org/10.14746/ssllt.2017.7.1.4.

Marshall, K. 2013. Burnout hits one in four teachers. *The Age*, http://theage.com. au/victoria/burnout-hits-one-in-four-teachers-20131005-2v13y.html.

McMillan, J. H., & Schumacher, S. 2010. *Research in Education: Evidence-Based Inquiry* (Seventh Ed.). Upper Saddle River, NJ: Person Education.

Mercer, S., Oberdorfer, P., & Saleem, M. 2016. Helping language teachers to thrive: Using positive psychology to promote teachers' professional well-being. In D. Gabrys-Barker & D. Galajda (Eds.), *Positive Psychology Perspectives on Foreign Language Learning and Teaching* (pp. 213–229). New York: Springer.

Ritter, J. T., & Hancock, D. R. 2007. Exploring the relationship between certification sources, experience levels, and classroom management orientations of classroom teachers. *Teaching and Teacher Education*, 23, 1206–1216.

Seligman, M. 2011. *Flourish: A Visionary New Understanding of Happiness and Well-Being*. New York: Free Press.

Tsui, A. S., and Gutek, B. 1999. *Demographic Differences in Organizations: Current Research and Future Directions*. Lanham: Lexington Books.

Wright, T. 2005. *Classroom Management in Language Education*. New York: Palgrave MacMillan.

Zainudin Awang. 2015. *A Handbook on SEM*. MPWS Publisher.

*The Social Sciences Empowered – Lumban Gaol, Hutagalung & Fong Peng (eds)*
*© 2020 Taylor & Francis Group, London, ISBN 978-0-367-00090-5*

# Gratitude feeling and cervical cancer preventive behavior in adolescents

Ambar Sulianti*, Fenti Hikmawati, Dadang Sahroni & Imam Sunardi
*UIN Sunan Gunung Djati, Bandung, Indonesia*

ABSTRACT: Indonesia is the country with the largest number of cervical cancer patients in the world. The guarding of reproductive health should be emphasized starting from adolescent girls. The aim of this study is to understand the relationship between feeling gratitude and the behavior of maintaining the health of sex organs. This research is a cross-sectional quantitative research. A total of 74 female psychology faculty students were used as research subjects. Measurement of the gratitude scale was taken by modifying the Western scale by adding a Godhead element. Data were analyzed using Pearson correlation test. The results showed there was a relationship between feeling gratitude and reproductive health behavior (p: 0.02). This study shows that being grateful for the healthy benefits of the reproductive system can lead to the behavior of women maintaining the health of their sex organs.

*Keywords*: gratitude, reproduction, health, prevention, behavior

## 1 INTRODUCTION

The number of incidences of cancer is increasing and cancer is a leading cause of death worldwide (Nurcahyanti, 2016). Although the disease can be prevented and cured at some stages, the incidence increases every year. Indonesia has a population of 93.15 million women aged 15 years and over who are at risk for cervical cancer. Current estimates show that every year 20,928 women are diagnosed with cervical cancer and 9,498 die from this disease. Cervical cancer ranks as the second most common cancer among women in Indonesia and cancer is most common among women between the ages of 15 and 44 (Bruni et al., 2017; Centre, 2017).

One of the efforts undertaken to reduce the number of cervical cancer victims is by increasing prevention programs. Cervical cancer is caused by the Human Papilloma Virus (HPV), especially types 16 and 18. There are several conditions that cause the cervix to be susceptible to HPV attacks, including infection and irritation of the vaginal and cervical areas. Guarding the health of reproductive organs becomes very important for women to avoid cervical cancer.

H. L. Blum (in Foller, 1992) explains there are four main factors that affect the degree of public health. These four factors are the determinants of health problems. The four factors consist of behavioral factors/life style, environmental factors (social, economy, politics, and culture), health service factors (type of coverage and quality), and genetic factors (heredity). Among these factors, human behavior is the largest determinant and the most difficult factor to overcome, followed by environmental factors (Follér, 1992).

## 2 STATEMENT OF THE PROBLEMS

Research shows that gratitude predicts psychological subjective well-being (Sadeghi & Pour, 2015), while well-being is a positive predictor of physical health (Hill, Allemand, & Roberts, 2013). Thus, gratitude is related to physical health.

Health behavior is influenced by a person's attitude towards health phenomena. Attitude towards health is influenced by three components, namely cognitive, affective, and behavioral. In other words, someone's attitude involves a function of cognition, emotionality, or action (Shahed, 2008).

Gratitude is specifically perceived as a positive emotion (Fredrickson, 2013; Tugade, Fredrickson, & Feldman-Barrett, 2004). Gratitude is associated with satisfaction (Datu & Mateo, 2015; Puente-Díaz & Meixueiro, 2016), happiness, positive pride (not arrogance), hope, life satisfaction, optimism, empathy, outlook, and positive feelings (Meherunissa, 2016).

Watkins et al. defines gratitude as an affective nature, referring to how much the individual experiences feelings of appreciation for the gifts he obtains. An individual experiences many favors, among them an intact body, and this should be appreciated by the individual by maintaining his body if the individual has a high level of gratitude (Watkins, Khathrane Woddward, Stone, & Kolts, 2003). Positive feelings toward

*Corresponding authors: ambarsulianti@uinsgd.ac.id

life is one component in the gratitude proposed by Watskin et al. 2003; that is, not feeling lack in his life, or in other words, having a sense of abundance. A person who does not feel deprived will have a positive feeling in himself. He will feel affluent for what he has, satisfied with the life he lived.

Research on the relationship of gratitude with both physical and psychic health has been studied (Datu & Mateo, 2015; Dixit & Malhotra, 2017; Emmons & Stern, 2013; Hasemeyer, 2013; Shao, Gao, & Cao, 2016; Sulianti, Darmalaksana, Sulaeman, & Rahman, 2017). However, the relationship of one component of gratitude (a positive feeling on life) that is acknowledged by health care behavior has never been reported. This study aims to analyze the relationship between positive feelings of gratitude for life with the behavior of maintaining the health of female sex organs.

## 3 CONCEPT OF REPRODUCTIVE HEALTH PREVENTIVE BEHAVIOR

Reproductive health is a state of complete physical, mental and social well-being and not merely the absence of disease or infirmity, including in matters relating to the reproductive system and its functional processes (World Health Organization, 2006). Reproductive health involves all of the reproductive processes, functions, and systems at all stages of human life (Wang, 2003).

Lawrence identifies factors that can influence the behavior of adolescent reproductive health, namely knowledge and attitudes of the adolescents, access to information, and reinforcing factors that strengthen including family, teachers, and peers (House, Bates, Markham, & Lesesne, 2010). Lawrence also argues that cognitive, social, and behavioral subcontracts affect adolescent sexual and reproductive health

The indicator subset for health promotion covers major areas of health-related behavior that are typically targeted by health education and outreach campaigns, one of which is reproductive behavior (Marshall, Leatherman, Mattke, & Panel, 2004). The process and social behavior associated with health refers to a series of phenomena related to the interaction between a person and the environment that has a direct or indirect influence on health. Social behavior and processes are very often seen as aspects of lifestyle. Some dimensions of dynamic health concepts can be considered. So, from a positive perspective, health balance often means biopsychosocial well-being. A person's health is influenced by disease preventive behavior (Noack, 1988).

One dimension of health is the spiritual dimension. To achieve growth in the spiritual dimension of health, many people do a serious doctrinal study related to the set religious group. Research shows that life lasts longer for people who regularly participate in religious practices, especially for women. In today's world, work and productivity play an increasingly important role. The way we see ourselves, how we see "good" in the world in which we live, and how we are grateful for things in life, all interact with health (Larsson, Sundler, & Ekebergh, 2012; "Shaping Your Health," 1979).

Therefore, this study is aimed to determine the influence of feeling gratitude on reproductive health preventive behavior.

## 4 RESEARCH METHODS

This research was conducted by using quantitative method of cross sectional study. Subjects in this study were 74 female students not physically or spiritually suffering from serious illness. The sampling technique used in this study was double sampling from four groups of entry-year college students. The population of this study was 368 women who were actively registered as students of the psychology faculty of UIN Sunan Gunung Djati.

There are several scales developed by Western researchers in order to measure gratitude, including Gratitude Questionnaire-6 (Emmons & Stern, 2013; Froh, Fan, Emmons, Huebner, & Watkins, 2004) Gratitude Adjective Checklist (Froh et al., 2004), and Gratitude Resentment and Appreciation Test (GRAT) short form (Duran, 2017). Unfortunately, if we examine further, the gratitude tools that exist and were developed in other countries, generally eliminate aspects of the Godhead. Therefore, measuring the scale of positive feelings of gratitude to God based on Indonesian culture was taken from the scale developed by Listiyandini, Nathania, Syahniar, Sonia, and Nadya (2017). This test measures the components of positive feelings toward the life they have.

In this questionnaire, the components of gratitude are divided into three, namely:

a) Have a sense of appreciation towards others or God and life, a feeling of warm appreciation of someone or something. And clarified by Watkins (2003) with the characteristics of second- and third-level gratitude for people, that is, to appreciate the contribution of others to one's well-being, and have a tendency to appreciate simple pleasures.
b) Positive feelings for life owned. This component is derived from the characteristics of people grateful according to Watkins (Watkins et al., 2003), namely, not to feel lack in life or, in other words, have a sense of abundance.
c) The tendency to act positively as an expression of positive feelings and appreciation.

The measuring tool for the behavior of maintaining the health of intimate organs used a questionnaire developed by researchers from the indicators. The following are indicators of mantaining the health of female intimate organs (Access, 2011; Health, 2011; Kaur, Kaur, & Kaur, 2018; Marcell, Wibbelsman, & Seigel, 2011; Wang, 2003):

a) Mantaining hygiene of the body, especially during menstruation.
b) Cleansing the genitals after defecation and urination.
c) Methods and habits of cleaning the vagina.
d) Maintaining the cleanliness of underwear.
e) Selection, frequency of change, storage, and disposal of sanitary napkins (House, 2012).
f) Paying attention to the symptoms of vaginal discharge.
g) Mantaining healthy sexual behavior.

## 5 FINDINGS

The average scores of gratitude and preventive behavior are shown in Table 1.

Correlation between gratitude feeling and reproductive health preventive behavior is shown in Table 2.

Based on data in Table 1, we divided gratitude feeling data into three groups: Low (11–18), Average (-19–25), and High (26–28). In the same way, reproductive health preventive behavior data is also divided into three groups, namely Low (22–27), Average (28–35), High (36–39).

The results of correlation analysis in Table 2 showed that there was a significant positive correlation between gratitude feeling and reproductive health preventive behavior.

Table 1. *Descriptive statistics of gratitude feeling and reproductive health preventive behavior.*

| | N | Minimum | Maximum | Mean | Std. Deviation |
|---|---|---|---|---|---|
| Gratitude Feeling | | 11 | 28 | 21.76 | 3.491 |
| Reproductive Health Preventive behavior | 74 | 22 | 39 | 31.57 | 3.716 |
| Valid N (listwise) | 74 | | | | |

Table 2. Analysis of gratitude feeling to reproductive health preventive behavior.

| | | Gratitude feeling | Reproductive health preventive behavior |
|---|---|---|---|
| Gratitude feeling | Pearson Correlation | 1 | 0.334* |
| | Sig. (2-tailed) | | 0.004 |
| | N | 74 | 74 |
| Reproductive health preventive behavior | Pearson Correlation | 0.334* | 1 |
| | Sig. (2-tailed) | 0.004 | |
| | N | 74 | 74 |

* Correlation is significant at the 0.01 level (2-tailed).

## 6 DISCUSSION & INTERPRETATION

Results of the data analysis suggested that gratitude for what is owned, for the gift from God in the form of an intact physical body, in this case an intimate organ, could be a factor leading to maintaining the health of a person's intimate organs, guarding the gift from God. According to Green, behavior is influenced by three main factors, one of which is the knowledge and attitude of society toward health, including traditions and beliefs of the society to health matters, society's value system, education level, socioeconomic level, and so on.

Gratitude is one of the "healing effects" that changes health behavior in a better direction. A positive feeling of gratitude is part of a broader sense of positive transformation categories. Other directional healing effects of gratitude, love, and tenderness arise in an interpersonal context and facilitate the process of bringing closure to old wounds which, in turn, alleviates emotional suffering. The nature of gratitude has its aspect of fulfillment, both in terms of present and previous benefits. This encourages the emergence of energy from within the body in the form of positive motivation that is directed towards the achievement of future goals (Emmons & Stern, 2013).

Being grateful makes one have a more positive view and a broader perspective on life – the view that life is a gift (Seligman, Steen, Park, Peterson, & Report, 2005) Being grateful will cause a person to gain an emotional and interpersonal advantage, improve his or her new coping skills both consciously and unconsciously, triggering a sense of self that will lead a person's life in a more positive direction (Krause, 2006). Being grateful can also help improve a person's ability to deal with problems and find the best solution to a problem. A person who is grateful has more control over his or her environment, personal development, purpose in life, and self-acceptance (Dixit & Malhotra, 2017). Grateful people also have positive coping in the face of life's difficulties, seeking social support from others, interpreting experiences with different perspectives, and having plans for solving problems (McCullough, Tsang, & Emmons, 2004). In an effort to prevent cervical cancer, people who are grateful have a positive plan for preventive efforts in the form of maintaining the cleanliness of sex organs.

## 7 CONCLUSION

Overall, based on the results, it can be concluded that there is a positive relationship between positive feelings on life owned and preventive reproductive health behavior. The greater the feeling of gratitude, the better the behavior of prevention of reproductive health. Thus people who do not feel deprived will have positive feelings in themselves, feel affluent about what they have and satisfied with the life lived, and then take care of it, especially in terms of health. Based on the value of the belief that the health is a positive thing to be grateful for.

## REFERENCES


Access, O. (2011). Evaluation of a reproductive health awareness program for adolescence in urban Tanzania-A, 1–9.

Bruni, L., L, B.-R., Albero, G., Serrano, B., Mena, M., Gómez, D., & De, S. S. (2017). *Human Papillomavirus and Related Diseases Report*, (July). http://www.hpvcentre.net/statistics/reports/MYS.pdf

Centre, I. I. (2017). *Human Papillomavirus and Related Cancers, Fact Sheet 2017* (2017-07-27) ICO Information Centre on HPV and Cancer, India, 2017, 6–7.

Datu, J. A. D., & Mateo, N. J. (2015). Gratitude and life satisfaction among Filipino adolescents: The mediating role of meaning in life. *International Journal for the Advancement of Counselling*, 37(2), 198–206. Doi: 10.1007/s10447-015-9238-3

Dixit, S., & Malhotra, M. (2017). Role of happiness, hope and gratitude on health and wellbeing among young adults. *International Journal of Science and Research*, 6 (6), 1779–1784. https://www.ijsr.net/archive/v6i6/ART20174630.pdf

Duran, N. O. (2017). The Revised Short Gratitude, Resentment, and Appreciation Test (S-GRAT): Adaptation for Turkish college students Minnettarlık Gücenme ve Takdir Ölçeği Gözden Geçirilmiş Kısa F ormu (K-MGTÖ) : *Journal of Hapiness & Wellbeing*, 5(1), 23–37.

Emmons, R., & Stern, R. (2013). Gratitude as a psychotherapeutic intervention. *Journal of Clinical Psychology*. http://onlinelibrary.wiley.com/doi/10.1002/jclp.22020/full

Follér, M.-L. (1992). Social determinants of health and disease: the role of small-scale projects illustrated by the Koster health project in Sweden and Ametra in Peru. *Cadernos de Saúde Pública*, 8(3), 229–239. Doi: 10.1590/S0102-311X1992000300002

Fredrickson, B. L. (2013). *Positive Emotions Broaden and Build. Advances in Experimental Social Psychology* (1st ed., Vol. 47). Elsevier Inc. doi:10.1016/B978-0-12-407236-7.00001-2

Froh, J. J., Fan, J., Emmons, R. A., Huebner, E. S., & Watkins, P. (2004). *Measuring Gratitude in Youth*, 1–38.

Hasemeyer, M. D. (University of S. F. (2013). The Relationship between Gratitude and Psychological, Social, and Academic Functioning in Middle Adolescence. Graduate Thesis and Dissertations, (January).

Health, E. (2011). Introduction to Reproductive Health and the (Draft for review) Training Module 1 Children â €TM s Environmental Health Basic hormone and endocrine functions Reproductive physiology Importance of environmental exposures.

Hill, P. L., Allemand, M., & Roberts, B. W. (2013). Examining the pathways between gratitude and self-rated physical health across adulthood. *Pers Individ Dif*, 54 (1), 92–96. doi: 10.1016/j.paid.2012.08.011.

House, L. D., Bates, J., Markham, C. M., & Lesesne, C. (2010). Competence as a predictor of sexual and reproductive health outcomes for youth: A systematic review. *Journal of Adolescent Health*, 46(3 SUPPL.), S7–S22. doi: 10.1016/j.jadohealth.2009.12.003

House, S., Mahon, T., & Cavill, S. (2012). *Menstrual hygiene matters*, 1–354.

Kaur, R., Kaur, K., & Kaur, R. (2018). Menstrual hygiene, management, and waste disposal: Practices and challenges faced by girls/women of developing countries. *Journal of Environmental and Public Health*, 2018. doi: 10.1155/2018/1730964

Krause, N. (2006). Gratitude toward God, stress, and health in late life. *Research on Aging*, 28(2), 163–183. Doi: 10.1177/0164027505284048

Larsson, M., Sundler, A. J., & Ekebergh, M. (2012). The influence of living conditions on adolescent girls' health. *International Journal of Qualitative Studies on Health and Well-Being*, 7(May 2017). doi: 10.3402/qhw.v7i0.19059

Listiyandini, R. A., Nathania, A., Syahniar, D., Sonia, L., & Nadya, R. (2017). Mengukur Rasa Syukur: Pengembangan Model Awal Skala Bersyukur Versi Indonesia. *Jurnal Psikologi Ulayat*, 2(2), 473. Doi: 10.24854/jpu22015-41

Marcell, A. V., Wibbelsman, C., & Seigel, W. M. (2011). Male adolescent sexual and reproductive health care. *Pediatrics*, 128(6), e1658–e1676. doi: 10.1542/peds.2011-2384

Marshall, M., Leatherman, S., Mattke, S., & Panel, M. of the O. H. P. P. and P. C. (2004). Selecting indicators for the quality of health promotin, prevention and primary care at the health systems level in OECD countries. *Primary Care*. Doi: 10.1787/164604735126

Meherunissa, S. (2016). An analysis of gratitude and hope in relation to happiness. *The International Journal of Indian Psychology*, 4(1).

Noack, H. (1988). Measuring health behaviour and health: Towards new health promotion indicators. *Health Promotion International*, 3(1), 5–11. doi: 10.1093/heapro/3.1.5

Nurcahyanti, A. D. R. (2016). Cervical cancer : The case in Indonesia and natural product-based therapy. *Journal of Cancer Biology & Research*, 4, 1–7.

Puente-Díaz, R., & Meixueiro, G. (2016). An exploration of the relationships between gratitude, life satisfaction, and importance of helping others among a representative sample of the adult population of Mexico. *Cogent Psychology*, 3(1), 1–7. doi: 10.1080/23311908.2016.1160558

Sadeghi, A., & Pour, S. B. (2015). The Effect of Gratitude on Psychological and Subjective Well-being among Hospital Staff, 3.

Seligman, M. E. P., Steen, T. A., Park, N., Peterson, C., & Report, P. (2005). *Positive Psychology Progress Empirical Validation of Interventions*.

Shahed, S. (2008). *Health Locus of Control, Health Beliefs, and Health Related Behaviors: A Study of Urban Females*. University of the Punjab.

Shao, D., Gao, W., & Cao, F. L. (2016). Brief psychological intervention in patients with cervical cancer:

A randomized controlled trial. *Health Psychology*, 35 (12), 1383–1391. Doi: 10.1037/hea0000407

Shaping Your Health. (1979). *Focus on Health*, 1(1), 43–52.

Sulianti, A., Darmalaksana, W., Sulaeman, D., & Rahman, A. A. (2017). Happiness in suffering of systemic lupus erythematosus: Does it make sense? *Icse*, 64–68. file:///C:/Users/Salma/Downloads/ICSE_2017_12(1).pdf

Tugade, M. M., Fredrickson, B. L., & Feldman-Barrett, L. (2004). Psychological resilience and positive emotional granilarity: Examining the benefits of positive emotions on coping and health. *Journal of Personality*, 72(6), 1161–1191.

Wang, C. (2003). Reproductive health indicators for China. *Social Science*, 57(2), 217–225. Doi: 10.1016/S0277-9536(02)00341-6

Watkins, P. C., Khathrane Woddward, Stone, T., & Kolts, R. L. (2003). Gratitude and Happiness. Pdf. Social Behavior and Personality.

World Health Organization. (2006). Defining Sexual Health: Report of technical consultation on sexual health 28-31 January 2002, Geneva. Sexual Health Document Series, (January), 0–30. doi: http://www.who.int/reproductivehealth/publications/sexual_-health/defining_sexual_health.pdf

*The Social Sciences Empowered – Lumban Gaol, Hutagalung & Fong Peng (eds)*
*© 2020 Taylor & Francis Group, London, ISBN 978-0-367-00090-5*

# The conceptual framework for understanding roles of person factors and listening strategy in listening comprehension

Guohui Du*, Fong Peng Chew & Zuwati Hasim
*Faculty of Education, University of Malaya, Kuala Lumpur, Malaysia*

ABSTRACT: English listening as an important part of language learning is challenging for most L2/EFL learners. Although many researches have demonstrated that some factors influenced listening success, few researches provided empirical evidence for two or more factors concertedly contributing to effective listening comprehension based on some listening comprehension theories. This study explores how person factors (L2 vocabulary, metacognitive knowledge and self-efficacy) interact, predict listening success directly, and, through strategy, are used indirectly in the listening process. In the present study, the participants are 400 English major students at a college in Anyang, China. Data will be collected through survey questionnaires, aural vocabulary test, and listening level test. Data analysis will be done by SPSS and SEM (structural equation modelling) to figure out their contribution to listening success in the future.

*Keywords*: EFL, listening strategy, listening comprehension theories, metacognitive knowledge, self-efficacy

## 1 BACKGROUND

Since the implementation of reform and opening policy in 1978, China has seen great improvements in a variety of fields: in particular, China has become the second-largest economy in the world. With the ascent of economic status, an all-round diplomatic layout is necessary for Chinese development. For example, "one belt and one road" policy, China-Africa forum, or the BRICS summit, which effectively promote economic development and cooperation between China and other countries.

Diplomatic policies can't be carried out without foreign language talents. "Outline of national medium and long term educational reform and development plan (2010–2020)" suggests "To accommodate our nation's economic and social open policy, a great number of internationalized talents with international view, familiarity with international rules and capability of participating international business and competition need to be trained.:

Therefore, English, as one of the most popular foreign languages in the world, takes up a vital position in Education in China. For instance, it is stipulated as compulsory that pupils learn English from grade four in elementary school to college for almost 11 years; before enrollment in universities, students have to participate in entrance examinations including English proficiency tests.

However, even if much time and energy are invested in English teaching and learning, the result is not good. There still exists a large gap between the market's demand and English talent training: most graduates have qualified reading, but they are poor at speaking, listening, and writing.

Given that the learner could access information from listening input to construct the knowledge necessary for language use, listening, as a crucial skill in foreign language learning, has a significant influence on language communication and acquisition (Anderson & Lynch, 1988; Rost, 1990). In spite of the importance of listening, it gained less attention than other language skill, and even was often ignored by L2/foreign language teachers and researchers (Moyer, 2006). Therefore, English listening research is very necessary, especially in China with 0.26 billion English learners (Liu Yandong, 2014).

Nowadays, English listening teaching is still based on exam-orientation in China: for example, in listening class, students are required to complete listening exercises as teachers play the record; then the right answers are read out by teachers without instructions or modeling on the listening process. Moreover, after class, students are advised to do after-class exercises, or listen to English songs and broadcasts, watch more English movies, and practice more listening activities without strategic support. The focus of their listening teaching has been getting students to comprehend, on

*Corresponding authors: duguohui2013@sina.com

their own and with little support, the meaning underlying the spoken text. Few considerations for learners' person factors are made in listening lessons, thus, more could be done to engage students directly in improving their listening comprehension and managing their own learning.

## 2 STATEMENT OF PROBLEM

Listening is a complex cognitive skill. Listeners must be able to process aural input quickly and, simultaneously, attend to new incoming messages. Processing at a rapid speed in our first language is mostly implicit, effortless, and automatic, with little attention to what we are doing as we comprehend. However, L2/EFL listening is more demanding and most of people think that listening is the most challenging language skill.

According to Rubin (1994), there are five groups of factors hindering or facilitating L2/EFL listening success: text characteristics, interlocutor characteristics, task characteristics, listener characteristics, and process characteristics. Of them, listener characteristics play a vital role in improving listening competence and can be developed by listeners themselves.

The listener characteristics or person factors include: (1) cognitive factors: vocabulary knowledge, syntactic knowledge, discourse knowledge, pragmatic knowledge, metacognitive knowledge, prior knowledge, L1 listening ability, sound discrimination ability; and (2) affective factors: anxiety, motivation, and self-efficacy.

There is some previous research on listener characteristics or person factors directly impacting listening success: for example, L2 vocabulary size. Wang and Treffers-Dellar (2017) and Mattews and Cheng (2015) stated that L2 vocabulary size explained 13% and 55% of the variance respectively in listening comprehension; metacognitive knowledge: Goh and Hu (2014) and Zeng and Goh (2015) showed that metacognitive awareness could predict 22% and 13–22% of the variance in listening performance respectively by MALQ; self-efficacy: Chen (2007) stated that compared with listening anxiety, English listening self-efficacy could predict English listening performance better with 277 Taiwanese college-level English learners. Mills et al. (2006) also stated that self-efficacy influenced listening performance, but only for females. However, these results are mixed and not clear-cut; for instance, how much of the variance in listening performance could be explained by L2 vocabulary size with the aural vocabulary size test in different native language context? Ghapanchi and Taheryan (2012) stated that metacognitive knowledge was not significant in predicting listening proficiency; Doç and Alci (2012) reported that there was a correlation between self-efficacy, metacognition, and academic success, moreover, only metacognitive awareness predicted academic performance through the regression analysis.

Besides, Coutinho (2008) and Yerdelen-Damar and Peşman (2013) showed that self-efficacy was a mediator between metacognition and performance. However, Mango (2009) stated that self-efficacy affected metacognition by path analysis. The results are mixed. Meanwhile, Vandergrift and Baker (2015) found that L2 vocabulary appeared to be impacted by metacognition (0.29) by the path analysis.

In fact, the influence of listening strategy on listening success is also a research focus. Lau (2017) stated that effective listening requires effective used and skillful orchestration of some cognitive and metacognitive strategies to control the listening process so as to gain better comprehension. However, others supported the lack of clear evidence that listening strategy was significantly efficient (Graham, Santos & Vanderplank, 2011; Renandya & Farrell, 2010). Some researches revealed that both EFL learners' self-efficacy and metacognitive awareness could predict learning strategies use in a way, but it was only in this study population without being generalized to other contexts (Nosratinia, Saveiy, & Zaker, 2014). However, Bonyadi, Nikou, and Shahbaz (2012) found that there was no significant relationship between self-efficacy and language learning strategies use. Another, language proficiency affected participants' use of strategies (Lee & Cai, 2010). Therefore, L2 vocabulary size may affect subjects' use of strategies, as vocabulary knowledge is an important predictor of language proficiency (Nasir, Manan, & Azizan, 2017).

However, few researches discuss the relationship between person factors, listening strategy, and listening success according to Imhof and Janusik's (2006) adapted Systems Mode of Listening from (Vandergrift & Goh, 2012).

## 3 RESEARCH OBJECTIVE

The aim of this study is (1) to identify whether listening strategy is a mediator of the relationship between person factors and listening success: (a) to examine whether person factors (L2 vocabulary size, metacognitive knowledge and self-efficacy) are significant predictors of listening success; (b) to examine whether person factors are significant predictors of listening strategy; (c) to examine whether listening strategy is a significant predictor of listening success; (2) to identify the relationship between L2 vocabulary size, self-efficacy, and metacognitive knowledge.

## 4 CONCEPTUAL FRAMEWORK

The current study discusses the cognitive processing of listening comprehension by exploring the factors that can impact the quality of that processing and result in different outcomes for different learners.

The Systems Model of Listening adapted from Imhof and Janusik (2006) was adopted in this study, which constructed the process of aural information processing by modifying a systems model of study processes (Biggs, 1999). The model includes three interdependent stages: person related, context related, and process and results. It helps us more clearly understand the listening construct and visualize the interdependent relationships between person factors, listening context, and different processes. Vandergrift and Goh (2012) stated that the process is a comprehensive and integrated system in which person factors and listening context can impact the process (quality of the processing) and the results (listening success, learning or affective factors, e.g. self-efficacy). Vice versa, the result (e.g. comprehension or miscomprehension) can impact the factors that affect the listening process (e.g. self-efficacy) and listeners' further efforts at processing subsequent input. That is to say, person factors and context factors will impact on the processing quality and strategies listeners may employ, which will influence the quality and nature of the outcome. The quality of this outcome will, in turn, affect some person factors, especially affective factors that will influence the strength of continued efforts to listen to texts, for instance, or a more joint effort to recall prior knowledge to elucidate the texts.

The person factors are categorized into two groups: cognitive factors and affective factors. Among cognitive factors are what listeners bring to their language learning, such as prior knowledge, metacognition, L1 listening ability, working memory capacity, and sound discrimination ability; others are developed as a result of the language learning process: L2 vocabulary, syntactic knowledge, discourse knowledge, and pragmatic knowledge. The latter factors may also develop differently as a function of the former.

On the basis of sample size, population demography, individual factors' importance, and the availability of reliable and valid instruments to gauge a given variable, the person factors researched in this study are limited to three variables: L2 vocabulary size, metacognitive knowledge, and self-efficacy. For example, first language listening ability is scarcely transferred to L2 listening ability because of completely different typology. Background knowledge is thought to play an important role in L2 listening comprehension (e.g., Long, 1990), but gauging this variable would be very complicated, as, for example, we have to create a new listening test with a limited number of topics to be measured and then evaluate student knowledge of all the topics covered in this test. It is also difficult to measure learners' pragmatic knowledge, discourse knowledge, syntactic knowledge, and sound discrimination ability, which possibly contribute a little to listening competence in terms of existing literature. Then, we try to figure out the degree to which these important factors affect L2/EFL listening success in terms of cognitive factors, affective factors. and listening strategy.

(Vandergrift & Goh, 2012) Among cognitive factors, L2 vocabulary is categorized into factors that are developed from the language learning process, and it seems that L2 vocabulary size plays a critical role in listening achievements (Mattews & Cheng, 2015). Metacognitive knowledge is defined by Flavell (1979) as "that segment of your (a child's, an adult's) stored world knowledge that has to do with people as cognitive creatures and with their diverse cognitive tasks, goals, actions, and experiences." Metacognitive knowledge is subsumed in the factors that listeners bring to their language learning. Vandergrift et al. (2006) determined that metacognitive knowledge, as employed by respondents, was capable to contribute about 13% of the variance to L2 listening performance of university students.

The affective factors also have a certain impact on listener engagement with a speaker or the listening/learning environment. In particular, self-efficacy is defined by Bandura (1986) as "people's judgment of their capabilities to organize and execute courses of action required to attain designated types of performance. It is not related to the skills one has but to the judgments of what one can do with whatever skills one possesses" (p. 391). It is the basis for self-confidence and motivation, and plays a certain role in L2 listening proficiency. Chen (2007) stated that compared with listening anxiety, the self-efficacy had a stronger prediction of English listening proficiency. Therefore, we limited person factors to L2 vocabulary size, metacognitive knowledge, and self-efficacy.

Other listening strategies include methods of managing personal mental and observable behavior to accomplish a listening activity (Graham & Macaro, 2008; Richards, 2008). Yeldam (2009) stated that low-level learners benefit most from top-down strategy instruction and most probably incorporate these strategies into their repertoire.

Moreover, some researches revealed that both foreign language learners' self-efficacy and metacognitive awareness could predict use of learning strategies in a way, but only in this study population without being generalized to other contexts (Nosratinia, Saveiy, & Zaker, 2014).

Lee and Cai (2010) reported that language proficiency affected participants' use of strategies: that is to say, L2 vocabulary size may affect subjects' use of strategies, for vocabulary knowledge is an important predictor of language proficiency (Nasir, Manan & Azizan, 2017).

Finally, L2 vocabulary and self-efficacy could be influenced by metacognition in some researches (Coutinho, 2008; Vandergrift & Baker, 2015).

Thus, in our research, the context (listening assessment) is fixed, and L2 vocabulary size, metacognitive knowledge, and self-efficacy are chosen as independent variables to predict the dependent variable (listening success) through the mediator variable (listening strategy) as per Imhof and Janusik (2006)'s adapted Systems Mode of Listening from Vandergrift and Goh (2012).

The following is the conceptual frame in this study.

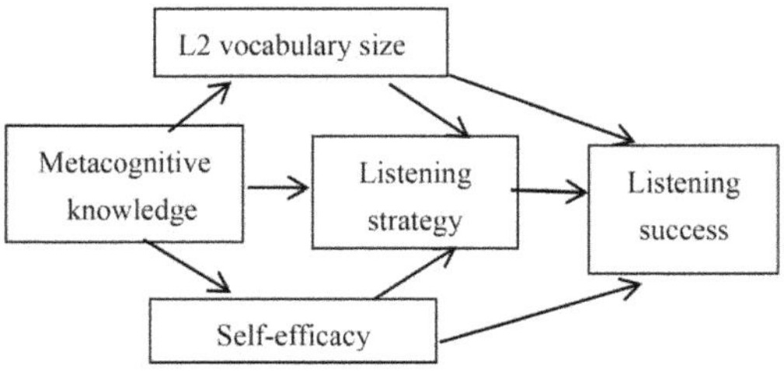


Figure 1. The conceptual framework adapted from Imhof and Janusik's (2006) adapted Systems Mode of Listening.

## 5 METHODOLOGY

### 5.1 *Participants*

Our participants are 400 English major students from school of foreign languages: freshman, sophomore, junior students in a university, Anyang City, China. They have at least nine years' English learning experiences. However, most time and energy are spent in reading and written exams without focused listening teaching. Only when they are enrolled in the university as English majors do they have two hours' English listening course per week, but few teachers teach them how to perceive the aural vocabulary and process aural information strategically.

### 5.2 *Instruments*

#### 5.2.1 *L2 vocabulary size*

The listening vocabulary levels test (LVLT) was designed to measure aural English vocabulary knowledge from the first five 1,000-word frequency levels and the Academic Word List (AWL). The 1,000 to 5,000 word-frequency levels consist of 24 items per level with 30 items from the AWL. A 150-item test is completed within 30 minutes. The LVLT uses the multiple-choice format with each item having four options, from which participants must choose the Chinese meaning closest to the target word. The target word is read, then following it is the non-defining context sentence with the target word. The LVLT items come from Nation's (2012) British National Corpus (BNC)/Corpus of Contemporary American English (COCA) word lists. The listening vocabulary levels test in this study derives from (McLean et al., 2015).

#### 5.2.2 *Metacognitive awareness listening questionnaire*

The MALQ is a self-report instrument with robust psychometric property to measure students' metacognitive knowledge. It is widely used to measure L2/foreign listeners' metacognitive awareness and strategic competence of processing aural input. MALQ has five distinct factors – person knowledge, directed attention, problem-solving, planning and evaluation, mental translation (Vandergrift et al., 2006) – with 21 items. A six-point Likert scale is used to measure 21 items in the MALQ with validity and reliability, ranging from "strongly agree" to "strongly disagree." The MALQ in this paper is adapted from Vandergrift et al.'s (2006) MALQ.

#### 5.2.3 *Listening self-efficacy questionnaire*

The English listening self-efficacy scale comprised 18 items that requested participants to estimate their capability to complete various listening comprehension tasks typically embedded in their listening exams and listening class. This was done to make sure that the self-efficacy instruments match performance measures closely, as suggested by Bandura (1986) and Pajares (1997) to improve the accuracy of the prediction. The strength of participants' self-efficacy is measured on the 11-point scale ranging from 0 (not confident at all) to 10 (highly confident). This questionnaire is adapted from Chen (2007).

#### 5.2.4 *Listening strategy questionnaire*

The ELLSI (EFL Listening Strategy Inventory) adapted from Nix (2016) is employed to measure students' listening strategy use. It is concerned with trait strategy use (habitual strategic use in experiential memory), not state strategy use (related to specific language tasks and typically investigated in treatment studies). The inventory consisted of 23 items based on a hierarchy of four aspects of listening: setting (interaction/conversation, self-study, academic), channel (face-to-face, VOIP/telephony, webcam, audio-visual media), strategies (metacognitive, cognitive, socio-affective), and processing (interactive, top-down, bottom-up). These items are integrated with this larger structural model which could explain the multivariate interactions which determine L2 learners' possibility to listen with comprehension. A six-point Likert scale is used to measure each item.

#### 5.2.5 *Listening comprehension level test*

The listening comprehension level test in this study is the listening part of Test for English Majors (TEM4). TEM4 is a formal and standard English proficiency test generally employed in colleges and universities of China, to examine English major students' English proficiency in listening, reading, writing, and translation. The listening section of TEM4 consists of two parts: (1) Dictation: students are required to note down the paragraph word by word; and (2) listening comprehension with two sections:

section A: mini-lecture with filling in ten blanks when listening to it; section B: two conversations with ten multiple choices. An advantage of employing TEM4 is that participants are familiar with the types of listening sections of TEM4 and the topics present in it are associated with university students' daily life and learning. Besides, it is a standardized test nationwide with high validity and reliability.

## 6 CONCLUSION

English listening is a complex cognitive skill, especially for L2/EFL learners, which is also the most difficult part of English learning. Thus, listeners have to simultaneously and comprehensively process the input strategically. This study explores the conceptual framework of how person factors and listening strategy interact and influence listening comprehension. In the future, we will establish a structural equation model (SEM) by data analysis to gain an insight for their relationship, which will provide a new venue for L2/EFL listening research.

This study will be conducive to L2/EFL listening teaching and learning. By metacognition or discussing listening process, students will clearly know how they understand listening processes and confidently process information input through strategy use in class or after class.

## REFERENCES


Anderson, A., & T. Lynch.1988. *Listening*. Oxford University Press.

Bandura, A. 1986. *Social Foundations of Thought and Action*. Englewood Cliffs, NJ: Prentice-Hall.

Biggs, J. 1999. *Teaching for Quality Learning at University*. Buckingham, UK: Society for Research into Higher Education.

Bonyadi,A., Nikoi, F. R., & Shahbaz, S. 2012. The Relationship between EFL Learners' Self-efficacy Beliefs and Their Language Learning Strategy Use. *English Language Teaching*, *5*(8).

Chen, H.-Y. 2007. The Relationship Between EFL Learners' Self-Efficacy Beliefs and English Performance. Unpublished Ph.D. papers.

Coutinho, Savia. 2008. Self-efficacy, metacognition and performance. *North American Journal of Psychology*. 10(1), pp. 165–172.

Doç, Y., & Alci, B. 2012. An examination into self-efficacy, metacognition and academic performance of pre-service ELT students : Prediction and difference. *Kalem Eğitim ve İnsan Bilimleri Dergisi*, 2(1), 143–165.

Flavell, J.H. 1979. Metacognition and cognitive monitoring: A new era of cognitive-developmental inquiry. *American Psychologist*, 34, 901–911.

Ghapanchi, Z., & Taheryan, A. 2012. Roles of linguistic knowledge, metacognitive knowledge and metacognitive strategy use in speaking and listening proficiency of Iranian EFL learners. *World Journal of Education*, *2* (4), 64–75.

Goh, C. C. M., & Hu, G. 2014. Exploring the relationship between metacognitive awareness and listening performance with questionnaire data. *Language Awareness*, *23*(3), 255–274.

Graham, S., & MacAro, E. 2008. Strategy instruction in listening for lower-intermediate learners of french. *Language Learning*, *58*(4), 747–783.

Graham, S., Santos, D., & Vanderplank, R. 2011. Exploring the relationship between listening development and strategy use. *Language Teaching Research*, *15*(4), 435–456.

Imhof, M., & Janusik, L. A. 2006. Development and validation of the Imhof-Janusik Listening Concepts Inventory to measure listening conceptualization differences between cultures. *Journal of Intercultural Communication Research*, *35*(2), 79–98.

Lau, K. L. 2017. Strategy use, listening problems, and motivation of high- and low-proficiency Chinese listeners. *Journal of Educational Research*, *110*(5), 503–514.

Lee & Cai. 2010. The effects of language proficiency on unfamiliar word processing in listening comprehension. *Hong Kong Journal of Applied Linguistics*, 12(2) 61–82.

Liu Yandong. 2014. Speech in the opening of the World Language Conference.

Long, D. R. 1990. What you don't know can't help you. *Studies in Second Language Acquisition*, 12, 65–80.

Magno, C. 2009. Investigating the effect of school ability on self-efficacy, learning approaches, and metacognition. *The Asia-Pacific Education Researcher*, 18(2), 233–244.

Matthews, J., & Cheng, J. 2015. Recognition of high frequency words from speech asapredictor of L2 listening comprehension. *System*, *52*, 1–13.

McLean, S., Kramer, B., & Beglar, D. 2015. The creation and validation of a listening vocabulary levels test. *Language Teaching Research*, *19*(6), 741–760.

Mills, N., Pajares, F., & Herron, C. 2006. A reevaluation of the role of anxiety: Self-efficacy, anxiety, and their relation to reading and listening proficiency. *Foreign Language Annals*, *39*(2), 276–295.

Moyer, A. 2006. Language contact and confidence in second language listening comprehension: A pilot study of advanced learners of German. *Foreign Language Annals*, *39*(2), 255–275.

Nasir, N. F. W. M., Manan, N. A. B., & Azizan, N. 2017. Ex amining the relationship between vocabulary knowledge and general English language proficiency. *ESTEEM Journal of Social Sciences and Humanities*, 1(11), 15–22.

Nation, I. S. P. 2012. The BNC/COCA word family lists (17 September 2012). Unpublished paper.

Nix, J. M. L. 2016. Measuring latent listening strategies: Development and validation of the EFL listening strategy inventory. *System*, *57*, 79–97.

Nosratinia, M., Saveiy, M., & Zaker, A. 2014. EFL learners' self-efficacy, metacognitive awareness, and use of language learning strategies: How are they associated? *Theory and Practice in Language Studies*, *4*(5), 1080–1092.

Pajares, F. 1997. Current directions in self-efficacy research. In M. L. Maehr, & P. R. Pintrich (Eds.), *Advances in Motivation and Achievement*, Vol.10, Greenwich, CT: JAI Press.

Renandya, W. A., & Farrell, T. S. C. 2011. "Teacher, the tape is too fast!" Extensive listening in ELT. *ELT Journal*, *65*(1), 52–59.

Rost, M. 1990. *Listening in Language Learning*. Harlow: Longman.
Rubin, J. 1994. A review of second language listening comprehension research. *The Modern Language Journal*, *78*(2), 199–221.
Vandergrift, L., & Baker, S. 2015. Learner variables in second language listening comprehension: An exploratory path analysis. *Language Learning*, *65*(2), 390–416.
Vandergrift, L., & Goh, C. C. M. 2012. *Teaching and Learning Second Language Listening*. New York and London: Routledge Taylor & Francis Group.
Vandergrift, L., Goh, C. C. M., Mareschal, C. J., & Tafaghodtari, M. H. 2006. The metacognitive awareness listening questionnaire: Development and validation. *Language Learning*, *56*(3), 431–462.
Wang, Y., & Treffers-Daller, J. (2017). Explaining listening comprehension among L2 learners of English: The contribution of general language proficiency, vocabulary knowledge and metacognitive awareness. *System*, 65, 139–150.
Yeldham, M.A. 2009. Approaches to second language listening instruction: Investigating thetop-down/bottom-up debate. Unpublished doctoral dissertation, University of Melbourne, Australia.
Yerdelen-Damar, S., & Peşman, H. 2013. Relations of gender and socioeconomic status to physics through metacognition and self-efficacy. *Journal of Educational Research*, 106(4), 280–289.
Zeng, Y., & Goh, C. 2015. Learners' level of metacognitive awareness and its relationship with listening performance. Proceedings - IEEE 15th International Conference on Advanced Learning Technologies: Advanced Technologies for Supporting Open Access to Formal and Informal Learning, ICALT 2015, 345–349.

*The Social Sciences Empowered – Lumban Gaol, Hutagalung & Fong Peng (eds)*
*© 2020 Taylor & Francis Group, London, ISBN 978-0-367-00090-5*

# Relationship with the community and social–emotional competence of youth in Sarawak, Malaysia

Keezrawati Mujan Yusuf, Abd Razak Zakaria, Abdul Muhsien Sulaiman & Fonny Dameaty Hutagalung
*Faculty of Education, University of Malaya, Malaysia*

ABSTRACT: The development of individuals is largely affected by external factors around them. This includes the development of social and emotional competence, which encompasses the ability of a person to be effective in social interaction, including different aspects describing the social and emotional skills necessary for day-to-day life. Based on Social Learning Theory by Albert Bandura, the researchers propose that the relationship individuals have with members of their community plays an important role in developing their social and emotional competence. The study is aiming to study the level of relationship between individuals and community and how this relationship influences their level of social and emotional competence in the context of Sarawak, East Malaysia.

*Keywords*: Social competence, emotional competence, family well-being, community involvement, community relationship

## 1 INTRODUCTION

The relationship of individuals with the community around them is one of the key aspects of Malaysian society. According to research on family well-being conducted by the National Population and Family Development Board (NPFDB) Malaysia (2016), Malaysian families scored 7.835 of 10 in the domain of community involvement. This score indicates the high value that Malaysian families put on the community they live in.

School community is the closest community for an adolescent for an extended period of time. Community involvement in school usually happens through extracurricular activities. There are many activities involved in extracurricular programs such as community service projects, internships, involvement in associations, organizations, sports events, and representing the school for programs or competition.

Besides school-based activities, community involvement also includes activities conducted within the local communities such as the village and residential park. According to a research on the participation of Malaysian youth in community activities conducted by Krauss et al. (2013), programs organized by religious groups show prevalence over other activities, probably due to the central role that religion plays in Malaysian society in general. There are also informal community programs, referred to as *gotong royong*, which are often conducted in a smaller and closer community such as a village and are less common in a big city.

The significance of community compels many scholars to study how it may impact the development of a person, especially during adolescence, known to be the most significant period of one's life in terms of character development. This includes social and emotional competence – a concept that is made up of various aspects of social, emotional, and cognitive skills necessary for day-to-day life. It has gained attention from many researchers, mainly due to its significant influence in many other aspects of development and is seen as a critical aspect of academic and social success.

## 2 STATEMENT OF THE PROBLEM

Much research conducted in Malaysia considers it one large group of people and generalizes the findings from one part of the country to the whole nation. Although Malaysia is a relatively small country, this practice is problematic because Malaysia is a very diverse society. There are significant differences in terms of demography, culture, living condition, family and community characteristics, and educational attainment between people in different parts of the country, especially between Peninsular Malaysia, Sabah, and Sarawak.

As part of the richness and diversity in terms of culture, as the biggest state in Malaysia with a large rural area, Sarawak is also facing a more notable urban–rural gap as compared to other states. While urban areas of Sarawak are progressing as rapidly as other major cities in Malaysia, the rural areas of Sarawak have limited access to basic services such as tarred roads, schools, electricity, water supply, and healthcare (Regina Garai Abdullah, 2017). These differences in physical development may lead to major differences in lifestyle, especially among the younger generation as technologies such as Internet and smart devices play a big role in shaping the day-to-day life of those with access to them (i.e. the urban residents). This may influence their choice of activities and subsequently their relationship with people and community around them. This makes Sarawak a very interesting context to study.

The current research is seeking to describe:

I. the strength of the relationship of adolescents in Sarawak and their community
II. the level of social–emotional competence of adolescents in Sarawak
III. the correlation between the relationship with the community and the level of social and emotional competence of adolescents in Sarawak

## 3 CONCEPTUAL FRAMEWORK

### 3.1 *Concept of relationship with the community*

According to NPFDB (2016), an indicator of community involvement is the relationship one has with the community. There may be correlation between community involvement and community relationship as active participation in community programs can create a closer bond between the members through frequent interaction. In studies on youth or children, two major types of relationship are discussed: the relationship between young individuals and non-familial adults, and with their peers.

In terms of Bronfenbrenner's model (1979), trusted adult relationships may occur where young people extend their relationships across their micro- and meso-systems (immediate/extended family, school, peers), exo-systems (communities), and macro-systems. These important non-parental adults include extended family, teachers, friends' parents, family friends, neighbors, coaches, or tutors (Ahrens et al., 2011). Scholars propose that the relationship with a non-familial adult is an influential developmental relationship because it empowers the developing person while at the same time providing scaffolding, showing empathy, and allowing open dialogue that benefit the children from the higher degree of control (Li & Jullian, 2012). A positive and supportive relationship with a non-familial adult is characterized by trust, respect, and reciprocity (Krauss et al., 2013). Subsequently, children who receive support from adults are more likely to be participative in the community, show greater confidence and independence, and greater and community connections. A recent study by Meltzer, Muir, and Craig (2018) on Australian youth indicates that supports given by trusted adults that are deemed as useful for youth include genuine conversation, encouragement, role modeling, and practical assistance.

Apart from adults, the relationship with peers is also very significant for a child. The degree to which children are accepted by their peers has important implications for their well-being. Children who experience peer rejection tend to fall into loneliness, a cognitive and affective state of feeling disconnected and lacking in supportive relationships (Rotenberg, 1999). This leads to distorted social and emotional development, and in some extreme cases, it can cause serious psychological condition such as depression. Children who are rejected by their peers are mostly associated with high levels of aggression, low prosocial skills, poor academic skills, low academic attainment, criminality, and poor relationships in adulthood (Krause et al., 2010). On the other hand, peer acceptance is associated with many positive indicators of social–emotional competence, such as low levelss of aggression and high levels of prosocial skills. Peer relationships characterised by mutual warmth, affection, shared experiences, reliability and loyalty promote the feeling of security and trust (Erdley et al., 2001). Saarni (2000) lists several benefits of close peer relationships, or more commonly referred to as friendship, including developing self-esteem, protecting oneself against stress, and improving learning by exploration of ideas. However, it should be noted that close peer relationships can lead to both positive and negative development. Negative development happens when a person tries to gain acceptance of their peer groups through imitating the behavior of their peers even if it is morally wrong.

### 3.2 *Concept of social and emotional competence*

Consortium on the School-Based Promotion of Social Competence (CASEL) (1994) outlines five domains of social-emotional competence namely, self-awareness, self-management, social awareness, relationship skills, and responsible decision-making.

Self-awareness refers to one's understanding of his or her own personal goals, emotions and values. This capacity marks a significant step in one's development as one learns to connect thoughts, emotions, and actions. This capacity allows a person to have a well-grounded positive mindset, self-efficacy, and optimism due to the ability to assess one's own strengths and limitations accurately. Competence in this domain is very crucial as people who are aware of their own strengths and emotions are more reflective of their state of being and clear about the reasons for their emotional responses. A high level of self-awareness leads to better self-control and eventually better decision making. Research has shown that children with better self-awareness tend to regulate their behavior using pro-social self-schemas (Froming, Nasby & McManus, 1998).

Self-management encompasses the skills and attitudes that facilitate the ability to regulate emotions and behaviors. This regulation happens on the individual, relational, and contextual level: on an individual level such as when a student regulates his own behavior in class; on a relational level when a child follows the behaviors that are emphasized by his parents; and on a contextual level such as when immigrants try to recognize and adopt the behaviors that are considered culturally important in the new society they are living in. Good self-management results in delayed gratification, effective management of stress, controlled impulses, and perseverance through challenges to achieve one's goals. It is often assessed through one's ability to remain calm in the midst of a stressful or changing situation. Similar to self-awareness, the ways people manage themselves are also affected by the culture of the society, which determines the behaviors and emotions, or the degree to which they can be expressed, appropriate to be expressed in a specific situation. Those who successfully manage themselves in accordance to their respective societal norms reportedly have lower levels of depression and higher levels of satisfactions in relationships (Matsumoto et. al, 2008).

Social awareness is the ability to read and understand other person's cues and to appropriately respond to their feelings (Frey, Hirschstein, & Guzzo, 2000). In other words, it is the ability to take the perspective of others, including those of different background or culture. Social awareness also includes a good understanding of social norms for behavior. People with strong social awareness more easily feel empathy and compassion towards others, which shows the awareness of sensitivity of complex issues. These will eventually lead to more altruistic and pro-social behaviors. The level of social awareness is often assessed through the ability to interpret emotions through facial expressions and to guess the reasons behind people's emotions accurately.

The next domain of social-emotional competence is relationship skills. These skills include the ability to communicate clearly, listen actively, cooperate, resist inappropriate social pressure, negotiate conflict effectively, and seek help when necessary. All of these skills are crucial to provide people with the tools needed to establish and maintain healthy and satisfying relationships, according to the social norms. Relationship skills play significant roles in people's life, especially school students. Studies found that students who are distant from their friends due to poor relationship skills are more likely to become disengaged from academic activities (Wentzel, 1999).

The last domain, responsible decision-making, refers to the ability to consider safety and ethical and societal factors in making decisions that will contribute to the well-being of the community. This helps people to make constructive choices and realistic evaluations of consequences in terms of social behavior and social interactions in diverse settings, taking the well-being of oneself and others into consideration. Those who are competent in this domain will ensure that there are more positive outcomes when making a choice by carefully weighing the strengths and weaknesses of each available option.

## 4 THEORETICAL FRAMEWORK

***Social learning theory.*** Social learning theory was developed by Albert Bandura (1977) as an extension to the behaviorist explanation of development. A key aspect of this theory is that behavior is not just the outcome of direct internal influences and external influences, but rather the complementary interaction between them which Bandura (1977) refers to as reciprocal determinism.

Social learning theory provides an explanation on ways immediate surroundings shape an individual as it emphasises the interaction between individuals and their significant others. This theory emphasizes that the environments surrounding a person are not random; those environments are often chosen and altered by the person through their own behavior. The core of social learning theory is modeling where most people learn through selective observation and remembering the behavior of others. This theory predicates that social interactions, which include role modeling and verbal instruction, influence the acquisition of behavior. Social learning theory also posits that significant others, especially adults, are responsible for modeling emotional skills and positive social behaviors, not just observable behaviors (Elias, Zins, Graczyk & Weissberg, 2003).

The main concept of Albert Bandura's learning theory is modeling, vicarious learning, and self-regulation. Modeling is the basic concept of social learning theory. Bandura classifies four phases of learning from modeling: the attentional phase of giving attention to a model, a retentional phase where the behaviors are coded and kept in the long-term memory, a reproduction phase where codes stored in the memory are reflected in observed behavior, and lastly the motivational phase where the observers will be motivated to imitate the model because they feel that, by doing so, they will gain reinforcement.

Vicarious learning, commonly known as observational learning, occurs when an individual learns something through observation without direct reinforcement or punishment of the behavior (Friedman & Schustack, 2012; Nicholle, Symmonds, & Dolan, 2011). Bandura suggests that vicarious learning depends on the consequences or the possible anticipated consequences of the modeled behavior. For example, behaviors observed to be punished or to have a strict rule in place against them are likely to be imitated.

This theory also suggests that human behavior is largely self-regulated: a continuously active process in which individuals monitor their own behavior, both in terms of influences and of consequences: judge their behavior in relation to their personal standards and broader standards; and lastly, react to their own behavior. Self-regulation is the central of causal processes because it mediates the effects of most external influences and provides the basis for purposeful action.

## 5 CONCLUSION

In conclusion, the research is proposing that the relationship with the community may influence the development of social and emotional competence, with the support of evidence from past studies and Albert Bandura's social learning theory. The research is still in the planning stage. The data collection stage is expected to proceed in January 2019. The findings of this study are hoped to improve our understanding of links between relationship with community and the development of adolescents' social and emotional competence. This will be useful for policy makers, educators, NGOs, or anyone in the community interested in organizing youth development programs in both urban and rural areas in the state of Sarawak, or other states with similar populations and culture such as Sabah. Hopefully, the findings of this research will help understand the context of local society and therefore be able to deal with the subjects appropriately, especially in guiding them towards character development.

## REFERENCES


Ahrens, E. N, Lerman, D. C., Kodak, T., Worsdell, A. S., & Keegan, C. (2011). Further Evaluation of Response Interruption and Redirection as Treatment for Stereotypic Behavior. *Journal of Applied Behavior Analysis, 44*, 95–108.

Bandura, A. (1977). *Social Learning Theory*. New York, USA: General Learning Press.

Bronfenbrenner, U. (1979). *The Ecology of Human Development*. Cambridge, MA: Harvard University Press.

Consortium on the School-Based Promotion of Social Competence (1994). The school-based promotion of social competence: Theory, practice, and policy. In R. J. Haggerty, L. R. Sherrod, N. Garnezy, & M. Rutter (Eds.), *Stress, Risk, and Resilience, in Children and Adolescents: Processes, Mechanisms, and Interventions* (pp. 268–316). New York: Cambridge University Press.

Elias, M. J., Zins, J. E., Graczyk, P. A., & Weissberg, R. P. (2003). Implementation, sustainability, and scaling up of social–emotional and academic innovations in public schools. *School Psychology Review, 32*, 303–319.

Erdley, C. A., Nangle, D. W., Newman, J. E., & Carpenter, E. M. (2001). Children's friendship experiences and psychological adjustment. In D. Nangle & C. Erdley (Eds.), *The Role of Friendship in Psychological Adjustment: New Directions in Child and Adolescent Development* (Vol. 91, pp. 5–24). San Francisco, CA: Jossey-Bass.

Frey, K. S., Hirschstein, M. K., & Guzzo, B. A. (2000). Second Step: Preventing aggression by promoting social competence. *Journal of Emotional and Behavioral Disorders, 8*(2), 102–112.

Friedman, H. S., & Schustack, M. W. (2012). *Personality: Classic Theories and Modern Research* (5th ed.). Boston, MA: Allyn & Bacon.

Froming, W. J., Nasby, W., & McManus, J. (1998). Prosocial self-schemas, self-awareness, and children's prosocial behavior. *Journal of Personality and Social Psychology, 75*(3), 766–777.

Krause, K., Bochner, S., Duchesne, S., & McMaugh A. (2010*). Educational Psychology for Learning and Teaching* (3rd ed). Southbank, VIC, Australia: Cengage Learning.

Li, J., & Jullian, M. (2012). Developmental relationships as the active ingredient: A unifying working hypothesis of "what works" across intervention settings. *American Journal of Orthopsychiatry, 82*(2), 157–186.

Matsumoto, D., Yoo, S. H., Fontaine, J., Anguas-Wong, A. M., Ariola, M., & Ataca, B.(2008). Mapping expressive differences around the world: The relationship between emotional display rules and individualism v. collectivism. *Journal of Cross-Cultural Psychology, 39*, 55–74.

Meltzer, A., Muir, K,. & Craig L. (2018). The role of trusted adults in young people's social and economic lives. *Youth and Society, 50*, 575–592.

National Population and Family Development Board Malaysia (2016). *Laporan indeks kesejahteraan keluarga Malaysia 2016*. Retrieved from http://www.lppkn.gov.my/Laporan_Indeks_Kesejahteraan_Keluarga_Malasia_2016.pdf

Nicholle, A., Symmonds, M., & Dolan, R. J. (2011). Optimistic biases in observational learning. *Cognition, 119* (3), 394–402.

Regina Garai Abdullah. (2017). Accessibility and development. A case study from rural Sarawak, Malaysia, *International Journal of Business and Society, 18* (4), 791–799.

Rotenberg, K. J. & MacKie, J. (1999). Stigmatization of social and intimacy loneliness, *Psychological Reports, 84* (1), 147–148.

Saarni, C. (2000). Emotional competence: A developmental perspective. In R. Bar-On & J. D. A. Parker (Eds.), *The Handbook of Emotional Intelligence: Theory, Development, Assessment, and Application at Home, School, and in the Workplace* (pp. 68–91). San Francisco, CA, US: Jossey-Bass.

Wentzel, K. R. (1999). Social-motivational processes and interpersonal relationships: Implications for understanding motivation at school. *Journal of Educational Psychology, 91*(1), 76–97.

*The Social Sciences Empowered – Lumban Gaol, Hutagalung & Fong Peng (eds)*
*© 2020 Taylor & Francis Group, London, ISBN 978-0-367-00090-5*

# The effectiveness of thinking maps, mind maps, and concept maps toward cognitive abilities among year 1 indigenous learners

Yeoh Sun Wei, Fonny Demeaty Hutagalung & Chew Fong Peng
*Department of Psychology & Counselling, Faculty of Education, University of Malaya, Kuala Lumpur, Malaysia*

ABSTRACT: The role of thinking skills becomes vital in the information era. Thinking tools help learners to think systematically and effectively. In the ever-changing world, we need to embrace change in order to remain competitive in the global market. Endeavoring to equip learners with critical and innovative thinking skills is an inevitable action to prepare them for the future world. Metacognition helps educators to understand learners' thinking and learning process. This awareness is the core to ensuring learners master the method of learning. There are many different types of thinking tools to guide us in the thinking process. This research focuses on three visual thinking tools: thinking maps, mind map and concept map. The study is to research the effectiveness of using different visual thinking tools to enhance learners' cognitive ability. This research will focus on indigenous learners because they are the at-risk group in the trend of modernization.

*Keywords*: Childcare, transition, secure base, separation individuation, attachment

## 1 INTRODUCTION

Educators should embed metacognition strategies in the teaching and learning process because it enables educators to maximize pupils' learning outcomes. Metacognition has a positive relationship with learners' cognitive ability. If we think of cognitive ability as a car; metacognition is the skill to drive it. A learner gifted with high cognitive ability must also be equipped with the skill to execute it (Sternberg, 1984; Abrami, Bernard, Borokhovski, Wade, Surkes, Tamim, & Zhang, 2008).

Metacognition helps educators to understand learners' thinking and learning process. This awareness is core to ensuring learners master the method of learning. Much research has been conducted regarding how thinking can be taught to children to improve their learning. These researches provide information about the process of "thinking about thinking." The awareness of metacognitive and the regulation of metacognitive improve learners' problem solving process (Flavell, 1979; Alikhan, 2014).

Metacognition and cognition can overlap in the same task. Cognition helps learners to achieve their goals but metacognition ensures the goals were achieved. Cognition provides strategies in problem-solving tasks whereas metacognition evaluates the best strategies to solve the problem. In other words, metacognition is used to monitor cognition (Roberts & Erdos, 1993; Anderson, Nashon, & Thomas, 2009).

There are many different types of thinking tools to guide the thinking process. In this research, researchers focused their study on three visual thinking tools: thinking maps, mind maps, and concept maps. Neuropsychology research shows that the thinking maps are the visualization of our thinking process. It helps thinkers transform abstract thought to visible image. This process provides deeper comprehension regarding a topic for the thinker (Hyerle, 1996; Alikhan, 2014).

The mind map was invented by Tony Buzan. It can be used in wide range of daily activity including education. This graphic technique expands thinkers' minds and improves creativity. Mind maps develop a variety of cortical skills, such as literacy, pictures, numeracy, logical thinking, art, and spatial awareness. Thinkers are free to expand ideas on the related subject. Understanding is strengthened through mind maps (Buzan, 2002).

A concept map is used as a learning and teaching tool. It illustrates the connection between concepts and ideas in a visual form. The concepts are connected by words and phrases to explain their relationship. Concept maps help learners to organize their thoughts in a systematic order. It deepens learners' understanding of the topic. It also helps learners to explore new concept connections (Eppler, 2006).

Cognitive ability is always related to education. Learners with high academic achievement have always been perceived as having high cognitive ability. The interpretation of abilities is varied and

*Corresponding author: ktcj85@siswa365.um.edu.my

controversial. It is culturally manipulated, therefore we must be open-minded when assessing the abilities of learners from different cultures. We should not restrict the evaluation criteria to academic ability only (Gardner, 2000).

Metacognitive skills improve learners' learning process, especially independent learning. Metacognition makes learners aware of their own thinking process. This awareness helps them to develop an effective learning strategy. According to Gardner's (2000) multiple intelligences theory, metacognitive skills enable learners to discover their own intelligence and select the most effective learning strategy to maximize learning outcomes.

## 2 RATIONALE OF STUDY

The theory of metacognition in processing information and cognitive development changed traditional learning instruction. Educators are encouraged to focus on learners' awareness of thinking and their ability to control the process of thinking. Metacognitive skills enable learners to develop essential skills such as critical thinking, creative thinking, and problem-solving skills. These skills are crucially important to the learning process (Baker, 2008).

Metacognition produces independent learners. Learners who are aware of their own thinking and thinking process are able to plan and select the most suitable strategy to complete the learning task. (Baker, 2008). Unfortunately, which thinking tool is the most effective metacognitive strategy to develop indigenous learners' cognitive ability is still undetermined.

The changing of curriculum in the Malaysian educational system is redirecting educators' endeavor to foster independent learning instead of passive learning. Transformation of the curriculum is a problematic period because most of the educators assigned to teach the new curriculum (KSSR) are educated under old curriculum (KBSR) (Ministry of Education, 2012).

KSSR curriculum encourages learners to plan their learning independently. Learners are engaged and active in learning if given the chance to make decisions in the learning process. They are able to customize learning objectives according to their own needs and interests. They have the freedom to pursue plans and strategies consistently over a long period of time. Independent learning also increases learners' self-esteem and motivation to learn as they are committed in their learning process (Brown, 1987; MOE, 2012).

Metacognition is divided into two parts: awareness of cognition and regulation of cognition. Awareness of cognition is awareness about one's own cognitive process. It includes the strengths and limitations of one's thinking abilities. The cognition can be influenced by internal and external factors. Education is important to nurture one's metacognitive ability (Schraw, 1998; Schneider, 2008).

Much research has been conducted on indigenous populations, but the development of metacognitive skills among indigenous learners and the effect of metacognitive skills to improve indigenous learners' cognitive ability remain an unfamiliar topic. Exploration on this topic helps educators to select the appropriate strategies to improve indigenous learners' learning outcomes.

## 3 STATEMENT OF THE PROBLEMS

There are numerous and longstanding literatures proving the significant effectiveness of visual thinking tools in the thinking and learning process (Flavell, 1979; Forrest-Pressley et al., 1985; Brown, 1987; Sugden, 1989; Schunk & Zimmerman, 1994; Zimmerman & Schunk, 2001). However, the pedagogy to introduce visual thinking tools to the indigenous population in Malaysia is yet largely under researched.

Stakeholders contributed persistent effort to upholding the educational quality of indigenous populations. However, the achievement gap between mainstream learners and indigenous populations remains huge (Abdullah & Mat, 2012).

The goal to provide equal education for all children in Malaysia becomes the first endeavor of the Ministry of Education. This endeavor was started long ago, even in Second Malaysia Plan developed in 1971 to 1975. It was continued in the Pelan Induk Pembangunan Pendidikan (PIPP) in 2006–2010 as the key plan of educational development in Malaysia (Kementerian Pelajaran Malaysia, 2007).

Unfortunately, all these goals have achieved success for the mainstream population only. Educational development among disadvantaged populations, particularly indigenous pupils, lags behind. MOE is determined to reduce the dropout rate of indigenous pupils (Kementerian Pelajaran Malaysia, 2007).

Statistics from JHEOA shows that the majority of indigenous people only have primary education. Many indigenous pupils do not continue their study in secondary schools. Furthermore, only 2% of the indigenous population successfully achieve higher education in tertiary institutions (Ali, 2012).

The government aspires to improve the indigenous population's welfare in terms of education. They targeted to reduce the indigenous pupil dropout rate to 15% by 2016. Besides, the government also targeted 305 indigenous pupils who will be able to continue their study to higher education institutions. They also promise to send at least 10 indigenous pupils to study abroad in the year (JAKOA, 2011).

Indigenous pupils' academic achievement is tremendously low compared to urban pupils. The percentage of the indigenous pupils passing public examinations such as UPSR, PMR, and SPM is small. One consequence of this situation is that indigenous pupils fail to continue their study in tertiary education. Only 880 indigenous pupils have succeeded in completing their studies in tertiary education between 1971 and 2010 (Razaq, 2011).

Despite efforts to improve education levels in the indigenous population, the education level of the indigenous population in Malaysia is still at a low level compared to other populations in the country. The dropout rate for indigenous students is significantly high: 29% in 2011. In 2012, the dropout rate for indigenous pupils decreased to 26.16%. In 2015, the dropout rate for indigenous pupils is 25%. Data shows improvement in indigenous pupils' dropout rate but it does not satisfy stakeholders. The rate is still a consistently large percentage in Malaysia (JAKOA, 2011).

Research from JHEOA shows that for every 100 indigenous pupils who enroll in year 1 primary school, only six of them successfully complete the study in form 5 secondary school. The findings show 94% of indigenous pupils drop out from the educational system. The enrollment of indigenous pupils to secondary school is lower than primary school enrollment (JAKOA, 2011).

Much research has been conducted on the educational system in regard to the indigenous population. These researches focused on the pedagogy and facility in indigenous schools to narrow the educational gap between indigenous learners and mainstream learners. Research on indigenous learners' metacognition development still remains obscure in the educational domain. Research on developing metacognitive skills should be conducted in order to achieve higher learning outcomes for indigenous learners.

Researchers were determined to find a statistically significant difference between the effectiveness of thinking maps, mind maps, and concept maps towards cognitive level among year 1 indigenous learners .

## 4 THEORY OF THE STUDY

The effectiveness of visual thinking tools will be measured by changes on learners' cognitive abilities. McCarthy Scale of Abilities (MCSA) serves as the measurement tool to assess learners' cognitive abilities in this research.

The inventory is commonly used to assess children's verbal abilities, quantitative abilities, perceptual performance, and memory skills. These cognitive abilities are essential for young learners to acquire further knowledge in future. MCSA was selected because it fulfills the definition of cognitive abilities (Erika, Luisa, Lizbeth, Stephen, & Lourdes, 2017).

MSCA focuses on cognitive abilities instead of IQ level of children. Cognitive abilities can be nurture by providing a constructive environment to the children (Erika et al., 2017). Unlike IQ level, which is natural ability, researcher believes that appropriate implementation of visual thinking tools are able to strengthen learners' cognitive abilities.

The definition of cognition becomes controversial in many domains of psychology. Researchers defined cognition as a mental process to transform, reduce, elaborate, store, recover, and use external and internal input of information (Cunha & Heckman, 2008).

Therefore, cognitive skills involves a wide range of activities, including perception, attention, memory coding, retention and recall, decision-making, reasoning, problem solving, imaging, planning, and executing (Cunha & Heckman, 2008).

These skills were assessed in MSCA with a wide range of interesting activities to attract children's attention. It provides a systematic assessment to observation of children's cognitive and motor abilities. The six scales included in MSCA are verbal, perceptual-performance, quantitative, general cognitive, memory load, and motor (McCarthy, 1972). Motor scale will be removed from this research because this research only focuses on cognitive abilities.

MSCA is an assessment to measure different cognitive abilities, including kinetic abilities. The assessment is suitable to evaluate cognitive level for 2½–8½-year-old children. Overall, the MSCA consists of thoroughly invented individual tasks to measure children's ability. McCarthy provides a set of scores and divergent compounded scores (William 2008).

The MSCA has a wide range of activities to measure children's abilities in five different scales. Variety of activities in MSCA includes xylophone, building blocks, and flash cards. These activities successfully retain children's attention during the long period of assessment (William 2008).

Presumably, the index reflects how well the child has integrated prior learning experiences and adapted them to the demands of the scales. The concept of combining various subtests to form composite scores is an important idea in testing. The verbal scale, consisting of five subtests, assesses comprehension and use of language. The quantitative scale consists of three subtests, which measure mathematical ability. The perceptual-performance scale consists of seven subtests, which evaluate a child's ability to conceptualize and reason without words. The memory scale consists of four subtests on short-term recall of words, numbers, pictures, and tonal sequences (Cohen, Libby, & Loraine 1994).

Verbal, perceptual performance, and quantitative scales are considered as general cognitive. The first scale is the verbal scale. Verbal scale assesses children's ability to express themselves. Visual thinking tools provide children an alternative way to receive and convey verbal information. Verbal scales assess mental processes such as divergent thinking and deductive thinking after implementing visual thinking tools in learning process (Nores & Barnett, 2010).

Verbal scale shows the maturity of children's verbal concept. Verbal ability is defined as the ability to understand and use language. It serves as a predictor for school achievement. It is the key abilities to achieve high performance in MSCA because children need to acquire verbal ability to understand examiner's instructions (Serpell & Jere-Folotiya, 2008; McCarthy, 1972).

Verbal scales consist of five items. Pictorial memory and word knowledge can be established by a mind map. Children need to recall names of objects shown on a card (Nores & Barnett, 2010). Researchers train children to memorize the names of objects using mind maps, which combine picture and verbal information in a meaningful and attractive manner.

Mind maps help to establish verbal fluency of young children. In MSCA verbal scale, children are required to name as many articles as they can in a category within 20 seconds. Spread branches in mind map develop divergent thinking. This establishes not only verbal ability but also higher order thinking skill (Nores & Barnett, 2010).

In MSCA verbal scale, children are required to complete sentences with antonyms. Bridge map can be used to introduce opposite analogies. Antonyms arranged in pairs are easier for young children to recognise and remember (Nores & Barnett, 2010).

Word knowledge is one of the items tested in MSCA verbal scale. It requires children to identify common objects and words. Circle maps can be used to introduce vocabulary for young children. Pictures can be included in the map to facilitate children's memory (Nores & Barnett, 2010).

The second scale is the perceptual performance scale. Participants' perceptual performance is assessed by several game activities. These activities assess children's reasoning skills and manipulation of materials. Children demonstrate this ability via imitation, classification, and organization of spatial, visual perceptual, and conceptual tasks (Serpell & Jere-Folotiya, 2008; McCarthy, 1972).

Visual thinking tools play an important role in developing visual organization. Logical classification can be trained using tree maps. Relationships of ideas can be developed using concept maps (Nores & Barnett, 2010).

The third scale is the quantitative scale, which assesses children's facility with numbers and understanding of quantitative words. The aim of the scale is to measure children's number aptitude, not their upper limit of computation. Number aptitude is related to the interest children have in numbers in daily life (Serpell & Jere-Folotiya, 2008; McCarthy, 1972).

Visual thinking tools can used to present numbers in an attractive manner. Attractive presentation helps to strengthen children's memory of numbers (Nores & Barnett, 2010).

The forth and the last scale is the memory scale. The tests in MSCA assess children's short-term memory. Verbal and non-verbal modalities are used to stimulate children's memory in this scale. Children's abilities to deal with specific content to be memorized have significant influences to their score in this scale. Therefore, test of memory has been located on verbal, perceptual performance, and quantitative scales (Serpell & Jere-Folotiya, 2008; McCarthy, 1972).

Visual thinking tools are used to stimulate children's memory ability. Memory ability is vital in cognitive development and learning processes (McCarthy, 1972). Researcher observes the impact of different visual thinking tools on children's memory ability.

Research uses modified MSCA as the instrument to measure indigenous learners' cognitive abilities. This is because the modified MSCA eliminated the language barriers for indigenous learners. The validity of modified MSCA will be determined later.

## 5 RESEARCH DESIGN

Researcher selected a pretest and posttest experiment approach to study the effectiveness of different visual thinking maps on development of indigenous children's cognitive ability. Statistical experiments are selected in this research because the researcher is allowed to manipulate different variables of the experiment. Besides, the experiment approach also provides more control for the researcher to the minimize influence of irrelevant factors for this research (McNiff & Whitehead, 2011).

This is a true experiment. Researcher has full control over the allocation and scheduling of the treatment. Samples involved are randomly assigned to different experimental groups (Imai, Tingley, & Yamamoto, 2013).

Random allocation of samples increases the statistical power of the research. It helps the researcher to control nuisance and confounding variables. Random allocation also increases the external validity of the research (Imai, Tingley, & Yamamoto, 2013).

Every sample has an equal opportunity to be allocated to any intervention, spreading the potentially nuisance variables among samples evenly across different interventions. The purpose is to minimize systematic differences other than the intervention. It is important to allocate the samples randomly because some of the nuisance variables are difficult to identify (Imai, Tingley, & Yamamoto, 2013).

In this research, the researcher endeavored to improve the cognitive level of indigenous children by using visual thinking maps such as thinking maps, mind maps, and concept maps. From a literature review on indigenous populations, the researcher is also aware that different living environments in indigenous village require educators to use different teaching and learning approaches.

Researcher used different groups of samples for each intervention. Therefore there are four groups involved in this research. Division of sample for different intervention is necessary to avoid order effects of pupils' previous knowledge from other interventions (Baldwin, 2012).

Cognitive levels will be measured using McCarthy Scale of Children Abilities (MSCA) in the pretest. Assessment resultd from the pretest will serve as an entry point for each respondent. Entry point is the baseline to measure the value-added in a new intervention, in this case, the cognitive level (Dynarski & Berends, 2015).

The learning session will be held once a week after school. Duration of a session is two hours. The first hour, researcher introduces the maps and shows the pupils how to construct the map. The second hour, respondents are given a task to construct their own visual thinking maps.

Researcher serves as a facilitator while respondents construct their maps. A variety of materials, such as plain paper, colored paper, pencils, colored pencils, magic colors, crayons, and stickers are provided in the session. Researcher provides variety of materials for respondents to attract them and encourage their creativity.

Cognitive levels will be measured again using MSCA after three months as new interventions were implemented in different groups. Value-added from the interventions is measured by comparing posttest and pretest results (Dynarski & Berends, 2015).

Changes will be compared among each group. The greater the distant between pretest and posttest, the greater the impact of the intervention (Dynarski & Berends, 2015). The highest value-added intervention is the most suitable visual thinking tool to improve indigenous children's cognitive level.

Researcher defines details of the research before it is conducted to ensure accuracy of findings. Improvement in designing the experiment is an ongoing process to ensure the usefulness of findings to improve quality of learning among indigenous populations (Schildkamp, Lai, & Earl, 2013).

### 5.1 *Selection of participants*

Researcher uses random sampling to identify samples for this research. Qualified respondents will be randomly assigned to a factor that it may affect the variables to be measured.

Sample is one of the subsets selected from the population of interest. In this research, 100 indigenous children, aged 7 and 8, from Pulau Carey were selected to serve as samples. Sampling process is vital for the research because time and cost consumed to conduct an experiment for whole population is too high (Barreiro & Albandoz, 2001).

Researcher determines sample size to 100 samples. They are divided into three experimental groups and one control group. Each group consists of 25 samples.

Randomized design is used in this research. Researcher randomly assign samples into four groups. Children from primary 1 and 2 are mixed, regardless of age, gender, and academic performance. This process is important to ensure all interventions were tested without prejudice or discrimination (Anderson & Shattuck, 2012).

The indigenous population in Carey Island is Mah Meri, the indigenous group inhabitanting Banting Port Klang, Selangor. They are recognized by the Malaysia government as Orang Asli from 18 indigenous groups. They are the subgroup from the Senoi population.

They also called as Orang Laut because they live along the coast and worked as fisherman. Their main villages stretch from Sungai Pelek to Carey Island. Their native language is Besisi. It is an Austroasiatic language or Mon-Khmer. The native language is a popular native language in the Mainland of Southeast Asia.

## 6 CONCLUSION

Researcher was aware that metacognitive skill plays an important role in students' early learning process, especially in indigenous education. To ensure indigenous learners do not get left behind in the mainstream education, we should emphasize customized education to fulfill variety needs and interests.

This research will help to identify the most effective visual thinking tool to develop indigenous learners' cognitive abilities. It is an important step to solve student-learning problems at an early stage for indigenous learners whose language is a barrier to coping with the current syllabus.

This research can be conducted in variety of circumstances for demographic comparison, such as gender, race, and SES.

Based on the methodology used, researchers can conduct more structured and systematic research. The methodology of this study can also ensure that the study is progressing to meet the objectives and provide guidance to researchers about the procedures implemented in the study.

After this research, researcher found the most appropriate visual thinking tool to develop learners' cognitive abilities, which are verbal abilities, perceptual performance, quantitative, and memory.

The findings from this research will serve as a guideline for future research on a combination of different visual thinking maps to develop young children's cognitive abilities.

## REFERENCES:

Abdullah, R., & Mat, F. B. (2012). Educational potency of the Orang Asli students in the State of Kelantan: Perspectives of the Department of Orang Asli Development

(JAKOA). In Abdullah et al. (Eds.), *Education and Orang Asli in the Mainstream*. Kuala Terengganu: Unversity of Sultan Zainal Abidin.
Alikhan, N. (2014). *Thoughts on Thinking Maps: A New Way to Think*. Los Angeles: New Horizon School.
Anderson, D., Nashon, S. M., & Thomas, G. P. (2009). Evolution of research methods for probing and understanding metacognition. *Research in Science Education*, 39 (2),181–195.
Anderson, T., & Shattuck, J. (2012). Design-based research: A decade of progress in education research. *Educational Researcher*, 41(1),16–25.
Baker, L. (2008). Metacognitive development in reading: Contributors and consequences. In K. Mokhtari & R. Sheorey (Eds.), *Reading Strategies of First and Second Language Learners: See How They Read*. Norwood, MA: Christopher Gordon.
Baldwin, M. (2012). Participatory action research. In Gray M., Midgley J. and Webb S. (Eds.), *The SAGE Handbook of Social Work* (pp. 467–481). London: SAGE.
Barreiro, P. L., & Albandoz, J. P. (2001). Population and sample sampling techniques. *Management Mathematics for European Schools*. R.
Buzan, T. (2002). *The Power of Social Intelligence*. London: Thompsons- HarperCollins Publisher.
Cunha, F., & Heckman, J. (2008). Formulating, identifying and estimating the technology of cognitive and noncognitive skill formation. *Journal of Human Resources*, 43(4),738–782.
Curriculum Development Division. (2014). *Primary School Standard Curriculum*. Putrajaya: Malaysian Ministry of Education.
Dynarski S and Berends M (2015) *Introduction to Special Issue: Research using longitudinal student data systems: Findings, lessons, and prospects*. Educational Evaluation and Policy Analysis 37(1): 3S–5S.
Eppler, M. J. (2006). A comparison between concept maps, mind maps, conceptual diagrams, and visual metaphors as complementary tools for knowledge construction and sharing. I, 5(3),202–210.
Gardner, H. (2000) *Intelligence Reframed: Multiple Intelligences for the 21st Century*. New York: Basic.
Hyerle, D. (1996). *Thinking Maps: Seeing Is Understanding*. Educational Leadership, 53(4),85–89.
JAKOA. (2011). *Development Strategic Planning for the Orang Asli 2011–2015*. Kuala Lumpur. Department of Orang Asli Development. Malaysia.
Schildkamp, K., Lai, M. K., & Earl, L. (2013) *Data-Based Decision Making in Education: Challenges and Opportunities*. Dordrecht: Springer.
Serpell, R., & Jere-Folotiya, J. (2008) Developmental assessment, cultural context gender, and schooling in Zambia. *International Journal of Psychology*, 43(2),88–96.
Zimmerman, B. J. & Schunk, D. H. (2001) *Self-Regulated Learning and Academic Achievement*, 2nd ed., Hillsdale, NJ: Lawrence Erlbaum.

*The Social Sciences Empowered – Lumban Gaol, Hutagalung & Fong Peng (eds)*
 *ISBN 978-0-367-00090-5*

# LINUS Program for Bahasa Melayu reading literacy among indigenous students

Md. Nasir bin Masran*, Nur Aida binti Che Aziz, Siti Rahaimah binti Ali & Farah binti Mukhtar
*Faculty of Human Development, Sultan Idris Education University, Malaysia*
*Institut Pendidikan Guru Kampus Raja Melewar, Negeri Sembilan*

ABSTRACT: This study aims to identify the effectiveness of the LINUS intervention program, which has been coordinated by the Ministry of Education (MOE) at the primary-school level, in improving the literacy level of Bahasa Melayu among indigenous students. This study used quantitative methods that involved surveys from primary schools around Batang Padang, Cemeron Highland, and Gua Musang. The instruments used were questionnaires, interviews, and document analysis. The questionnaires was divided into six sections: part A is on the demographic part; part B is on the level of achievement in Bahasa Melayu reading literacy; part C is on the implementation of Linus program; part D is on the teachers' teaching approach; part E is on the teaching and learning method; and part F is on the contributing factors. A total of 30 LINUS Bahasa Melayu teachers were selected as a sample for the study. The researchers used descriptive statistics to analyze respondent profile and for questionnaire analysis. The descriptive statistics used were mean, mean score, and standard deviation. The results showed the LINUS intervention program was effective in improving Bahasa Melayu reading literacy. The findings also showed high mean score for all the reading parts tested except for the contributing factors. This shows that there are other factors influencing the effectiveness of the Linus intervention program for the indigenous students. The implementation of the Linus program, the teacher's teaching approach, and the teaching and learning method used showed positive mean. In conclusion, the Bahasa Melayu teachers who had undergone the LINUS intervention program implementation stated that this program is very helpful in their teaching. The findings clearly demonstrated that teachers practice every construct and module provided in the program but their impact might be influenced by other factors.

*Keywords*: effectiveness, LINUS intervention program, Bahasa Melayu reading literacy, indigenous students

## 1 INTRODUCTION

As a basis for education, literacy skills or reading and writing skills are the ones that need to be given priority at the early stage of schooling. Without this foundation, pupils find it difficult to continue their everyday learning, which will become more challenging and also involve high level thinking skills (HOTS) in today's education.

Education opportunities are not limited to certain groups but rather are for all levels of society as everyone needs to acquire knowledge so that they could change their future. Therefore, indigenous students should also receive education in order to have the same education opportunities and no dropouts.

In extending the context of education to various levels and forms, the main emphasis at the early stage of education is the reading literacy of Bahasa Melayu. Bahasa Melayu is now viewed as a major and very important subject in school and is used at every level of learning and also as a medium of communication. This is consistent with the government's strategy since 2010, under the National Key Result Area (NKRA) mandate, to ensure students master literacy and numeracy. The Ministry of Education Malaysia has taken the initiative in providing literacy education from early childhood. According to Mahzan Arshad (2010), human capacity in pursuing science will change the pattern of society life.

## 2 PROBLEM STATEMENT

The problem of failing Bahasa Melayu, especially among indigenous students, needs serious attention.

*Corresponding author: md.nasir@fpm.upsi.edu.my

It is important for these students to pass the subject as it is related to the continuity of life and progress in navigating the flow of education in tandem with students in the mainstream schools. There are several studies focusing on the causes of failure in student literacy, such as health, family background, and emotional development. The study shows that the problem of literacy requires serious and deeper attention. The problem of lack of learning experience also has a negative impact on vocabulary development and the growth of students' thinking.

The previous studies always focused on the problem of dropouts among students (Punanesvaran, 2004), and student dropouts and their relationship with the rural socioeconomy (Mohd Eusoff, 1974). The dropout rate of indigenous students (Kumar, 2001) has shown that one of the factors that contributed to the problem of dropout was not being able to master reading literacy. Among the causes of inability to master reading literacy is that the level of teaching and learning the subject is too high (Tuzana, 2014).

According to Tabah (2008), the students' level of mastery and the difficulty of reading and writing are at a modest low level. In order to achieve a more advanced level of literacy, teachers need to master the LINUS teaching methodology specifically and professionally. Therefore, a study should be conducted on the LINUS program that has been implemented so that a finding explains why, after various programs have been implemented, the literacy rate is still low and for indigenous students still lack literary incompetence.

## 3 RESEARCH OBJECTIVES

The objectives of this study are:

a) to identify the level of Bahasa Melayu reading literacy among indigenous students
b) to identify the suitability of the syllabus or teaching and learning content with the level of Bahasa Melayu reading literacy among indigenous students
c) to identify the level of teachers' knowledge skills in developing reading literacy skills of Bahasa Melayu among indigenous students.
d) to identify the effective teaching and learning methods that enable the enhancement of the reading literacy skills of Bahasa Melayu among indigenous students.
e) to identify the contributing factors to the implementation of the LINUS intervention program towards reading literacy skills of Bahasa Melayu among the indigenous students.

## 4 RESEARCH QUESTIONS

a) What is the level of Bahasa Melayu reading literacy among indigenous students?
b) What is the suitability of the syllabus or teaching and learning content with the level of Bahasa Melayu reading literacy among indigenous students?
c) What is the level of teachers' knowledge skills in developing reading literacy skills of Bahasa Melayu among indigenous students?
d) What is the effective teaching and learning methods that enables the enhancement of the reading literacy skills of Bahasa Melayu among indigenous students?
e) What are the contributing factors to the implementation of the LINUS intervention program towards reading literacy skills of Bahasa Melayu among indigenous students?

## 5 RESEARCH METHODOLOGY

### 5.1 *Research limitation*

The limitations of the study are the location of the study, the sample of the study, and the contributing factors to be studied. The study also uses only two instruments to collect the data needed to answer the questions of the study.

### 5.2 *Research design*

This study uses quantitative methods and surveys to gather information on the effectiveness of LINUS intervention programs for Bahasa Melayu reading literacy for indigenous students. This method is chosen because researchers are able to gather accurate information and to focus more on the research objectives. According to Mohd. Majid Konting (2005), a survey aims to collect information on variables.

### 5.3 *Sample*

Sample is a case or subject extract from a population. According to Sekaran (2003), the sample size refers to the actual number of subject or material chosen as a sample to show population characteristics (N). In this study, the researchers chose to use the purposive sampling method. The researchers selected 30 respondents among the LINUS teachers who taught Bahasa Melayu subjects in primary schools located around the districts of Batang Padang, Cameron Highlands, and Gua Musang.

### 5.4 *Research instrument*

This study uses two instruments: questionnaires and interviews. The questionnaire is distributed to 30 respondents with six sections and each part has 10 items and 11 items with the aim of obtaining data. The interviews are conducted face-to-face with four respondents consisting of LINUS Bahasa Melayu teachers and LINUS primary school coordinator who are located around Batang Padang, Cameron Highland, and Gua Musang districts.

### 5.5 *Pilot study*

The pilot study was conducted on 30 respondents consisting of LINUS Bahasa Melayu teachers who taught at several Batang Padang and Kuala Kubu Baru district schools. This study was conducted before the actual study. From the findings of the pilot study, the questionnaire items used were positive and suitable for use as a research instrument. Cronbach alpha reliability for 51 items of this study was 0.8378. The researchers used the SPSS (Statistical Package for the Social Science) software version 2.0 to obtain the Alpha Cronbach's value.

## 6 RESEARCH FINDINGS

Below are the research results.

Based on Table 1, the overall mean for identifying the extent of literacy proficiency in Bahasa Melayu or achievements to be at a high level with high score and the mean value (M = 3.913). The highest mean is 4.07

Table 1. Bahasa Melayu literacy competency (achievement).

| No | Item | Mean | Mean Score | Standard Deviation (SD) |
|---|---|---|---|---|
| 1 | The LINUS class is very fun | 3.80 | High | 0.55 |
| 2 | LINUS test measures the ability of student achievement | 4.03 | High | 0.67 |
| 3 | My students like activities in LINUS class | 3.80 | High | 0.48 |
| 4 | The LINUS program improves the reading of Malay students | 3.97 | High | 0.61 |
| 5 | The LINUS program helps students to read easily | 4.07 | High | 0.58 |
| 6 | LINUS class activity allows students to answer verbal questions | 4.00 | High | 0.53 |
| 7 | Activity in LINUS class guides students to answer written questions | 3.83 | High | 0.53 |
| 8 | The LINUS program motivates students to learn | 3.87 | High | 0.68 |
| 9 | The LINUS module provided by the KPM complies with the student standard | 3.93 | High | 0.74 |
| 10 | Students acquire reading skills faster through the LINUS program | 3.83 | High | 0.70 |
| | **Overall Mean** | **3.913** | **High** | |

(SP = 0.58), at item no. 5: the five LINUS programs help students to read easily. The analysis showed that the level of proficiency or achievement in the LINUS program for Bahasa Melayu reading literacy is high.

Table 2. Implementation of LINUS program.

| No | Item | Mean | Mean Score | Standard Deviation (SD) |
|---|---|---|---|---|
| 1 | I know that LINUS is related to reading | 4.33 | High | 0.61 |
| 2 | Teaching and learning in LINUS class is fun | 3.87 | High | 0.51 |
| 3 | Planning lesson for LINUS class is mandatory | 4.27 | High | 0.52 |
| 4 | Teaching and learning activities in LINUS class is the basic introduction to students who are weak in Bahasa Melayu | 4.20 | High | 0.48 |
| 5 | The LINUS program makes students read | 4.10 | High | 0.48 |
| 6 | The LINUS program can enhance students' Bahasa Melayu language skills | 4.03 | High | 0.56 |
| 7 | Students are required to pass LINUS screenings | 3.97 | High | 0.67 |
| 8 | Students involved actively during the learning session | 3.70 | High | 0.47 |
| 9 | Students gain a lot of basic knowledge and techniques of reading Bahasa Melayu from LINUS class | 3.97 | High | 0.49 |
| 10 | Parents also know about the LINUS program at school | 3.57 | Moderate | 0.71 |
| | **Overall Mean** | **4.000** | **High** | |

The findings show that the LINUS intervention program is well implemented as it has a high mean score; the overall mean value (M = 4.00). A moderate mean score is 3.57 (SP = 0.71), which is the parents of the pupils also know about the LINUS program at school. The highest mean score is 4.33 (SP = 0.61), that is most of the teachers know that LINUS is related to reading. This means that the overall mean of the LINUS intervention program implementation on reading literacy among indigenous students is high and shows that it is well implemented.

Based on Table 3, the mean score for the level of teacher's knowledge skill in reading literacy is high (M = 4.213). The highest mean is 4.33 (SD = 0.55) for item no. 5: I use an appropriate tone of voice when teaching; and also item no. 7: I will correct the

Table 3. The level of teachers' knowledge skill.

| No | Item | Mean | Mean Score | Standard Deviation (SD) |
|---|---|---|---|---|
| 1 | I taught in a fun way | 4.03 | High | 0.61 |
| 2 | I often practice drilling following the LINUS construct to the students | 4.10 | High | 0.61 |
| 3 | I use simple and easy sentences | 4.30 | High | 0.47 |
| 4 | I always check LINUS student's training | 4.20 | High | 0.48 |
| 5 | I use the appropriate tone of sound when teaching | 4.33 | High | 0.55 |
| 6 | I am friendly with the students | 4.30 | High | 0.47 |
| 7 | I will correct the error of the student during the LINUS class | 4.33 | High | 0.55 |
| 8 | I am good at teaching LINUS | 3.93 | High | 0.64 |
| 9 | I will make LINUS class in good condition | 4.20 | High | 0.55 |
| 10 | I always praise the students during the lesson | 4.40 | High | 0.62 |
| | **Overall Mean** | **4.213** | **High** | |

Table 4. Teaching and learning methods.

| No | Item | Mean | Mean Score | Standard Deviation (SD) |
|---|---|---|---|---|
| 1 | I use a computer while teaching LINUS | 2.93 | Moderate | 0.83 |
| 2 | I use reading cards while teaching LINUS | 3.93 | High | 0.58 |
| 3 | I used the Hibur method while teaching LINUS | 4.03 | High | 0.57 |
| 4 | Cards and colorful materials | 3.90 | High | 0.48 |
| 5 | I use pictures and videos while teaching LINUS | 3.53 | Moderate | 0.90 |
| 6 | I diversify LINUS class teaching techniques | 4.27 | High | 0.58 |
| 7 | BBM that I provide is interesting | 4.07 | High | 0.58 |
| 8 | I encourage students to communicate during the learning session | 4.27 | High | 0.52 |
| 9 | I always praise every student achievement level | 4.27 | High | 0.45 |
| 10 | My PDP sessions are not just concentrated in the classroom | 3.93 | High | 0.58 |
| 11 | I apply every LINUS construct in PDP | 4.13 | High | 0.57 |
| | **Overall Mean** | **3.99** | **High** | |

mistakes of students while in class. The level of the teacher's knowledge skill is appropriate to the situation and acceptance of the learning and teaching of indigenous students. These indigenous students are sensitive and require teachers to use an appropriate tone when teaching. Nevertheless, each error should always be corrected at each level so that learning basic reading obtained with good practice.

Based on Table 4, overall mean of the Teaching and Learning method applied in the teaching Bahasa Melayu reading literacy is high at 3.99. A moderate mean score is 2.93 (SP = 0.83), that is item no. 1: I use a computer while teaching LINUS. Item no. 6: I diversify the LINUS class teaching technique, has the highest mean score of 4.27 (SP = 0.58). Item no. 8: I encourage students to communicate during the learning session, also has the highest mean score of 4.27 (SP = 0.52) and item no. 9: I always praise each level of student achievement, also has the highest mean score of 4.27 (SP = 0.45)

Table 5. Contribution factors of the LINUS implementation.

| No | Item | Mean | Mean Score | Standard Deviation (SD) |
|---|---|---|---|---|
| 1 | Students are ready to learn | 3.10 | Moderate | 0.75 |
| 2 | Parents are given exposure to the LINUS intervention program | 3.50 | Moderate | 0.78 |
| 3 | Parents always care about the lesson I teach at school | 2.37 | Moderate | 0.85 |
| 4 | Parents provide additional reading guidance at home | 2.13 | Low | 0.97 |
| 5 | I master the contents of LINUS Teaching and Learning Modules | 3.93 | High | 0.64 |
| 6 | I used the LINUS Teaching and Learning Module | 4.03 | High | 0.49 |
| 7 | Students know the LINUS program is important for improving reading skills | 3.47 | Moderate | 0.97 |
| 8 | Students enjoy learning in LINUS class environment | 3.73 | High | 0.78 |
| 9 | Nutrition is a contributor to students' learning | 4.40 | High | 0.72 |
| 10 | LINUS class learning time is suitable for every learning session | 4.13 | High | 0.78 |
| | **Overall Mean** | 3.479 | Moderate | |

The contributing factor to the success or failure of students to improve literacy skills by reading Bahasa Melayu, the LINUS intervention program, is at a moderate level with the overall mean value of (M = 3,479). The lowest mean score with the mean value = 2.13 (SD = 0.97) is item no.4: Parents give additional guidance at home reading. Whereas for the moderate mean score, mean = 2.37 (SD = 0.85) is item no. 3: Parents always care about the lesson I teach at school. While the highest mean score with mean = 4.40 (SD = 0.72) is item no. 9: Nutrition is a contributor to students' learning

## 7 CONCLUSION

Based on the interviews and the findings, the majority of the respondents supported the LINUS intervention program. According to them, the Linus program is good and useful. It has detailed content modules and is constructed in an orderly way. It has effective use for the Linus students and assists in teachers' teaching and learning process. However, this program is seen to be less effective due to other factors that affect all the environmental conditions of the program. Among these factors are the language, culture, self factor, lack of balance and continuity of learning at home, no encouragement from parents, teachers, and also dietary factors.

The findings also show low self-esteem among indigenous students, especially the first year students. This self-concept is shy, sensitive, and less self-esteem. However, this attitude decreases when they are in the second year and beyond. But the level of readiness is still moderate with the mean value finding of 3.10. The indigenous students perceive school as the place for them to get food and meet their intention of going to school but there is no readiness for them to learn. The findings from the interview also found that the indigenous students were easily bored and tired of the learning activities. They can only focus between 10 to 15 minutes (Mohamad Johdi, 2009). Then they will turn to other activities like playing and no longer focus on the learning. With this attitude, formal teaching activities can not be fully implemented. The indigenous students are more interested in learning in the form of playing and entertaining methods that require a lot of movement and music. The play method can enhance the interest of indigenous students to learn (Haslinda & Zanaton, 2015). This teaching method makes their learning fun.

In this study, it can also be seen that the mean score for additional reading guidance at home by parents is low. The mean score for parents caring about the lesson taught at school is only moderate. Therefore, it is hoped that future study uncover the details of the real problem of parents who are not concerned about their child's learning in school.

## REFERENCES


Ab. Halim Tamuri & Mohamad Khairul Azman Ajuhari, (2010). Amalan pengajaran guru Pendidikan Islam berkesan berterskan konsep Mu'allim. *Journal of Islamic and Arabic Education*, 2(1),43–56.

Chua, Y. P. (2014). *Kaedah penyelidikan: Kaedah dan statistik penyelidikan*. Edisi-3 .Shah Alam: McGraw-Hill Education.

Hasbullah Abd. Rahman, Azhar Abdul Aziz & Mohd Yadman Sarwan. (2010). *Pendidikan seks Menurut Pendekatan Islam: Satu Tinjauan*. Dlm Abd Rahim Abd Rashid. Pendidikan Seksual: Perspektif Kurikulum, Pengajaran dan Tingkahlaku Moral. ms. (53–61). Kuala Lumpur: Utusan Publications.

Kitila, A. M. (2012). *Teachers' Attitudes towards and Comfort about Teaching Schoolbased Sexuality Education in Urban and Rural Tanzania*. University of Dar es.

Salaam, Tanzania. *Global Journal of Health Science*. Vol. 4 Mahzan. A. (2010). *Literasi dalam Pendidikan Seks*. Dlm Abd Rahim Abd Rashid.

Pendidikan Seksual: Perspektif Kurikulum, Pengajaran dan Tingkah laku Moral. ms. (42–52). Kuala Lumpur: Utusan Publications.

Tara, T. S. et al. (2014). *Knowledge and Teaching Confidence of Educators about Sexual Health Topics*.Department of Health & Human Performance Texas A & M University. National Forum of Teacher Education Journal. Vol. 24.

*The Social Sciences Empowered – Lumban Gaol, Hutagalung & Fong Peng (eds)*


# Measuring the relationship of technical university student's English proficiency with their academic achievement and attitude

Norhatta Mohd* & Yasmin Yahya*
*Faculty of MIIT, Technical Foundation Section, University of Kuala Lumpur (UniKL), Malaysia*

Nurul Sharaz Azmanuddin*
*Faculty of MIIT, Software Engineering Section, University of Kuala Lumpur (UniKL), Malaysia*

Nurhadiana Nurulmatin*
*Faculty of MIIT, Student Development Section, University of Kuala Lumpur (UniKL), Malaysia*

ABSTRACT: Proficiency in the English language has vital implications on social integration and level of education. This study examined the level of English proficiency among students at Malaysian Institute of Information Technology (MIIT), Universiti Kuala Lumpur (UniKL). This study also observed the relationship between attitude and gender towards students' English language proficiency. A total of 116 UniKL MIIT undergraduates were randomly selected as samples. The aim was to find out the relationship between English proficiency and academic achievement as well as students' attitude. Descriptive and inferential statistical analyses were used to analyse the collected data. The research findings showed there was a significant relationship between attitude, academic achievement and gender towards English proficiency. Overall, the findings of this study will help stakeholders to have a better understanding of the relationship between English language proficiency and the factors that affect the students' academic achievement. It will also provide strong evidence and add up to the current literature to strengthen the strategies in helping the technical University students to become marketable employees in the future who are excel in their field. Some recommendations were suggested at the end of the study where future research analyses towards other factors such as their course selection and environment will be included.

*Keywords*: Academic achievement, English proficiency, attitute, effect, factors

## 1 INTRODUCTION

The English language has become an international language in today's global work life. Since English is the language for many scientific and technological written information sources, it is crucial for higher learning institutions to internationalise their education systems to meet the needs of students who will be working in the global market. (Anadolu 2001). In addition, the increase in global interaction has stimulated the growing demand for proficiency in English language. The demand for English proficiency in the academia is on the increase because most of interactions either for studies or works in many countries in the world are done in English. As a result, most of the countries in the world today use English language as either a foreign or second language. In many English medium tertiary institutions, English is taught to improve the students' proficiency and communicative competence from the very beginning of their studies to ensure they can follow their academic study courses successfully (Emmanual, Yakubu & Yusof, 2014).

Most of the methods adopted in teaching English were not geared towards achieving specific purpose nor tailored to the needs of the learners. Such methods range from Grammar Translation method; the Direct Method, the Audio-Lingual Method; the Cognitive - Code Method; situational language teaching, communicative language teaching among others (Nwogu & Nwoke 2005). Bachman (1990) defines language proficiency as the language ability or ability in language use. Oller (1983, as cited in Fakeye & Yemi, 2009) avers that language proficiency is not a single unitary ability, but that it consists of several distinct but related constructs in addition to a general construct of language proficiency.

Early detection of the factors that contribute to the decrease of English language proficiency among

*Corresponding author: norhatta@unikl.edu.my; yasmin@unikl.edu.my; nsharaz@unikl.edu.my; nurhadiana@unikl.edu.my

university students is essential since the English proficiency is crucial for their educational achievement as well as for the students' future career development. This is in line and parallel with the advancement of technology where the use of English language has become prominent and it is important for someone to acquire and master it. Nonetheless, some issues arise when the language has been incorrectly used especially among the second language learners who take this language acquisition lightly. Present studies to measure the link between students' attitude, gender and academic achievement towards English language proficiency have not shown much evidence on how does it affect the technical students in a university. Thus, it has been a growing need to do in-depth studies to determine whether the factors such as gender, academic achievement and attitude do really affect the students especially among University Kuala Lumpur (UniKL MIIT) undergraduates who can be considered as technical students.

The research questions posed in this study were as below:

1. Is there any significant difference between students' proficiency in English language based on gender?
2. Is there any relationship between students' academic achievement and English proficiency?
3. Is there any relationship between English proficiency and students' attitude?

Hypotheses:

1. There is no significant mean difference between gender and English proficiency
2. There is no significant relationship between English proficiency and academic achievement.
3. There is no significant relationship between English proficiency towards students' attitude among MIIT students.

These three null hypotheses were tested through descriptive and inferential analyses. The first hypothesis was tested using t-test and was presented in the frequency table. The second hypothesis was tested using Pearson correlation and the third hypothesis was tested using one-way ANOVA. The findings of the research may help the university to identify the problems in the early semester. This will help to find suitable solutions for students to get a good academic achievement. The findings of this study may also help to provide better strategies to improve the students' level of English proficiency and academic achievement.

## 2 LITERATURE REVIEW

### 2.1 *Causes that contribute to the undergraduates' academic performance*

Based on the existing research, there are a number of factors in multiple dimensions that contribute to the undergraduates' academic performance. In general, these factors fall into the following four categories: academic, psychosocial, cognitive, and demographic (McKenzie & Schweitzer, 2001). All these factors have been widely explored and studied by previous researchers.

For example, among academic factors, prior academic achievement (e.g., McKenzie & Schweitzer, 2001; McKenzie, Gow, & Schweitzer, 2004), learning skills and habits (e.g., AbbottChapman, Hughes, & Wyld, 1992), learning strategies (i.e., general learning strategies, subject-matterspecific strategies) and approaches (e.g., Duff, Boyle, Dunleavy, & Ferguson, 2004; Pokay & Blumenfeld, 1990; Sadler-Smith, 1996; Watkins & Hattie, 1981) were explored as variables influencing academic performance.

According to Farisa Jazima, Nurul Ashikin and Syamaizar (2017), environmental factors such as parents or family, teacher and peers were the least contributors that affect the English language proficiency level among technical students. Earlier studies by Abdul & Hanifah (2010) did not see a significant effect on the English language proficiency among students across gender. Meanwhile, Alsheikh, Parameswaran & Elhoweris (2010) argued that parents have a challenging role in influencing the performance of their children in education. According to them, family cohesion and effort were remarkably important factors that help with the children's GPA scores.

### 2.2 *Factors that contribute to the undergraduates' English proficiency*

Fresh graduates are facing with the problems that their level of proficiency especially in English is deteriorating. This has contributed for the unemployment issue as the employers have seen them as not having the quality needed in order to be a global player in a company or organisation. Shamsuddin (2015, as cited by Yuen, 2015) believes the main cause to this issue comes from the tendency of the current Generation Y graduates who are more into their gadgets and very technology-savvy. As a result, the use of proper language has been abandoned as they constantly communicate with one another using short forms, acronyms and slang words. However, there have been cases where the language-learning situation and the environment can be associated immediately with the learner's attitudes as a whole. It was generally agreed that although attitudes did not necessarily determine behavior, positive attitudes did facilitate the language learning process (Sayeedur Rahman, 2005).

Lack of interest in learning English can contribute to the low proficiency in English among students. Norrizan (1992) stated that in regard of speaking in language learning, there were many interrelated factors that can influence the interest of the students in learning the target community's language such as in terms of political, cultural and technical way. Many Malaysian students exhibit a lack of interest in learning English due to the fact that Malaysian societies are different socially or culturally from the English-speaking Western societies. This somehow may contribute to the negative attitudes towards English, low proficiency, lack of confidence and the feeling of anxiety and apprehension.

## 3 MATERIAL & METHOD

### 3.1 *Study design*

This study applied a quantitative approach by means of cross-sectional research design among UniKL MIIT students. It was conducted for two semesters between January to December 2018. A total of 116 questionnaires were distributed through random sampling technique for third semester degree students. Random sampling was used as it is considered as the suitable method to collect data since it is a quick and easy method to give equal chance of anyone being picked within one large target group (Tuckman, B.W. 1978). The programs that participated in this study are Bachelor in Computer and Entrepreneur Management (BCEM), Bachelor in Networking System (BNS) and Bachelor in Software Engineering (BSE). Respondents were given a set of questionnaires to be answered within 30 minutes during their class hours and the response rate was hundred percent.

### 3.2 *Study instrument: questionnaire and data analysis*

The instrument used was adapted from B. Yushau and M. Hafidz (2015); Latu (1994); Atetwe (2013); Marian; Blemenfeld and Kaushanskaya (2007). All the questionnaires were adapted to suit the scope of this study. There were two sections in the questionnaire: the first section is seeking for students' demographic background where closed ended questions were asked. The second section consists of questions regarding students' attitude towards the use of English. For the purpose of the participants to understand the items better, some amendment, enhancement and improvement have been made for several phrases and sentence construction of the questionnaire. Likert Scale items were used for the rating from 1 to 5 with 1 for totally disagree, 2 for disagree, 3 for agree and 4 for totally agree.

This study applied the IBM Statistical Package for Social Science (SPSS) version 20 to analyze the data. For obtaining demographic information, the data was analyzed using descriptive analysis to get the frequency and the percentage of the respondents. For other variables the data was analyzed to obtain the mean, standard deviation, minimum and maximum values of the variables. For research question 1, data was analyzed using independent sample t-test to investigate whether there is a significant difference in mean between two independent groups. Pearson Correlation was used to analyze data for obtaining the results for research question 2. The correlation coefficient ranging from -1 to 1 indicates significant strong negative correlation to strong positive correlation between two continuous variables. For research question 3, the data for both variables; English proficiency and students' attitude were recoded to three category levels; high, medium and low. The values for these three levels are demonstrated in Table 1. The Chi-square analysis was used to examine the association among the levels of English proficiency and students' attitude.

Table 1. The level values of English proficiency and students' attitude towards the use of English.

| Level | Value |
|---|---|
| High | 3.0 – 4.0 |
| Medium | 2.0 – 2.9 |
| Low | Below 2.0 |

### 3.3 *Reliability of instruments for questionnaires*

Reliability is the condition in which measures are free from error and therefore produce consistent outcomes. The instrument is considered reliable if a measurement design or method consistently assigns the same score to individuals or objects with equal values (Lakshmi & Mohideen, 2013).

The use of coefficient Alpha is the most common technique of assessing internal consistency reliability estimation. There are three different measures of coefficient Alpha but the most commonly used measure is Cronbach's coefficient Alpha. Cronbach's Alpha is in fact an average of all the possible reliability estimates of an instrument (Cronbach, 1951).

The value for coefficient starts from 0 to the maximum of 1. In general, a value of 0.6 or less indicates unacceptable internal consistency reliability. Mostly in the social sciences, acceptable reliability estimates ranges from 0.70 to 0.80.

In this study, the questionnaire for attitude towards the use of English with nine items were tested through Cronbach's Alpha reliability analysis. The result for the Alpha coefficient is 0.759 which is in the range of acceptable reliability estimates (Lakshmi & Mohideen, 2013).

## 4 RESULTS AND DISCUSSION

Table 2 shows the demographic and descriptive statistics of respondents from UniKL MIIT students. The distribution of the sample by gender found 68 of the students are males and 48 are females. Meanwhile, majority of the them are originated from city (83.6%) and only 16.4% of them are from village.

Table 2. Demography and descriptive statistics of respondents.

| Variable | n (%) | Mean | Std. Dev | Min | Max |
|---|---|---|---|---|---|
| Male | 68 (58.6) | | | | |
| Female | 48 (41.4) | | | | |
| Residency: City | 97 (83.6) | | | | |
| Residency: Village | 19 (16.4) | | | | |
| Stay in hostel | 36 (31.0) | | | | |
| English Proficiency | | 3.31 | 0.60 | 2.00 | 4.00 |
| Students' achievement | | 3.04 | 0.47 | 1.89 | 3.90 |
| Students' attitude | | 3.09 | 0.38 | 2.27 | 4.00 |

Most of the students do not stay at hostel where approximately only one third of them stay in hostel. English proficiency is defined as the achievement of English subject taken during semester 2; and students' achievement is the value of their cumulative grade point average (CGPA). The values for English proficiency and students' achievement ranging from 0 to 4 where 0 means they are totally not performed and 4 indicates the highest achievement. The mean score for English proficiency is 3.31 (SD = 0.60) with the minimum of 2.0 and maximum of 4.0. These values show that all students in this sample acquired a good English proficiency. Moreover, some students manage to obtain the highest score. As for the students' achievement and students' attitude, the mean score is approximately similar, 3.04 (SD = 0.47) and 3.09 (SD = 0.38) respectively. The minimum score for students' achievement is 1.89 which is below 2.00 and the maximum value is 3.90. The score value which is less than 2.00 could suggest that other academic subjects or environment factors might influence the result. In contrast to students' attitude, the minimum value is 2.27 and maximum value is 4.00. These values imply that majority of students have excellent attitude towards the use of English.

The result for the independent sample t-test showed that there is a significant difference between male and female students in English proficiency. The t-value of 2.373 with p-value of 0.019 clearly shows that the test is significant at 0.05 level; therefore, the first hypothesis is rejected. Table 3 displays the full results of the t-test.

Table 4 presents the results of matrix correlation between students' achievement, students' attitude and English proficiency. The correlation is significant at 0.01 level between students' academic achievement and English proficiency and between students' attitude and English proficiency. Therefore, the score hypotheses is rejected. However, there is no correlation between students' achievement and students' attitude. Table 5 demonstrates the number and percentage of the levels of students' English proficiency and the level of students' attitude towards the use of English. There is no student in the category of high English proficiency and low level of attitude, medium level of English

Table 3. T-test analysis of English proficiency based on gender.

| | t-value | Mean diff | Std. error | df | p-value |
|---|---|---|---|---|---|
| English Proficiency | 2.373 | 0.265 | 0.111 | 114 | 0.019* |

* Significant at 0.05 level

Table 4. Pearson correlation matrix of students' achievement and attitude towards their English proficiency.

| | | Students' achievement | Students' attitude | Students' English proficiency |
|---|---|---|---|---|
| Students' achievement | Pearson Correlation | 1 | 0.033 | 0.243 |
| | p-value | | 0.727 | 0.009* |
| | N | 116 | 116 | 116 |
| Students' attitude | Pearson Correlation | 0.033 | 1 | 0.330 |
| | p-value | 0.727 | | 0.000* |
| | N | 116 | 116 | 116 |
| Students' English proficiency | Pearson Correlation | 0.243 | 0.330 | 1 |
| | p-value | 0.009* | 0.000* | |
| | N | 116 | 116 | 116 |

* Significant at 0.01 level

Table 5. The crosstabulation of the English proficiency level and attitude level.

| Level of Students' English Proficiency | Level of Students' Attitude | | | |
|---|---|---|---|---|
| | High | Medium | Low | Total |
| High | 61 (52.6%) | 22 (19.0%) | 0 | 83 |
| Medium | 15 (13%) | 18 (15.5%) | 0 | 33 |
| Low | 0 | 0 | 0 | 0 |
| Total | 76 | 40 | | 116 |

Table 6. The Chi-Square test between the level of students' English proficiency and the level of students' attitude towards the use of English.

| | Chi-Square value | Fisher's Exact Test | p-value | N |
|---|---|---|---|---|
| Level of English Proficiency and Attitude | 8.144 | 7.981 | 0.004* | 116 |

* Significant at 0.01 level

proficiency and low level of students' attitude, low level of English proficiency and high level of students' attitude, low level of English proficiency and medium level of students' attitude and low level of English proficiency and low level of students' attitude. Out of the total sample, there are 52.6% of the students in the category of high level of English proficiency and high level of attitudes towards the use of English. Where else, for other categories the percentage is less than 20%. This means that those who are excellent in English proficiency are influenced by their good attitude towards the use of English. As shown in Table 6, the Chi-square test proves that there is significant association among the level of students' English proficiency and the level of students' attitude towards the use of English. Therefore, the third hypothesis is rejected.

## 5 LIMITATIONS AND RECOMMENDATIONS FOR FUTURE RESEARCH

We recommend that future research should be conducted into other variables that have not been tested in this study. Future studies may consider investigating the factors that might contribute to the students' English proficiency such as students' performance on the use of English language and understanding of technical subjects in English language. Furthermore, the small size of sample did not permit generalization and this study should be extended to all programmes offered in UniKL MIIT in order to obtain more accurate results. A mixed method of approach of qualitative and quantitative data will be implemented to obtain a more accurate results.

Overall, the findings of this study will benefit lecturers and the university in determining the factors that might influence students' English proficiency to become more globally marketable employees who are not only competence and skilful in their area of specialisation but also able to communicate and use English fluently. As such, these elements could contribute to the global competitiveness and the country's strong development after they graduate from the university.

## ACKNOWLEDGEMENT

The authors would like to thank the Universiti Kuala Lumpur on Short Term Research Grant (STRG), the lecturers and respondents involved in this study.

## REFERENCES

Abbott-Chapman, J et al. (1992). Monitoring student progress: A framework for improving student performance and reducing attrition in higher education. *National Clearinghouse for Youth Studies*. Hobart: Australia.

Abdul, L. A. & Hanifah, A. A (2010). Penguasaan dan Penggunaan Bahasa Inggeris dalam Kalangan Pelajar PKPG Kemahiran Hidup Universiti Teknologi Malaysia. Retrieved February 20, 2014, fromhttp://eprints.utm.my/id/eprint/ 10799/1/penguasaan_dan_penggunaan_bahasa_ing geris_dalam_kalangan_pelajar.pdf

Alsheikh, N., Parameswaran, G., & Elhoweris, H. (2010). Parenting Style, Self-Esteem and Student Performance in the United Arab Emirates. Current Issues in Education, 13(1). Retrieved October 13, 2013 from http://cie.asu.edu/

Atetwe J.A. (2013). Analysis of the Influence of Mother Tongue on Students' Performance in English in KCSE in Public Day Secondary Schools in Gatundu District, Kiambu County, Kenya. Thesis*: Master of Education in Curriculum Studies and Instruction*. Nairobi: Kenya.

B. Yushau & M. Hafidz Omar. (2015). Mathematics Performance and its Relation to English Language Proficiency Level of Bilingual Arab University Students. *Indian Journal of Science and Technology 8(13)*.

Bachman, L. F. (1990). Fundamental consideration in language teaching. Oxford: *Oxford University Press*.

Emmanual, C.S, Yakubu,S. & Yusof, I.T. (2014) A Tool for Enhancing Students' Proficiency in English Language Skills. *International Journal of English Language Teaching.1(2)*.

Cronbach, L. J. (1951). Coefficient alpha and the internal structure of tests. *psychometrika*, *16*(3), 297–334.

Duff, A. et al. (2004). The relationship between personality, approach to learning and academic performance. Personality and Individual Difference, *36*,1907–1920.

Fareen, J.A.M. (2013). English Communication Today: Problems and Perspectives on Communicating in English in the Technical Colleges of Tamilnadu. *Journal of Advances in English Language Teaching*. 1(14):106–116.

Farisa Jazima Mat Yusof, Nurul Ashikin Abu Bakar & Syamaizar Bilah (2018). English Language Proficiency among Technical Students of Kolej Kemahiran

Tinggi. *Human Sustainability Procedia*.Retrieved from http://ocs.uthm.edu.my/uthmocr/index.php/hsp/article/view/153

Lakshmi, S., & Mohideen, M. A. (2013). Issues In Reliability and Validity Of Research. *International journal of management research and reviews*, *3*(4), 2752.

Latu.M.F. (1994). Factors affecting the learning of English as a second language macroskills among Tongan secondary students. *Theses*: *Doctorates and Masters*, Edith Cowan University.

Marian, V., Blumenfeld, H. K., & Kaushanskaya, M. (2007). The Language Experience and Proficiency Questionnaire (LEAP-Q): Assessing Language Profiles in Bilinguals and Multilinguals. Journal of Speech, Language, and Hearing Research, 50 (4),940–967.https://doi.org/10.1044/1092-388(2007/067).

McKenzie, K. & Schweitzer, R. (2001).Who succeeds at university? Factors predicting academic performance in the first year Australian university students. *Higher Education Research and Development* (20): 21–33.

McKenzie, K., Gow, K., & Schweitzer, R. (2004). Exploring first year academic achievement through structural equation modelling. *Higher Education Research and Development* (23) 95–112.

Nwogu K. N., & Nwoke A. (2005). A Hand Book on the Teaching of English Language in Secondary Schools. *Jos, Mono Expressions LTD*.

Norrizan Razali. (1992). *ESL in Malaysia: Looking Beyond the Classroom*. Retrieved May 7 2007. Site: http://www.melta.org.my/ET/1992/main8.html.

Fakeye,D.O., & Yemi,O. (2009). English Language Proficiency as a Predictor of Academic Achievement among EFL students in Nigeria. *European Journal of Scientific Research, 37(3),490-495.*

Pokay, P., & Blumenfeld, P. (1990). Predicting achievement early and late in the semester: The role of motivation and use of learning strategies. *Journal of Education Psychology*, *82*, 41–50.

Sadler-Smith, E. (1996). Approaches to studying: age, gender and academic performance. *Educational Studies*, *22*, 367–380.

Sayeedur Rahman. (2005). *Orientations and Motivation in English Language Learning: a Study of Bangladeshi Students at Undergraduate Level*. Retrieved May 5 2007. Site: https://www.asian-efl-journal.com/1236/quarterly-journal/orientations-and-motivation-in-english-language-learning-a-study-of-bangladeshi-students-at-undergraduate-level/#squelch-taas-tab-content-0-3

Secil, S. (2001). The Role of English Medium Instruction in Engineering And Architecture Courses at Anadolu University. *Thesis*: *Master of arts in teaching English as a foreign language*. Anadolu: Turkey.

Springer, S. & Collins, L. (2008). Interacting inside and outside of the language classroom. *Language Teaching Research*, *12*, 39–60.

Tuckman, B.W. (1978). Conducting educational research. New York: Harcont Brace Jovanovich Inc.

Warschauer, M. (2000). The changing global economy and the future of English teaching. *TESOL Quarterly, TESOL in the 21st Century*, *34*(3), 511–535.

Yuen, M. (2015). Poor English a Major Handicap. The Star Online.Retrieved from https://www.thestar.com.my/news/ nation/2015/11/15/poor-english-a-major-handicap-its-not-just-young-doctors-workers-in-various-sectors-too-have-a-weak/

*The Social Sciences Empowered – Lumban Gaol, Hutagalung & Fong Peng (eds)*
*© 2020 Taylor & Francis Group, London, ISBN 978-0-367-00090-5*

# Improving quality of life and psychological well-being and reducing work stress in inclusive-education teachers using *kebersyukuran* training (gratitude training)

Mira Aliza Rachmawati*, Indahria Sulistyarini, Prapti Leguminosa & Dwiva Aditya Putri
*Universitas Islam Indonesia, Indonesia*

ABSTRACT: This research examined the effect of *kebersyukuran* training (gratitude training) on improving the quality of life and psychological well-being and reducing the stress of teachers in inclusive schools. The hypothesis of this study is that quality of life and psychological well-being in the group treated will be higher than in the untreated group, while the level of work stress in the group treated with *gratitude training* will be lower than the untreated group. The design of this research is using the untreated control group design with pretest and posttest dependent by adding the follow-up. Survey respondents as many as 14 people, divided into 7 in the experimental group and 7 in the control group. This study uses the quality of life measurement tool developed by WHO, a psychological well-being scale with reference to the theory developed by Ryff (1989), as well as Fimian's (1988) Teacher Stress Inventory (TSI). Gratitude training modules are based on three aspects (Al-Munajjid, 2006; Al-Jawziyya, 2010): intention (liver), speech (oral), and action (limbs). Data were analyzed using mixed design analysis variance to see the difference in scores in the group (pretest, posttest and follow-up), as well as the difference in scores between groups (experimental and control). The results of data analysis showed a significant difference in scores at pretest, posttest and follow-up between the two groups for the quality of life and work stress variable, and nonexistent differences for psychological well-being variable.

## 1 INTRODUCTION

Statistically, the birth rate of children with special needs in Indonesia has not been accurately and specificly recorded. BPS Inter-Census Survey of 2005, the National Population and Family Planning (Wimbarti, 2016), puts the number of school-age children (5–14 years) at 42,870,041 inhabitants. Teachers are one of the most important components in improving the quality of both regular education and inclusive education. The quality of education will improve when teachers have good competence, based on preconditions set by the government. Competencies required by teachers for children with special needs are, of course, different when compared to regular teachers.

Various problems experienced by teachers in inclusive schools are associated with prosperity. In this case, the low wage and poor working conditions contribute to the stress faced by teachers and, of course, will impact the teachers' quality of life (Hasstrup & Adenike in Chadha & Pandey, 2015). Sultana (Chadha & Pandey, 2015) adds that the stress experienced by teachers is closely related to several dimensions of life, including professional, personal, and financial. When this dimension cannot be fulfilled properly, it will affect the quality of life of the teacher.

Results of research conducted by Frederickson (Gavian, 2011) demonstrated that positive emotions have an important role in the mental and physical health of individuals, including the fight against stress. Psychological welfare is a term closely associated with mental health. Bardburn, Campbell, Warr, and Wall (War, 1978) state that psychological well-being in essence refers to a person's feelings about the activities of daily living.

Stress will also affect quality of life, and motivation and encouragement to teachers will affect the performance of teachers (Chadha & Pandey, 2015). All aspects related to well-being come from quality of life (Damasio, Malo, & Silva, 2013; Shabani & Shamir, 2014). That is, people will have high quality of life when they perceive themselves well with all that they have. According to the World Health Organization (WHO, 1997), quality of life is individuals' perception of their position in life in the context of culture and value systems in which they live. Quality of life is related to goals, expectations, standards, and individual attention, and is influenced by physical health, psychological state, level of independence, social relationships, personal beliefs, and

*Corresponding author: miraaliza@uii.ac.id

relationship with important situations in their environment (WHO, 1997).

Emmons and McCullough (2003) mentioned that feeling grateful has many positive emotional, physical, and interpersonal effects. People who have gratitude tend to have higher levels of positive emotions and low levels of negative emotions such as anxiety and envy (McCullough, Emmons, & Tsang, 2002). Some other studies also prove that gratitude effectively lowers daily stress (Emmons & Stern, 2013) and work stress (Cahyono, 2014), as well as improving the welfare and quality of life of the poor (Howells, 2007; Anggarani, Andayani, & Karyanta, 2013; Tricarico, 2013).

Thankfulness plays an important role in how a person perceives an event (Fitch-Martin, 2015). Gratitude interventions can be used to decrease irrational beliefs because of the appreciation contained in gratitude intervention allows changes in cognitions that later will affect the irrational belief (Tricarico, 2012). By lowering their irrational beliefs, teachers will be better prepared for interpreting various situations experienced with students in a more rational way and react accordingly. Wood, Joseph, and Linley (Fitch-Martin, 2015) also mention that gratitude encourages someone to interpret events more positively and this will minimize the emergence of stress. Gratitude intervention will also improve the teacher's relationship with students (Tricarico, 2012).

This was confirmed by studies conducted by Emmons and Shelton (Sulistyarini, 2010): when an individual is able to feel and express gratitude, that person will likely be able to enjoy his job, have a sense of optimism, and be more energetic and better able to help another individual than someone who has never felt grateful. Furthermore, results of research by Park, Peterson, and Seligman (2004) showed that 18.5% of individual happiness is determined by how much gratitude people feel. Based on this variety of explanations, it is not surprising gratitude has a close link with happiness and life satisfaction of individuals (Emmons & McCullough, 2003; Watkins et al., 2003; McCullough et al., 2004; Wood et al., 2007; Lambert et al., 2009).

Based on the above, the research questions posed in this study are whether gratitude training can improve quality of life and psychological well-being, and can reduce work stress on teachers in inclusive schools?

The purpose of this study is to find proof that gratitude training improves the quality of life and psychological well-being, and reduces work stress, on teachers in inclusive schools.

## 2 LITERATURE REVIEW

WHO (1997) defines quality of life as individuals' perception of their position in life in the context of the culture and value systems in which they live. Quality of life is related to goals, expectations, standards, and individual attention. Quality of life is a concept that is influenced by physical health, psychological state, level of independence, social relationships, personal beliefs, and individuals' relationship with important situations in their environment.

There are four domains of quality of life, according to WHO (1997), namely:

1. Physical Health, covering aspects of energy and fatigue, pain and discomfort, sleep and rest, mobility, activities of daily living, and dependence on substances or drugs
2. Psychological, covering aspects of appearance, negative feelings, positive feelings, self-esteem, thinking, learning, memory and concentration, and spirituality/religion/beliefs.
3. Social Relations, covering aspects of personal relationships, social support, and sexual activity.
4. Environmental, covering aspects of financial resources, liberty and physical security, health and social care (including accessibility and quality), the home environment, the opportunity to acquire new information and skills, participation and opportunities for recreation, and physical environment (including pollution, noise, congestion, and climate as well as transportation).

Psychological well-being, according to Ryff (1989), is a psychological condition characterized by individual healthy functioning of the psychological dimension in the process of self-actualization. Dimensions of positive psychology consist of six aspects: the acceptance of self, positive relationships with others, autonomy or independence, mastery of the environment, purpose in life, and personal development/growth.

Fimian (1988) states that there are two things that are measured to identify the stress of work: events that became a source of job stress and events that are manifestations of work stress. In the model of teacher work stress, Fimian (1988) mentions ten factors that measure the level of work stress experienced by teachers. Sources of stress include time management, work stressors, professional distress, discipline and motivation, and professional investment. Manifestations of stress include emotional manifestations, fatigue manifestations, cardiovascular manifestations, gastronomic manifestation of, and behavioral manifestations.

Gratitude, as delivered by Al-Jawziyya (2010) and Al-Munajjid (2006), emphasizes that all the good an individual receives is a gift given by God, and can come directly or in the form of intermediary humans (others), nature, and so forth. That is, every blessing people experience should be returned to God. Second, the expression of gratitude should be in line with religious rules and in order to bring as well as seek the pleasure of Allah. Gratitude to the heart (heart), oral (speech), and limbs (actions) are aspects of the grateful (Al-Jawziyya, 2010; Al-Munajjid 2006).

a. Gratitude to the heart at the recognition that all goodness and grace comes from God to His servants. Humans do not have the resources and efforts to bring favors; only God can confer favor, without expecting a dime of His servants.
b. Oral gratitude is flattery and praise to the Lord for His blessings, as well as recognition of His gifts and the need for it, not because of riya ', showing off, or arrogance. Oral gratitude is a means to express what is in a person's heart. If a person's heart is filled with gratitude to God, then he would utter verbal praise and thanksgiving to God.
c. Gratitude to the members of the body indicates that the body is used to worship God, included using His favors positively, and not to commit adultery or do things that are forbidden by God.

## 3 METHODS

### 3.1 *Data collection*

Collecting data in this study is done through observation, interviews, and measurement using the Teacher Stress Inventory (Fimian, 1988) to measure the level of teacher work stress; WHOQOL (WHO, 1997) scale to measure teachers' quality of life, and the Psychological Welfare (Ryff, 1989) scale to measure the level of psychological well-being of teachers.

a. Quality of life scale
For measuring quality of life, the scale used is based on the WHOQOL scale that includes dimensions of quality of life, physical health, psychological health, social relationships, and the environment. The blueprint scale of quality of life, according to WHO (1997) is as in Table 1:

Table 1. Blueprint quality of life scale.

| Domain | Favorable | Unfavorable | Total |
|---|---|---|---|
| Quality of Life and Overall Health | 1, 2 | | 2 |
| Physical Health | 10, 15, 16, 17, 18 | 3, 4 | 7 |
| Psychological Health | 5, 6, 7, 11, 19 | 26 | 6 |
| Social Relationships | 20, 21, 22 | | 3 |
| Environment | 8, 9, 12, 13, 14, 23, 24, 25 | | 8 |
| Total | 23 | 3 | 26 |

Validity is defined as how well a test measures what it intends to measure (Anastasia & Urbina, 1997). A gauge rated higher validity will have a small measurement error. The validity of the instrument is said to be a good contribution when the discrimination power index ≥ 0.30 (Anwar, 2005). Validity of the scale in this study was tested using the technique of content validity. Content validity refers to the extent to which the items on the scale cover the entire area of the contents of the object to be measured, or the extent to which the test content reflects characteristics of the attributes to be measured.

WHOQOL scale has been tested by WHO in 20 countries and has been translated into many languages, including Indonesian. This instrument has been tested on 11,830 people worldwide. Subjects consisted of diverse age, occupation, marital status, and state of health. Cronbach alpha values obtained in each of these aspects, among others: physical health (0.82), psychological (0.81), the environment (0.80), and social relationships (0.68). Cronbach alpha values in all aspects range from 0–1, which means that this measure has good reliability. Item–domain correlations ranged from 0.48 to 0.7 for the physical health domain, from 0.5 to 0.65 for the psychological domain, from 0.45 to 0.57 for the domain of social relations, and from 0.47 to 0.56 for the environmental domain. This scale has been used in various studies: the research of Chadha and Pandey (2015) concerning the quality of life of government and nongovernmental teachers; Yuliati research, Baroya and Ririanty (2014) regarding the quality of life of the elderly; Daughter research, Wati and Ariyanto (2014) concerning the quality of life of menopausal women; research by Alves, Salim, Martinez, Passos, Carlo, and Scarpelini (2009) regarding the quality of life of the victims of trauma; and research by Damasio, Melo, and Silva (2013) concerning the quality of life of teachers.

b. Psychological Well-Being Scale
The psychological well-being scale is used to reveal the extent of psychological well-being of inclusive-school teachers. The scale used in this study is an adaptation of the existing measuring devices discovered by Ryff (1989) with reference to the six aspects of psychological well-being; namely, self-acceptance, positive relationships with others, autonomy, mastery of the environment, purpose in life, and personal development/growth, which has been used by Pinem (2014) on high school teachers with a reliability coefficient of 0.809. The scale consists of 42 items, grouped into five levels of frequency: Never (1) Rarely (2) Sometimes (3) Often (4) Always (5).

Each answer means, never get a score of 1, rarely gets a score of 2, sometimes gets a score of 3, often gets a score of 4, and always gets the job stress score of 5. Score is obtained by adding an overall score of the scale items. The higher the job stress scale score, the higher the stress of the work of teachers. Conversely, the lower the job stress scale score, the lower the stress of the work of teachers. Blueprint scale teacher job stress before the test can be seen in Table 2:

Table 2. Blueprint teachers job stress scale.

| Aspect | Factors | No items | Total items |
|---|---|---|---|
| Source of stress | Time management | 1, 2, 3, 4, 5, 6, 7, 8 | 8 |
| | Work-related stress | 9, 10, 11, 12, 13, 14, | 6 |
| | Professional distress | 15, 16, 17, 18, 19 | 5 |
| | Professional investment | 20, 21, 22, 23 | 4 |
| | Discipline and motivation | 24, 25, 26, 27, 28, 29 | 6 |
| Manifestation of stress | Emotional manifestation | 30, 31, 32, 33, 34 | 5 |
| | Fatique manifestation | 35, 36, 37, 37, 38, 39 | 5 |
| | Cardiovascular manifestation | 40, 41, 42 | 3 |
| | Gastronomic manifestation | 43, 44, 45 | 3 |
| | Behavioral manifestation | 46, 47, 48, 49 | 4 |
| Total | | | 49 |

The teacher job stress scale that had been developed tested for validity and reliability of measuring instruments. Test validation is performed to determine the extent of provision and accuracy of the measuring instruments. A measuring instrument is said to have high validity if the instruments could explain the measuring function in accordance with the purpose of the measurement (Anwar, 2010).

### 3.2 *Sampling*

The method used in the determination of the subjects in this study is by using purposive sampling (i.e. teachers who teach in inclusive schools in the city of Jogjakarta, both men and women). Subjects in the study consisted of teachers who have high job stress scores, teachers who have lower quality of life scores, as well as teachers who have lower psychological well-being scores using scales of job stress, quality of life, and psychological well-being. Subjects were divided into two groups: the experimental group and the control group.

### 3.3 *Procedures*

Before gratitude training is granted, some steps being taken in this study are:

1. Training needs analysis
   Training needs analysis is carried out through observation and interviews conducted with teachers in inclusive schools. Training needs analysis obtained a description of the types of inclusion and whether the school teacher faced problems. The results of the needs analysis was to identify whether or not to provide gratitude training in terms of objectives, content, and methods.
   After needs analysis was conducted, the researchers conducted a meeting with the school to explain the research process and ask for a commitment of the inclusive school teachers to participate in this study. Researchers then sought approval from the inclusive-school teacher by explaining beforehand (informed consent) about the research to be carried out, including research goals, rights, and obligations.
2. Preparation of training modules and reading materials
   Gratitude training modules designed from the development of three aspects of gratitude according to Al-Jawziyya (2010) and Al-Munajid (2006), aim to help teachers optimize the inclusion of quality of life. The material in this gratitude training consists of a grateful heart, oral thankfulness, and deeds.
3. Preparation of the completeness of the training modules
   a) Observation sheet
      Training observation sheet used by the observer to record the behavior of the participants during the course. Recording method used in observation is the rating scale. Observations carried out on all training sessions to obtain a complete picture of the behavior of participants during the training.
   b) The end-of-training evaluation sheet
      Evaluation sheet completed by each trainee at the end of the training to see the ratings associated with the participants of the training session.
4. Test the feasibility of modules
   Before being used in research, modules of gratitude were first tested for feasibility through professional judgment: that is, discussing the module together with the supervisor and prospective trainers who provide inputs in order to improve.
5. Selection of participants, facilitator, co-facilitator, and observer
   Participants in the training were teachers working in inclusive schools in the city of Yogyakarta. Trainees were selected based on pretest results explaining the purpose of this study by completing a questionnaire scale on quality of life. Researchers also gave informed consent forms to

the trainees as evidence of consent to participate in the training process. After the pretest data was collected, the researchers made the selection of participants needed based on the following criteria: participants are teachers who have a quality of life scale score at the level of medium and low.

The facilitator in this study has the following criteria:

1) Psychologist
2) Have an interest or concern in the field of Islamic psychology
3) No previous research on gratitude

Criterion co-facilitators in this study have the following criteria:

1) Masters students in psychology
2) Never followed the program of activities of training and understand what needs to be done

Criteria observers in this research have the following criteria:

1) Students of S1 psychology who already took a course inobservation and interviews, or a masters student of psychology
2) Never followed the program of activities of training and understand what needs to be done

### 3.4 *Data analysis*

Hypothesis test used in this study is anova mixed design, which combines the two sub-analyses of within and between subject test. Within is testing the difference of scores in one group (pretest, posttest, follow-up), while between subject test is test score differences between groups (experimental and control) (Widhiarso, 2010).

## 4 DISCUSSION

This study aims to reduce the work stress of inclusive-school teachers, as well as improving the quality of life and psychological well-being of inclusion-school teachers. This study measured three times (pretest, posttest, and follow-up) to determine whether intervention by gratitude training can reduce stress, improve quality of life, as well as increase the psychological welfare of the inclusion-school teacher. Hypothesis test results show that gratitude training can reduce work stress and improve quality of life for the inclusion-school teachers, but does not improve the psychological well-being of teachers. This means two research hypotheses are accepted and one hypothesis is rejected.

This study proves that gratitude training can be instrumental in efforts to reduce the work stress of teachers in inclusive schools. This is in line with the opinion of Emmons and Stern (2013), which states that people who are grateful will be more effective in dealing with everyday stress and have higher resilience in the face of stress. Research by Cahyono (2014) also proved that gratitude training effectively lowers employee stress. In addition, the results of the data analysis on work stress are show gratitude training of contributions of 50.5% against a decrease in work stress in "source of stress" aspects and 53.1% against a decrease in "manifestation of stress" aspects. Analyses of these data prove that gratitude training can be quite effective to reduce work stress among inclusion-school teachers.

According to Robbins (2006), one of the variables that affect work stress is a person's perception. Fitch-Martin (2015) found that gratitude effectively reduces stress because gratitude plays an important role in how a person perceives an event. In the process, gratitude allows teachers to believe that every situation or event in life is the will of God as an altruistic act of God. This process encourages teachers to continue to have a positive perception in every situation including, when facing various sources of work stress.

Graitude also make teachers look for positive things behind the source of stress in their work (slow student progress, disruptive student behavior, meager salaries), rather than dwell on the source of stress itself. This means grateful teachers remain positive in difficult conditions. Gratitude encourages someone to interpret events in a more positive light and it will minimize the emergence of stress (Wood, Joseph, & Linley in Fitch-Martin, 2015).

The results also showed that the gratitude intervention made teachers feel positive emotions, feel calmer in the face of problems, and be more patient at work or facing the students. These findings are consistent with the findings of several previous studies about positive emotional impact (Emmons & McCullough, 2003) and encourage positive coping styles when facing a problem (Wood et al., 2007).

Another effect of the intervention of gratitude in this study was the improvement in the physical conditions of teachers. This can occur because of thoughts and emotions perceived positive when grateful make teachers more calm and more attuned to their physical health, so they avoid the various manifestations of stress manifested in physical symptoms such as fatigue, dizziness, increased blood pressure, and so on. According to Anderson, 2000; Bermejo-Toro & Prieto-Ursua, 2006; Lylte, 1991; Warren, 2010 (Tricarico, 2012) appreciation contained in gratitude training makes teachers avoid stress, fatigue, burnout, and a decrease in personal achievement.

Gratitude not only brings a positive impact for the individual but also can bring a positive climate to the work environment. Teachers who are grateful tend to be more eager to work, not to delay their work, encourage each other, and support each other. This is in line with the findings of McCullough et al. (2002) that a person who is feeling grateful is more prosocial, empathic, and helpful to others. In this study, we also found that after gratitude training, teachers can complete their tasks on time. This is because the teachers are more motivated to complete the task. Kaplan et al. (Fitch-Martin, 2015) found that gratefulness can increase the willingness to try.

Gratitude training can reduce stress, and improve the quality of life of teachers in inclusive schools. This is consistent with research by Froh, Seffick, and Emmons (2008) that individuals who often feel and express gratitude will be enjoying the job, be more optimistic and energetic, and be more help or support to others than those who did not experience gratitude associated with quality of life. Thankfulness fosters a positive experience of the situation at hand, so that one can have maximum satisfaction and enjoy their circumstances. Arief and Habibah (2015) also state that gratitude is a coping strategy. In gratefulness, someone interprets positively the problems of life.

The teachers feel the changes to a more positive perception so that they come to more enjoy their work and life, as well as being able to take the meaning of every event they experienced even in the event of a negative thought. When they were able to take a positive meaning of each event, their emotions will have a more positive impact, thereby increasing quality of life. Sulistyarini's (2010) research stated that gratitude can be sourced from a person's positive perception of the experience that happened. Favors given by God do not always come in the form of positive events for humans, but with gratitude one is taught to continue to take advantage or have a positive perception of events that seem negative. Romdhon (2010) also states that gratitude is useful not only when individual experience or is something fun. Gratitude could also serve as a coping mechanism when individuals experience pressing situations and unpleasant incidents.

This study has shown that gratitude training can reduce work stress and improve quality of life. However, it is not proven to improve psychological well-being. The results of this study reinforced research conducted by Timmons (2015) which states that gratitude training cannot improve the psychological well-being of the mothers of children with autism spectrum disorder. Timmons (2015) found that there was no difference in psychological well-being in either the experimental group or the group control when mothers are grateful. In fact, the research found that all groups had the same effect. According to David (Robinson, 1991), relating to the psychological well-being of income level, marital status, and social support, individuals with high income level are married, and gaining social support would have higher psychological well-being. This may explain why the inclusion school teachers (research subjects) had lower psychological well-being. They generally earn salaries below the minimum wage. The salary or income cannot meet their daily needs. Although they have been grateful with what they had, it nevertheless cannot be denied that being unable to meet one's needs (low income) will tend to result lower psychological well-being than those who have a high income.

## 5 CONCLUSIONS

According to the research done, it can be concluded as follows:

a. Gratitude training improves quality of life in their school teachers inclusion in Jogjakarta
b. Gratitude training has no effect on psychological well-being of inclusive school teachers in the city of Yogyakarta.

## REFERENCES


Al-Jauziyyah, I. Q. (2010) *Sabar dan Syukur: Menguak Rahasia di Balik Keutamaan Sabar dan Syukur*. Semarang: Pustaka Nuun.

Al-Munajjid, M. B. S. (2006) *Silisilah Amalan Hati: Ikhlas, Tawakkal, Optimis, Takut, Bersyukur, Ridha, Sabar, Introspeksi Diri, Tafakkur, Mahabbah, Taqwa, Wara'*. Bandung: Irsyad Baitus Salam.

Anggarani, F. K., Andayani, T. R., & Karyanta, N. A. (2013) Pengaruh Pelatihan Syukur terhadap *Subjective Well-Being* pada Penduduk Miskin di Surakarta. *Jurnal Ilmiah Psikologi Candrajiwa*. 2(2).

Arief, M. F., & Habibah, N. (2015) Pengaruh Strategi Aktivitas (Bersyukur dan Optimis) terhadap Peningkatan Kebahagiaan pada Mahasiswa S1 Pendidikan Guru Sekolah Dasar. *Seminar Psikologi dan Kemanusiaan UMM*. 198–205.

Brackenreed, D. (2011) Inclusive education: Identifying teachers' strategies for coping with perceived stressors in inclusive classrooms. *Canadian Journal of Education Administration and Policy*, 122, 1–37.

Cahyono, E. W. (2014) Pelatihan Gratitude (Bersyukur) untuk Penurunan Stres Kerja Karyawan di PT "X". *Jurnal Ilmiah Mahasiwa Universitas Surabaya*, 3(1),1–15.

Chadha, M., & Pandey, N. (2015) A study of quality of life among government and non-government secondary school teachers. *Indian Journal of Health and Well-being*, 6(1),37–41.

Damasio, B. F., Melo, R. L. P., & Siva, J. P. (2013) Meaning in life, psychological well-being and quality of life in teachers. *Paideia*. 23(54),73–82.

Dickerson, K. L. (2008) Stress and Self-Efficacy of Special Education and General Education Student Teachers During and After The Student Teaching Internship. *Dissertation*. Texas: Texas A & M University.

Emmons, R. A., & McCullough, M. E. (2003) Counting blessings versus burdens: An experimental investigation of gratitude and subjective well-being in daily life. *Journal of Personality and Social Psychology*, *84*, 377–389.

Emmons, R.A. & Stern, R. (2013) Gratitude as a psychotherapeutic Intervention. *Journal of Clinical Psychology*, 69(8),846–855.

Faragher, R., Broadbent, C., Brown, R. I., & Burgess J. Applying the principles of quality of life to education. *Int. Public Health J*, 6(2),119–132.
Fimian, M. J. (1988) *Teacher Stress Inventory*. USA: Clinical Psychology Publishing.
Fitch-Martin, A. (2015) Gratitude and health: A brief intervention to reduce undergraduate stress. *Dissertation*. Colorado: Colorado State University.
Froh, J. J., Sefick, W. J., & Emmons, R. A. (2008) Counting blessings in early adolescents: An Experimental study of gratitude and subjective well-being. *Journal of School Psychology*, 46, 213–233.
Gunay, G., & Boylu, A. A. (2014) Moderator effects of quality of life on job satisfaction of the academic staff in university. *International Journal of Arts & Sciences*, 7(6),659–669.
Huppert, F. A. (2009) Psychological well-being: Evidence regarding its causes and consequences. *Journal Compilation International Association of Applied Psychology: Health and Well-Being*, *1*(2), 137–164.
Jeyaraj, S. S. (2013) Occupational stress among the teachers of the higher secondary schools in Madurai District, Tamil Nadu. *IOSR Journal of Business and Management (IOSR-JBM)*, 7(5),63–76.
Kobayashi, H. (2007) Factores Associated ith Resilience, Quality of Life and Self-Esteem: Adolescents who Have Experienced Multiple Neurosurgeries and Chronic Pain. *Dissertations*. Doctor of Philosophy Wayne State University.
Kongchao, H., Min, L., Yongbo, Z., & Xiaoping, Y. (2015) Research on professional life quality of preschool teachers in Chinese poverty counties: job satisfaction' perspective. *Asian Social Science*, 11(9).
Lambert, N. M., Fincham, F. D., Stillman, T. F., & Dean, L. R. (2009) More gratitude, less materialism: The mediating role of life satisfaction. *Journal of Positive Psychology*, 4, 32–42.
McCullough, M. E., Emmons, R. A., & Tsang, J. A. (2002) The grateful disposition: A conceptual and empirical topography. *Journal of Personality and Social Psychology*, 82, 112–127.
McCullough, M. E., Tsang, J.-A., & Emmons, R. A. (2004) Gratitude in intermediate affective terrain: Links of grateful moods to individual differences and daily emotional experience. *Journal of Personality and Social Psychology*, 86, 295–309.
Park, N., Peterson, C., & Seligman, M. (2004) Strengths of character and well-being. *Journal of Social and Clinical Psychology*, *23*, 603–619.
Puspitasari, D. A., & Handayani, M. M. (2014) Hubungan tingkat *self-efficacy* guru dengan tingkat *burnout* pada guru sekolah inklusi di Surabaya. *Jurnal Psikologi Pendidikan dan Perkembangan*, 3(1), 59–68.
Shabani, E., & Shamir, A. S. High school teachers sensation seeking: Quality of life and hope of life. *International Journal of Education and Applied Sciences*. 1(6),287–296.
Robbins, S. P. (2006) *Perilaku Organisasi (terjemahan Benyamin Molan) edisi ke sepuluh*. Jakarta: PT. Indeks Kelompok Gramedika.
Robinson, J. P., Shaver, P. R., &Wrigthman, L. (1991) *Measure of Personality and Social Psychological Attitudes*. Academic Press.
Ryff, C. D. (1989) Happiness is everything, or is it? Explorations on the meaning of psychological well-being. *Journal of Personality and Social Psychology*, 57(6),1069–1081.
Ryff, C. D., & Singer, B. H. (2006) Know thyself and become what you are: A audaimonic approach to psychological well-being. *Journal of Happiness Studies*, 9(3),13–39.
Romdhon, A. (2010) Kebersyukuran sebagai Sebuah Strategi Coping. *International Converence and The 3rd Congress of Association of Islamic Psychology*. Malang: UIN Maliki Press. 69–77.
Shadish, W. R., Cook, T. D., & Campbell, D. T. (2002) *Experimental and Quasi-Experimental Design for Generalized Causal Inference*. USA: Houghton Mifflin Company.
Sultana, B., Bano, Y., Bano, F., & Shafa, M. D. (2012) The nature and impact of teacher stress in the private schools of Gilgit-Baltistan, Pakistan. *International Journal of Academic Research in Progressive Education and Development*, 1(2).
Sulistyarini, I. R. (2010) Pelatihan Kebersyukuran untuk Meningkatkan *Proactive Coping* pada Survivor Bencana Gunung Merapi. *Direktorat Penelitian dan Pengabdian Masyarakat Universitas Islam Indonesia*.
Timmons, L. N. (2015). The effectiveness of a gratitude at improving well-being for parents fo children with autism spectrum disorder. *Thesis*. Not Published. Texas: Faculty of College of Science and Engineering Texas Christian University.
Tricarico, H. L. (2012).The impact of a brief gratitude intervention for teachers. *Dissertation*. New York: UMI Dissertation Publishing.
Vinosa, L. (2012) *Cognitive Behavior Therapy* untuk Meningkatkan Kualitas Hidup pada Remaja yang Mengalami Luka Bakar dan *Post Traumatic Stress Disorder* akibat Erupsi Merapi 2010. *Tesis Program Magister Profesi Psikologi Universitas Islam Indonesia*.
Watkins, P. C., Woodward, K., Stone, T., & Kolts, R. L. (2003) Gratitude and happiness: Development of a measure of gratitude, and relationships with subjective wellbeing. *Social Behavior and Personality*, 31, 431–451.
Wehmeyer, M. L., & Schalock, R. L. (2001) Self-determination and quality of life: Implications for special education services and supports. *Focus on Exceptional Children*, 33(8).
Wimbarti, S. (2016) "Deteksi Dini dan Manajemen Kelainan Balita di Sekolah dan Keluarga". *Seminar dan Workshop Anak Berkebutuhan Khusus*. Fakultas Psikologi Universitas Gadjah Mada, Yogyakarta.
Wood, A. M., Joseph, S., & Linley, P. A. (2007) Coping style as a psychological resource of grateful people. *Journal of Social and Clinical Psychology*, 26, 1108–1125.
World Health Organization. (1997) *WHOQOL: Measuring Quality of Life*. Division of Mental Health and Prevention of Substance Abuse.

*The Social Sciences Empowered – Lumban Gaol, Hutagalung & Fong Peng (eds)*
*© 2020 Taylor & Francis Group, London, ISBN 978-0-367-00090-5*

# Family functioning, peer support, and student engagement among high school students in central Java Island, Indonesia

Hazhira Qudsyi, Siti Kholidiyatus Sa'diyah & Risa Mahara
*Department of Psychology, Islamic University of Indonesia, Indonesia*

ABSTRACT: This study aims to predict the relationship of family functioning and peer support to student engagement among high-school students in Indonesia. The hypothesis of this study is that both family functioning and peer support can predict student engagement among high-school students. Participants of this study were 201 high-school students, aged 14–21 years. The study was conducted with a quantitative approach and performed in Yogyakarta Province, Indonesia, and Central Java Province, Indonesia. Measurement of student engagement was done by using School Engagement Scale, developed by Fredericks et al. (2005); peer support variable was measured using Perceived Peer/Friend Academic Support Scale (PFASS) developed by Chen (2005); and family functioning variable was measured using Family Assessment Device (FAD), developed by Baldwin and Bishop (1983) and adapted by Putrie (2015). Results of data analysis showed that both family functioning and peer support can predict student engagement among high-school students in level of significance < 0.01 ($R = 0.456$; $p = 0.001$), with an effective contribution of 20.8%. Besides that, result of data analysis showed that peer support became independent predictors (stronger than family functioning) of student engagement in level of significance < 0.01 ($R = 0.399$; $p = 0.000$), with an effective contribution of 15.9%. Weakness and research implications will be discussed in this article.

*Keywords*: student engagement, peer support, family functioning

## 1 INTRODUCTION

The Law Act of the Republic of Indonesia Number 20 Year 2003 on National Education System, showing that Indonesian education serves to develop skills and character development and civilization of the nation's dignity in the context of the intellectual life of the nation, is aimed at developing students' potential in order to become people of faith and fear of God Almighty, noble, healthy, knowledgeable, skilled, creative, independent, and democratic and accountable citizensa. Seeing this, it becomes a necessity for all parties in the education system in Indonesia to be actively involved and engaged in achieving objectives of National Education, including the students.

Students who are actively involved in the educational process have their own expectations for all parties involved in the educational process. But in reality, there are many students who have not been sufficiently involved and engaged in the education process, and even tend to ignore it and are reluctant to be actively involved. In fact, news still shows the problems related to student inactivity in the educational process.

Fredricks, Blumenfeld, and Paris (2004) mention that student engagement is a form of student behavior of those who feel attached to activities in school, and manifests in behavioral, cognitive, and emotional engagement. Student engagement represents a multidimensional construct that includes feelings, thoughts, and behaviors of students with regard to school context (Appleton, Christenson, & Furlong in Griffiths, Sharkey, & Furlong, 2009). The existence of full student engagement in the process of education and learning in school would have a positive impact for students themselves. Student involvement in school activities has considerable impact for students and schools. Several studies have shown that involvement in school activities gives students better academic performance (Kuh, Cruce, Shoup, and Kinzie, 2008; Wefald & Downey, 2009; Van Ryzin, Gravely, & Roseth, 2009; Shernoff, 2010; Reeve, 2012; Dharmayana, Kumara, Masrun & Wirawan, 2012; Chase, Hilliard, Geldhof, Warren, & Lerner, 2014; Wonglorsaichon, Wongwanich, & Wiratchai, 2014; Baroody, Rimm-Kaufman, Larsen, & Curby, 2016; Green, Liem Martin, Colmar, Marsh, & Mclnerney, 2012), increases social competence (Baroody et al., 2016), reduces health risk behaviors (Dolzan, Sartori, Charkhabi, & DePaola, 2015), prevents students truant behavior (Virtanen, Lerkkanen, Poikkeus, & Kuorelahti, 2014), and reduces delinquency (Ludden, 2011). Students who are active in school activities tend to be more cooperative in every school policy. Obviously this will help schools achieve educational goals in accordance with Act (UU) No. 20 Year 2003.

High and low levels of student engagement are certainly influenced by various factors. Griffiths, Sharkey, and Furlong (2009) stated that individual and contextual affect student engagement. Various studies have shown the influence of individual factors on student engagement, such as academic self-efficacy (Anggraini, Setiyadi, & Sudirman, 2014; Kholid, 2015), self-concept (Rodriguez-Fernandez, Ramos-Diaz, Fernandes-Zabala, Goni, Esnaola, & Goni, 2016; Green et al., 2012), motivation (Green et al., 2012), a positive attitude toward school or study (Green et al., 2012; Omar, Hassan, & Atan, 2012), and learning experience (Miller, Rycek, & Fritson, 2011). Various studies have also explained the results with regard to contextual (environmental) factors that influence student engagement, such as peer support (Kholid, 2015; Furrer & Skinner, 2003), culture (DeWaelsche, 2015; Crede & Borrego, 2012), institution or school (DeWaelsche, 2015; Fernandez-Zabala, Goni, & Zuleika, 2015), family or parents (Fernandez-Zabala et al., 2015; Furrer & Skinner, 2003; Virtanen et al., 2014), teacher support (Furrer & Skinner, 2003; Virtanen et al., 2014; Klem & Connell, 2004), complexity of the environment (Shernoff, Kelly, Tonks, Anderson, Cavanagh, Sinha, & Abdi, 2016), regular communication between families and teachers (Kraft & Dougherty, 2012), and teacher-student relationship (DeLaet, Colpin, Leeuwen, Noortgate, Claes, Janssens, Goossens, & Verschueren, 2016).

Based on some of these findings, it can be seen that family and peers have a role in student engagement.

## 2 PROBLEM STATEMENT

As reported in Liputan6.com, dozens of students skipping school were caught by police officers, National Unity Politics and Public Protection (Kesbangpolinmas), and Department of Education Tegal. Raids were held in a number of places, such as shopping centers and internet cafes. Some students were caught playing games, or billiards with friends (www.news.liputan6.com, February 28, 2013). Raids were carried out by municipal police officers (Satpol PP), Department of Education, and police in Magelang also caught dozens of students who are playing games in internet cafes while school hours (www.kompas.com, 2013). In fact, one of the students who wascaught stated that he prefers skipping school rather than be late for school (www.kompas.com, 2013).

Student truant behavior also occurs in Manado, as reported in tribunmanado.co.id: some high school students are known to be skipping school to play online games (www.manado.tribunnews.co id, November 3, 2011). Some vocational school students were also caught by municipal police in Semarang during school hours. Schools also said that these students had not attended school since morning (www.daerah. sindonews.com, April 4, 2015). Cases of students who skip school occurred in many cities. Not least in the Student City, Yogyakarta. One of the residents said that almost every day residents find dozens of students who skip school and stop at the food stalls. The students spend time just to hang out from morning until noon (www.brilio.net, September 3, 2015). Not only that, as many as ten students who have missed classes during school hours were caught by officers and Order Department of Police of Yogyakarta at a number of online games place (www.antara news.com, October 20, 2015).

Student truant behavior is still going on in 2016. As provided in Tempo.co, the Department of Education along with Depok Police catch dozens of uniformed students who skip school and play on the internet cafes (www.m.tempo.co, February 12, 2016). In addition, Mataram City municipal police officers catch some truant students during school hours. The students were found in internet cafes and playing games (www.news.okezone.com, February 1, 2016). Some high school students in Ciamis were also caught in raids conducted by police and Koramil Ciamis. The students skipped school and were caught smoking and drinking (www.wartapriangan.com, January 23, 2016).

Reporting about student truancy during school hours showed that the students were reluctant to get involved or actively participate in learning activities at school. Truant behavior in students is not just an issue of morality or discipline. As stated by Wibowo (2009), the phenomenon of school truancy shows low involvement or participation in learning. Wibowo (2009), added that low student involvement may appear in various forms, such as students who talk in class, easy to ask permission to teacher, are passive and silent, do no homework, collect assignments late, or are not cooperative when teachers are teaching in the classroom. In scientific study, student involvement is often referred to as student engagement (Griffiths, Sharkey, & Furlong, 2009).

Education is not only the duty of schools but also of teachers who obviously become educators in formal institutions such as schools. Education is also the duty of parents and family as the smallest community of people and closest environment to the child. Parents and families also have the same responsibility in educating children. Al-Istanbuli (2006), stated that family was also partially responsible for children's education, because education is not just the task of teachers. According to Loutzenhiser (2001), family environment said to be a family function. Basically, a functional family is a family that can work and perform its functions properly (MacArthur, 2000). Walsh (2003), defines healthy families as a family that has the characteristics and properties that allow it to function optimally.

Family is an integral part of a child's educational process, including the involvement of students in school (student engagement). As explained previously, quite a lot of research proves the family's role in student engagement. As described in research by Havermans, Botterman, and Matthijs (2014), family is one of the most influential resources in student engagement in school. This is in line with the results of research by Fernandez-Zabala et al. (2015), that family has a significant correlation with student engagement in school.

In addition to family, a student's peers are another close environment. In adolescence, the peer support factor is very influential in the lives of many teenagers. Based on Erikson's stages of development (Chen, 2005), adolescents (ages 12–18 years) tend to be independent, and in order to be independent of their parents, they choose to rely on friends. This makes adolescents tend to consider the perspectives of peers to decide a case. This trend is also true in friendship patterns of students in school. The pattern of supportive friendships among students and a close attachment between students can help students pass through academic difficulties in school. This makes students more comfortable and cooperative towards school activities, and increasingly show the behavior of student engagement. Research by Kholid (2015), showed that there is a relationship between self-efficacy and peer support to student engagement, where peer support provides an effective contribution of 5.8%.

Based on the above explanation, it can be concluded that the effectiveness of teaching and learning activities is supported by active involvement of students in school. One thing that is capable of supporting it is the social environment; namely, cooperative friendship and family functioning This needs further research. Based on the previous explanation, this study will focus on empirical studies about the role of family functioning and peer support in student engagement.

Hence, this study aimed to answering the following research questions:

i. Can family functioning and peer support predict student engagement among high school students?
ii. Can those variables together predict student engagement, or predict independently?
iii. Which variables strongly predict student engagement?

## 3 PURPOSE OF STUDY

Purpose of this study is to empirically test whether family functioning and peer support can predict student engagement among high school students.

## 4 LITERATURE REVIEW

Fredricks et al. (2004) mentions that student engagement is a form of student behavior wherein the student feels attached to activities in school and manifests in behavioral engagement, cognitive engagement, and emotional engagement. Yonezawa, Jones, and Joselowsky (2009) also stated that involvement of students in school at the age of adolescence consists of aspects of behavioral engagement, cognitive engagement, and emotional engagement. Meanwhile, Appleton, Christenson, Kim, and Reschly (2006) define student engagement as behavior of entanglement by students who manifested four behavioral indicators: involvement in academics, participation in classes and extracurricular (behavior), strategy in the completion of academic tasks (cognitive), and feeling tied to school (psychological).

Some researchers suggest several terms that can be used as a reference in discussing family functioning. As described by Kelley and Sequeira (1997), the initial efforts of researchers and clinicians to understand family functioning was focused on the understanding of a variable that makes a family normal (normal family). However, according to Jackson (Kelley & Sequeira, 1997), a change in thinking made researchers prepare questions in the idea of normal families. In the end, according to Kelley and Sequeira (1997), researchers were no longer just focused on analysis of families experiencing distress or psychiatric disorders, but also looking at a component called healthy families.

Walsh (2003) revealed that normal family is a healthy family and ideal. Walsh (2003) defined healthy family as a condition of family that has ideal characteristics and properties so the family can function optimally. According to Smith, Elliott, and Lach (2004), family functioning is a broad term used to describe manifold characteristics of a family environment, such as the welfare of the elderly, the quality of marriage, the relationship between parents and children, cohesion, expression of feelings, conflicts, and so on. Meanwhile, Hartmann (2002) defines the functioning of a healthy family as having characteristics such as the nature of rules, boundaries, communication patterns, and roles. Additionally, Noller (Sheehan & Noller, 1998) also explained three important characteristics in the functioning of interpersonal relationships that exist in family system or important characteristics of family functioning: namely intimacy, conflict, and family control.

Peer group is a social context developed through the collective functioning of its members based on the norms and values of groups (Brown, 1990). Often meet, doing activities together, and their affective interpersonal attachments group members made the group could be a strong socializing influence in school (Kinderman, 2003). According to Rubin, Bukowski, and Parker (1998), connections and social networking with peers who formed and nurtured could be a major source of social support for adolescents in overcoming emotional distress and difficulties of adjustment which they experienced. In a networking group like this, it is possible to form support among individuals in group. This is often referred to social support.

Chen (2005) states that peer support is an endorsement of peers given to individuals to support the academic activities at school. Meanwhile, Malecki and Demaray (2002) define peer support as an individual's perception of aid to surrounding environment by encouraging other students in the form of emotional support, instrumental, information and others, so that people feel more useful and prevent others from things dangerous. The same thing also expressed by Mead, Hilton and Curtis (2001) which states that peer support is a system where people can give and receive help with the principles of mutual respect, trust and mutually beneficial sharing with each other.

## 5 METHODS

### 5.1 *Participants*

Participants of this study were high school students in Yogyakarta and Purbalingga (both in Indonesia), boys and girls, age 12–18 years old.

### 5.2 *Measures*

#### 5.2.1 *Student engagement*

The scale of student engagement in this study is an adaptation of the School Engagement Scale developed by Fredricks et al. (2005). Reliability coefficients of Cronbach alpha from the original School Engagement Scale is 0.75 to 0.80. This original scale included 15 items divided into 3 subscales: 4 items as behavioral engagement subscales, 6 items as emotional engagement subscales, and 5 items as cognitive engagement subscales. Cronbach alpha reliability coefficients of the Indonesian version are 0.859.

#### 5.2.2 *Family functioning*

Family functioning in this study was measured using Putrie and Qudsyi's (2015), scale, an adaptation of the Family Assessment Device (FAD) scale based on Epstein, Baldwin, and Bishop (1983). This scale included 40 items divided into 7 subscales: 4 items as problem solving subscales, 3 items as communication subscales, 5 items as roles subscales, 5 items as affective response subscales, 2 items as affective involvement subscales, 9 items as behavioral control subscales, and 12 items as general function subscales.

#### 5.2.3 *Peer support*

The scale of peer support in this study is an adaptation of Chen (2005), namely Perceived Peer Academic Support Scale (PFASS). Reliability coefficients of Cronbach alpha on PFASS original is 0.913. This original scale included 22 items and divided into 3 subscales: 6 items as emotional support subscales, 3 items as instrumental support subscales, and 13 items as cognitive support subscales.

### 5.3 *Statistical analysis*

The data analysis was conducted after research data were collected. This includes data of student engagement, family functioning, and peer support. This study will use regression analysis to analyze the predictive power of the independent variables in this study on the dependent variable, namely student engagement.

## 6 RESULT

Referring to the conditions of research, data were normally distributed, linear, and meet the assumptions of multicolinearity, then the analysis to test hypothesis in this study can be continued by using regression analysis techniques. The hypothesis proposed in this study is that family functioning and peer support can predict engagement in high school students. Based on the analysis, the results are as follows in Table 1:

In addition to testing the hypothesis, researchers also conducted additional analyses to strengthen the research results. Results of additional analyses are summarized in Table 2:

According to Table 2, peer support became the most powerful variable as predictor of student engagement. Seeing this, researchers then performed additional analyses with peer support variable conditions as dominant variable as a predictor.

Table 1. Regression analysis.

| Variable | F | p | R | $R^2$ |
|---|---|---|---|---|
| Family functioning & peer support * *Student Engagement* | 25.930 | 0.000 | 0.456 | 0.208 |

Table 2. Additional analyzes.

| Predictor of *student engagement* | F | p | R | $R^2$ |
|---|---|---|---|---|
| Family functioning + Peer support | 25.930 | 0.000 | 0.456 | 0.208 |
| Peer support | 37.684 | 0.000 | 0.399 | 0.159 |

Table 3. Additional analyzes (based on student engagement subscales).

| Independent Variable | Student engagement Subscales | p | R | $R^2$ |
|---|---|---|---|---|
| Peer support | Behavioral engagement | 0.000 | 0.279 | 0.078 |
| | Emotional engagement | 0.000 | 0.352 | 0.124 |
| | Cognitive engagement | 0.000 | 0.273 | 0.075 |

Table 4. Additional analyses (correlation between peer support and student engagement by gender).

| | Male | | | Female | | |
|---|---|---|---|---|---|---|
| Variable | p | R | $R^2$ | p | R | $R^2$ |
| Peer support * student engagement | 0.000 | 0.410 | 0.168 | 0.000 | 0.577 | 0.333 |

## 7 DISCUSSION AND RECOMMENDATIONS FOR FUTURE RESEARCH

This study aimed to empirically test whether family functioning and peer support could predict engagement in high school students. Based on previous analysis, obtained results showed a highly significant correlation between family functioning and peer support to student engagement in high school students. The results obtained from the value of R = 0456 and p = 0.000 ($p < 0.01$). Thus, it can be said that family functioning and peer support together can very significantly predict student engagement in high school students, with an effective contribution of 20.8%.

Based on additional analysis done anyway, the result obtained showed a very significant correlation between peer support and student engagement in high school students when the influence of family functioning is controlled. The results obtained from the value of R = 0.399 and p = 0.000 ($p < 0.01$). Based on these results, it can be said that peer support can significantly predict student engagement with an effective contribution of 15.9%. These results indicate that peer support can be a variable that independently predicts student engagement among high school students, without having to see the effect of family functioning variables. As in the results analysis, family functioning variables can not predict student engagement, otherwise see the effect of peer support variable. It can be said that family functioning variables do not become independent variables to predict engagement among high school students, and to be together with the peer support variable if it is to be a predictor for engagement among high school students.

Results of previous analysis show that, despite family functioning having the power to predict student engagement together with peer support, peer support is the only variable that can predict student engagement among high school students. These results are consistent with findings in previous studies that explain peer support is closely related to student engagement among high school students (Lynch, Lerner & Leventhal, 2013; Van Ryzin, 2011; Cappella, Kim, Jennifer, & Jackson, 2013).

Student engagement behavior is influenced by their peer support. Students' friendship environments are cooperative and supportive, so as to foster student engagement behavior that is also important to students. Chen (2005) suggests that peer support is an endorsement of peers given to individuals to support the academic activities at school. Peer support is given in emotional support, instrumental support, and cognitive support. Cooperative peer support makes students feel comfortable interacting with friends nearby. Peers are also able to provide support to students to get through difficulties encountered in learning activities at school. It makes students eager to complete tasks and challenges in the learning process. Peer support can also be in the form of material support, scuh as access to tools and materials that support learning.

When students have opportunity to interact with their friends as well, then students will easily share experiences and receive information related to a specific experience. Emotional support from friends at school also makes students have a passion for trying to pass obstacles in completing school assignments. This also applies when students have difficulty in completing tasks. Peer support comes in the form of a solution on how to complete tasks. Students can be helped by the support of their peers. Results of research by Wentzel, Rubin, Bukowski and Brett (2009) stated that good relations among friends will make students more actively involved in school activities. Taylor and Parsons (2011) state that supportive friendship environment can make students more actively involved in school activities.

Legault, Demers, and Pelletier (2006) proved that through emotional support provided by peers, students become more motivated and eager to work on school assignments. Meanwhile, Roseth, Johnson, and Johnson (2008) prove that friend relationships of mutual support emotionally enable students to become more active in school. Engagement with student activities has an impact on increasing student achievement. Furrer and Skinner (2003) also stated that emotional support provided by peers is able to make students more comfortable with being active in school activities, so that the level of student interest in school interaction increases.

Support from peers becomes a strategic solution to dealing with psychological problems experienced by adolescents. Yasin and Dzulkifli (2010), Glozah (2013), and Hashim (2007) showed that students who had support in school were better able to deal with psychological problems. Conflict resolution in adolescence has a psychological impact on students' performance in school. Students become more comfortable being actively involved in school activities. Li, Lynch, Calvin, and Liu (2011) stated that teenagers who get support from peers tends to be more actively involved in emotions and behavior in learning activities in the classroom.

Based on this research, the subscale of student engagement most influenced by peer support is an emotional engagement in class activities. The results showed that the level of contribution of peer support on these subscales is 12.4%. These results indicate that support of peers makes students feel comfortable, happy, and eager to participate in activities at school. Furrer and Skinner (2003) explain that peer support makes students become more engaged in classroom activities, especially emotional involvement. Support from peers make students become more appreciated and feel comfortable to attempt classroom activities. Support from peers also makes students feel part of the school neighborhood. It certainly makes students become more involved in school activities.

By gender, there are differences in the predictive level of peer support to student engagement in school. Women have a contribution rate of peer support that was higher (at 57.7%) against student engagement in school than male respondents who have peer support percentage of 41%. Research conducted by Pijl, Mjaavatn, and Frostad (2016), and Rose and Rudolph (2006) proved that female students had higher levels of peer support than men. This is because female patterns of interaction with friends are closer emotionally, while the friendship patterns of male students tend to be more a competition friendship. Friendship quality in female students makes them feel comfortable to interact with their friends and it makes the level of engagement of female students higher than male students. Research from Harper, Karini, Bridges, and Hayek (2004), Amir, Saleha, and Ahmad, and Hutkemri (2014) also showed that the level of student engagement in girls in school is higher than boys. This is due to female students more making preparations prior to attending classes than boys.

Based on the results of research and previous discussion, here are some thoughts for further research. For further researchers who want to conduct research on the same theme, it is expected to control every class when carrying out the questionnaires so that researchers can directly check and part or item not filled by respondents, and the respondents can be asked immediately to fill it in, which will be very helpful for research results. Moreover, it would improve research on the same topic as this research to increase the variable support of teachers. The addition of these variables could explain multidimensional student engagement behavior. Researchers can study student engagement behavior as influenced by peer support and teacher support.

## 8 CONCLUSIONS

Based on the analysis and previous discussion, it can be concluded that family functioning and peer support together can very significantly predict student engagement among high school students. Another conclusion of this study is that peer support has independently predictive power in terms of student engagement when compared to family functioning.

## REFERENCES


Al-Istanbuli, M. M. 2006. *Parenting Guide: Dialog Imajiner Tentang Cara Mendidik Anak Berdasarkan Al-Qur'an, As-Sunnah, dan Psikologi.* Jakarta: Hikmah PT Mizan Publika.

Amir, R., Saleha, A., Jelas, Z. M., Ahmad, A. R., & Hutkemri. (2014). Students' engagement by age and gender: A cross-sectional study in Malaysia. *Middle-East J. Sci. Res.*, *21*(10). DOI: 10.5829/idosi.mejsr.2014.21.10.85168

Anggraini, D., Setiyadi, A. B., & Sudirman. (2014). Correlation between students' academic self-efficacy and their engagement in speaking English class. http://download.portalgaruda.org/article.php?article=287414&val=7230&title=CORRELATION%20BETWEEN%20STUDENTS%20ACADEMIC%20SELF-EFFICACY%20AND%20THEIR%20ENGAGEMENT%20IN%20SPEAKING%20ENGLISH%20CLASS. Retrieved at June 26th 2016.

Appleton, J. J., Christenson, S. L., Kim, D., & Reschly, A. L. (2006). Measuring cognitive and psychological engagement: Validation of the Student Engagement Instrument. *Journal of School Psychology*, 44, 427–445.

Azwar, S. (2005). *Penyusunan Skala Psikologi. Edisi 1, Cetakan 7.* Yogyakarta: Penerbit Pustaka Pelajar. 10 pelajar Yogyakarta bolos sekolah terjaring razia. http://www.antaranews.com/berita/524543/10-pelajar-yogyakarta-bolos-sekolah-terjaring-razia. Retrieved at June 26th 2016.

Baroody, A. E., Rimm-Kaufman, S.E., Larsen, R.A., & Curby, R.W. 2016. A multi-method approach for describing the contributions of student engagement on fifth grade students' social competence and achievement in mathematics. *Learning and Individual Differences*, *48*, 54–60.

Bolos sekolah, puluhan pelajar main "game" di warnet.] http://regional.kompas.com/read/2013/11/12/1934216/Bolos.Sekolah.Puluhan.Pelajar.Main.Game.di.Warnet. Retrieved at June 26th 2016.

Bolos sekolah, siswa asik main game online. http://manado.tribunnews.com/2011/11/03/bolos-sekolah-siswa-asik-main-game-online. Retrieved at June 26th 2016.

Bolos sekolah, 8 pelajar dihukum push up. http://daerah.sindonews.com/read/985171/189/bolos-sekolah-8-pelajar-dihukum-push-up-1428143794. Retrieved at June 26th 2016.

Bolos sekolah malah asyik nongkrong di warung, tak baik-tak baik! https://www.brilio.net/news/bolos-sekolah-malah-asyik-nongkrong-di-warung-tak-baik-tak-baik-150903y.html. Retrieved at June 26th 2016.

Bolos sekolah, belasan pelajar Depok terjaring di warnet. https://m.tempo.co/read/news/2016/02/12/214744360/bolos-sekolah-belasan-pelajar-depok-terjaring-di-warnet. Retrieved at June 26th 2016.

Bolos sekolah, enam siswa ditangkap di warnet. http://news.okezone.com/read/2016/02/01/340/1301991/bolos-sekolah-enam-siswa-ditangkap-di-warnet. Retrieved at June 26th 2016.

Brown, B. B. (1990). Peer group and peer cultures. In S. S. Feldman & G.R. Elliot (Ed.), *At the Threshold: The Developing Adolescents* (pp. 171≠196). Cambridge, MA: Harvard University Press.

Cappella, E., Kim, H. Y., Neal, J. W., & Jackson, D. R. (2013). Classroom peer relationships and behavioral engagement in elementary school: The role of social network equity. *Am J Community Psychol, 52*, 367–379. DOI 10.1007/s10464-013-9603-5

Chase, P. A., Hilliard, L. J., Geldhof, L. J., Warren, D. J. A., & Lerner, R. M. (2014). Academic achievement in the high school years: The changing role of school engagement. *J Youth Adolescence, 43*, 884–896, DOI 10.1007/s10964-013-0085-4

Chen, J. J. (2005). Relation of academic support from parents, teachers, and peers to Hong Kong adolescents' academic achievement: The mediating role of academic engagement. *Geneti, Social, and General Psychology Monogrophs, 31*, 77–127.

Crede, E., & Borrego, M. (2012). International diversity and student engagement in graduate engineering research groups. *Procedia Social and Behavioral Sciences, 56*, 141–152.

DeLaet, S., Colpin, H., Leeuwen, K. V., Noortgate, W. V., Claes, S., Jansens, A., Goosens, L., & Verschueren. (2016). Teacher-student relationships and adolescent behavioral engagement and rule-breaking behavior: The moderating role of dopaminergic genes. *Journal of School Psychology, 56*, 13–25.

DeWaelsche, S.A. (2015). Critical thinking, questioning, and student engagement in Korean university English courses. *Linguistics and Education, 32*, 131–147.

Dharmayana, I.W., Kumara, A., Masrun., & Wirawan, Y.G. (2012) Keterlibatan Siswa *(Student Engagement)* sebagai Mediator Kompetensi Emosi dan Prestasi Akademik. *Jurnal Psikologi, 39*(1), 76–94.

Dolzan, M., Sartori, R., Charkhabi, M., & DePaola, F. (2015). The effect of school engagement on health risk behaviours among high school students: Testing the mediating role of self-efficacy. *Procedia Social and Behavioral Sciences, 205*, 608–613.

Eipsten, N., Baldwin, L., & Bishop, D. (1983). The McMaster family assesment device. *Journal of Marital and Family Therapy, 9*(2), 171–180.

Fernandez-Zabala, A., Goni, E., Camino, I., & Zulaika, L. M. (2015). Family and school context in school engagement. *European Journal of Education and Psychology* (article in press).

Fredricks, J.A., Blumenfeld, P., Friedel, J., & Paris, A. (2005). School engagement. In K.A. Moore & L. Lippman (Eds.), *What do children need to flourish?: Conceptualizing and measuring indicators of positive development*. New York, NY: Springer Science and Business Media.

Field, A. (2005). *Discovering Statistics Using SPSS for Windows*, London: Sage Publications.

Furrer, C., & Skinner, E. (2003). Sense of relatedness as a factor in children's academic engagement and performance. *Journal of Educational Psychology, 95*(1), 148–162.

Glozah, F.N. (2013). Effects of Academic Stress and Perceived Social Support on the Psychological Wellbeing of Adolescents in Ghana. *Open Journal of Medical Psychology, 2*, 143–150.

Green, J., Liem, G.A.D., Martin, A.J., Colmar, S., Marsh., H.W., & Mclnerney. (2012). Academic motivation, self-concept, engagement, and performance in high school: Key processes from a longitudinal perpective. *Journal of Adolecence, 35*, 1111–1122.

Griffiths, A.J., Sharkey, J.D., & Furlong, M.J. (2009). Student Engagement and Positive School Adaptation. *Handbook of Psychology Positive in School*, 1st edition, No 16, 197–212.

Harper, S.R., Karini, R.M., Bridges, B.K., & Hayek, J.C. (2004). Gender Differences in Student Engagement Among African American Undergraduates at Historically Black Colleges and Universities. *Journal of College Student Development, 45*(3), 271–284.

Hartmann, P.B. (2002). Family Functioning and Anorexia Nervosa: The Issue of Control. *Thesis*. School of Applied Psychology, Griffith University.

Hashim, I.H.M. (2007). Stress, Coping And Social Supports In The Adolescent Years. *Kajian Malaysia*. Jld. Xxv, No. 1.

Havermans, N., Botterman, S., & Matthijs, K. (2014). Family resources as mediators in the relation between divorce and children's school engagement. *The Social Science Journal, 51*, 564–579.

Kelley, D.L., & Sequeira, D.L. (1997). Understanding Family Functioning in A Changing America. *Communication Studies on Summer 1997*.

Kholid, A. (2015). Hubungan efikasi diri dan dukungan teman sebaya dengan keterlibatan siswa pada sekolah. *Publication Paper*. Muhammadiyah Surakarta University, Indonesia.

Kindermann, T. A. (2003). Development of children's social relationships. In J. Valsiner & K. Connolly (Ed.), *Handbook of developmental psychology* (pp. 407–430). Thousand Oaks, CA: Sage.

Klem, A.M., & Connell, J.P. (2004). Relationships matter: Linking teacher support to student engagement and achievement. *Journal of School Health, 74*(7), 262–273.

Kraft, M. & Dougherty, S. (2012). The Effect of Teacher–Family Communication on Student Engagement: Evidence from a Randomized Field Experiment. *Journal of Research on Educational Effectiveness*. *6*(3), 199–222.

Kuh, G.D., Cruce, T.M., Shoup, R., & Kinzie, J. (2008). Unmasking the Effects of Student Engagement on First-Year College Grades and Persistence. *The Journal of Higher Education, 79*(5), 540–563, *DOI: 10.1353/jhe.0.0019.*

Legault, L., Demers, I.G., & Pelletier, L. (2006). Why Do High School Students Lack Motivation in the Classroom? Toward an Understanding of Academic Amotivation and the Role of Social Support. *Journal of Educational Psychology American Psychological Association, 98*(3), 567–582. DOI: 10.1037/0022-0663.98.3.567

Li, Y., Lynch, A.D., Kalvin, C., & Liu, J. (2011). Peer relationships as a context for the development of school engagement during early adolescence. *International Journal of Behavioral Development, 35*(4), 229–342.

Loutzenhiser, L. (2001). Risk, family functioning, and child competence in head start families. *Thesis*. University of Saskatchewan.
Ludden, A.B. (2011). Engagement in School and Community Civic Activities Among Rural Adolescents. *J Youth Adolescence, 40*, 1254–1270, DOI 10.1007/s10964-010-9536-3
Lynch, A.D., Lerner, R.M., & Leventhal, T. (2013). Adolescent Academic Achievement and School Engagement: An Examination of the Role of School-Wide Peer Culture. *Journal Youth Adolescence, 42*,6–19. DOI 10.1007/s10964-012-9833-0.
MacArthur, J.D. 2000. *The Functional Family*. Utah: Brigham Young University.
Malecki, C.K., & Demaray, M.K. (2002). Measuring Perceived Social Support: Development Of The Child And Adolescent Social Support Scale (Casss). *Psychology In The Schools, 39*(1).
Mead, S., Hilton, D., & Curtis, L. (2001). Peer Support: A Theoretical Perspektive. *Psychiatric Rehabilitation Journal, 25*(2).
Miller, R.L., Rycek, R.F., & Fritson, K. (2011). The effects of high impact learning experiences on student engagement. *Procedia Social and Behavioral Sciences, 15*, 53–59.
Omar, N. D., Hassan, H., & Atan, H. (2012). Student engagement on online learning: Learners attitude toward e-mentoring. *Procedia Social and Behavioral Sciences, 67*, 464–475.
Pijl, S. J., Mjaavatn, P. E., & Frostad, P. (2016). Adolescents: Differences in friendship patterns related to gender. *Issues in Educational Research, 1*(26).
Priyatno, D. (2009). *SPSS untuk analisis korelasi, regresi, dan multivariat*. Yogyakarta: Penerbit Gava Media.
Puluhan siswa bolos sekolah terjaring razia. http://news.liputan6.com/read/523975/puluhan-siswa-bolos-sekolah-terjaring-razia. Retrieved at June 26th 2016.
Putrie, R. A., & Qudsyi, H. (2015). Hubungan antara keberfungsian keluarga terhadap efikasi diri dalam pengambilan keputusan karir akademik pada remaja. *Publication Paper*. Yogyakarta: Program Studi Psikologi Universitas Islam Indonesia.
Reeve, J. (2012). A self-determination theory perspective on student engagement. *Handbook of Research on Student Engagement, 149*. DOI 10.1007/978-1-4614-2018-7_7.
Rodriguez-Fernandez, A., Ramoz-Diaz, E., Fernandez-Zabala, A., Goni, E., Esnaola, I., & Goni, A. (2016). Contextual and psychological variables in a descriptive model of subjective well-being and school engagement. *International Journal of Clinical and Health Psychology, 16*, 166–174.
Rose, A. J., & Rudolph, K. D. (2006). A review of sex differences in peer relationship processes: Potential trade-offs for the emotional and behavioral development of girls and boys. *Psychological Bulletin*, January, *1* (132), 98–131. doi:10.1037/0033-2909.132.1.98.
Roseth, C. J., Johnson, D. W., & Johnson, R. T. (2008). Promoting early adolescents' achievement and peer relationships: The effects of cooperative, competitive, and individualistic goal structures. *Psychological Bulletin: American Psychological Association, 134*(2). DOI: 10.1037/0033-2909.134.2.223.
Rubin, K. H., Bukowski, W., & Parker, J. G. (1998). Peer interaction, relationship, and groups. In N. Eisenberg (Ed.), Handbook of Child Psychology: Vol 3. Social, Emotional, and Personality Development (pp. 619–700). New York: Wiley.
Santosa, P. B., & Ashari. (2005). *Analisis Statistik dengan Microsoft Excel dan SPSS*. Yogyakarta: Penerbit Andi Offset.
Sepuluh siswa siswi dai SMA ternama di Ciamis terjaring razia. http://www.wartapriangan.com/siswa-sma-ternama-di-ciamis-dirazia/ 1919963/. Retrieved at June 26th 2016.
Sheehan, G., & Noller, P. (1998). Adolescent's perception of parental favouritism: Links with family functioning. *Family Matters, 49* Autumn 1998. Australian Institute of Family Studies.
Shernoff, D. J. (2010). Engagement in after-school programs as a predictor of social competence and academic performance. *Am J Community Psychology, 45*, 325–337, DOI 10.1007/s10464-010-9314-0.
Shernoff, D. J., Kelly, S., Tonks, S. M., Anderson, B., Cavanagh, R. F., Sinha, S., & Abdi, B. (2016). Student engagement as a function of environmental complexity in high school classrooms. *Learning and Instruction, 43*, 52–60.
Smith, M. L., Elliott, I. M., & Lach, L. (2004). Cognitive, psychosocial, and family function one year after pediatric epilepsy surgery. *Epilepsia, 45*(6).
Taylor, L. & Parsons, J. (2011). Improving student engagement. *Current Issues in Education, 14*(1). Retrieved from http://cie.asu.edu/
Van Ryzin, M. J., Gravely, A. A., & Roseth, C. J. (2009). Autonomy, belongingness, and engagement in school as contributors to adolescent psychological well-being. *J Youth Adolescence, 38*, 1–12. DOI 10.1007/s10964-007-9257-4.
Van Ryzin, M. J. (2011). Protective factors at school: Reciprocal effects among adolescents' perceptions of the school environment, engagement in learning, and hope. *J Youth Adolescence, 40*, 1568–1580. DOI 10.1007/s10964-011-9637-7.
Virtanen, T. E., Lerkkanen, M. K., Poikkeus, A. M., Kuorelahti, M. (2014). Student behavioral engagement as a mediator between teacher, family, and peer support and school truancy. *Learning and Individual Differences, 36*, 201–206.
Walsh, F. 2003. *Normal Family Processes, Third Edition: Growing Diversity and Complexity*. New York: Guilford Publication.
Wefald, A. J., & Downey, R. G. (2009). Construct dimensionality of engagement and its relation with satisfaction. *Kansas State University The Journal of Psychology 143*(1), 91–111.
Wentzel, K. R., & Rubin, K. H. ; Bukowski, W. M. ; Laursen, B. (Ed). (2009). *Handbook of Peer Interactions, Relationships, and Groups. Social, Emotional, and Personality Development in Context*, (pp. 531–547). New York: Guilford Press, xvii, p. 654.
Wibowo, T. G. (2015). Anak "bolos" sekolah, ayo salah siapa? http://www.kompasiana.com/thomasgwibowo/anak-bolos-sekolah-ayo-salah-siapa_54fd6486a333119f1b50fefa. Retrieved at June 26th 2016.
Wonglorsaichon, B., Wongwanich., S., & Wiratchai, N. (2014). The influence of students school engagement on learning achievement: A structural equation modeling analysis. *Procedia: Social and Behavioral Sciences, 116*, 1748–1755.
Yasin, A. S., & Dzulkifli, M. A. (2010). The relationship between social support and psychological problems among students. *International Journal of Business and Social Science, 1*(3).
Yonezawa, S., Jones, M., & Joselowsky, F. (2009). Youth engagement in high schools: Developing a multidimensional, critical approach to improving engagement for all students. *J Educ Change* Vol 10, 191–209, DOI 10.1007/s10833-009-9106-1.

*The Social Sciences Empowered – Lumban Gaol, Hutagalung & Fong Peng (eds)*
*© 2020 Taylor & Francis Group, London, ISBN 978-0-367-00090-5*

# The subjective well-being of women as victims of infidelity: Lesson learned from forgiveness training

Muhammad Novvaliant*
*Faculty of Psychology and Social Cultural Sciences, Islamic University of Indonesia, Indonesia*

Subandi*
*Faculty of Psychology, Gadjah Mada University, Indonesia*

ABSTRACT: Infidelity is a major indicator for getting divorced. Consequences for the victims are the declining of subjective well-being, which is characterized by high negative affect, low positive affect and lower marital satisfaction. Forgiveness training is given to improve subjective well-being. This study used a qualitative approach to understand the psychological dynamics of each subject during and after forgiveness training. The subjects were three3 women who became victims of their husbands' infidelity. The results showed that the subjective well-being remained low, but some were able to improve their subjective well-being, which was characterized by low negative affect, higher positive affect, and higher marriage satisfaction.

*Keywords*: Infidelity, subjective well-being, forgiveness training, case studies

## 1 INTRODUCTION

Marriage involves two individuals with different habits, visions of life, and educational strata. These differences require adjustment that often leads to conflict. Conflict itself is a thing that cannot be avoided in an interpersonal relationship (Tsang, Mc Cullough, & Fincham, 2006). Even in the most excellent marriage, a couple sometimes hurt or injure another couple (Ripley & Worthington, 2002). Marital conflict could have adverse effects on mental health, physical health, and family. Marital conflict is associated with symptoms of depression, eating disorders, alcoholic behavior, poor health, and some specific diseases, such as cancer, heart disease, and chronic pain (Fincham, 2003). If the conflict worsens and there is no immediate solution, it could lead to divorce.

Most studies on divorce begin with the assumption that the divorce is a stressful transition for the adult couple, as well as their parents, spouses, and children too (Amato, 2000). Plummer and Kock-Hattem (1986) mentioned that a divorce is the end of a family relationship. The post-divorce impact is an adjustment to post-divorce family conditions that are no longer intact. Lasswell and Lasswell (1987) reveal that the divorce is the solution to the problem which cannot be solved because of the increasing frustration experienced by the couple. Amato and Previti (2003) stated that divorce is a complex event that can be seen from many points of view, including the psychological or sociological standpoint. The definition of divorce has evolved from the view that it is a single event to that it is a complex process (Frustenberg & Kathleen, 2001), particularly if spouses live together until after the divorce is official. This complexity appears along with many conflicts before getting divorce, even if both spouses are living separately and rarely interacting with each other (Stevenson & Black, 1995).

Divorce has multiple causes, one of which is partner infidelity. When a wife knows her husband is having an affair, there is a feeling of betrayal and marital satisfaction will decrease (Fan & Lui, 2004). Some studies also mention that infidelity tops the list of causes of divorce. Amato & Previti's (2003) study of 208 subjects who had just divorced in the United States, reported infidelity and physical and psychological violence as the most common causes of divorce. In Indonesia, a qualitative study of seven women who experienced domestic violence in Palu showed that the woman (wife) chose to divorce because the husband had an affair more than because of the violence perpetrated by her husband. Some women who are victims of domestic violence choose to stay in the marriage because of confidence that change is possible, as well as for the sake of the children (Sahi, 2009).

*Corresponding authors: muhanovva@gmail.com, masubandi@gmail.com

Similar results were also found in Yogyakarta. Another qualitative study conducted on three women who sued their husbands for divorce in Yogyakarta, shows that one wife sued her husband for divorce because the marital relationship had not lived up to expectations. Other factors that led to divorse included, among others, lack of loyalty; physical, verbal, and sexual violence; lazy at work so did not have a regular income to meet family needs (not responsible); the wife felt exploited; irrational jealousy; intervention from the husband's family; and the notion of women as 'maids' which can be treated as desired, or simply that the husband or wife does not appreciate the other, Khumas in Khumas (2012).

The decline of subjective well-being when victims of infidelity feel negative emotions such as anger and disappointment. Arthaud-Day, Rode, Mooney, and Near (2005) mentioned that the subjective well-being has three domains: cognitive evaluation, positive affect, and negative affect. Satisfaction in life is often considered in the cognitive evaluation of an individual's life. Cognitive evaluation can be in the form of feelings of alienation that usually appears at the beginning of the process of divorce in couples where these feelings grow stronger in intensity along with the development of perceived disappointment due to the conflict (Amato, 2000). Affective components are often measured by the balance between positive and negative emotions as felt by the individuals (Nieboer, Lindenberg, Boomsma, & van Bruggen, 2005).

Divorce may cause a feeling of pain and have a long-term effects on the individuals. Those individuals will try to heal that feeling via a coping strategy. Coates and LaCrosse (2003) mentioned that blaming the former partner and the idea of *I am correct* is one strategy to make one feel better. Unfortunately, this strategy does not overcome the problem and it cannot make the person feel better. Besides, blaming behavior needs an object, in this case, the other individual, to blame. Coates and LaCrosse (2003) recommended individuals forgive themselves and the others concerned in the problems that hurt them. Forgiving will release them from blaming and anger and allow them to move on with their lives constructively (Coates & LaCrosse, 2003)

Furthermore, this research is intended to explain the dynamics of subjective well-being of women as victim of infidelity after forgiveness training.

## 2 METHOD

This research used the qualitative method with case study approach to study the psychological dynamic deeply and comprehensively. The subjects were individuals who participated in the forgiveness training. Case study approach was chosen because it is appropriate for research which asks *why* and *how* questions and the researcher has less opportunity to control which events to investigate and when the research focus is on contemporary phenomena (Yin, 2008). The qualitative analysis used for the data was achieved from subject's workbook, assignments, and diaries in every session, and the follow-up interview.

### 2.1 *Participants*

There were three participants in this study. They were the clients of the Women and Children Protection Department. Criteria of participants were women who: (1) have been married over five years, and (2) have a report from Women and Children Protection about marriage problems, especially infidelity. The participants were initialized Mrs. M, Mrs. S, and Mrs. Z.

Mrs. M is a mother of two children. Age 41 years. Is a freelance marketing activities of an insurance company. Last education was Senior High School and has been married for more than 18 years. Overall, the problem is she experienced betrayal by her husband. Mrs. M can't accept this and stores an enormous rage against her husband.

Mrs. S is also a mother of two children. Age 39 years. Currently active as a housewife. Last education was Junior High School and has been married for more than 20 years. In general, experienced problems almost identical to Mrs. M, namely, betrayal by her husband. Mrs. S feels very sad, even feeling she no longer knows what is called happiness.

Mrs. Z has one child. Age 31 years. Active as a butcher. Last education was Senior High School and have been married for more than 12 years. Main problem is when her husband, who works as a security officer, had an affair with a teenage girl who was 16 years old. The husband then became detach from Mrs. Z and also his daughter. The husband prefers to be with his mistress rather than Mrs. Z. Mrs. Z tried to find a resolution by using mediation, including her husband also was reported to his supervisor and demoted. Despite mediation efforts having been successful and Mrs. Z reunited with her husband, Mrs. Z's sense of disappointment and anger is still high. This makes Mrs. Z often feel anger when thinking of the events that occurred in the past.

## 3 RESULT

The results indicated that there were different dynamics among the respondents. Infidelity happened to all respondents. But their cases were different because of personal and environmental factors. These differences influenced the score of each aspects. The following are the dynamics:

### 3.1 *Forgiveness*

The changes of respondent's scores were different. Forgiveness score showed the level of respondents to forgive their husbands who had hurt them through infidelity.

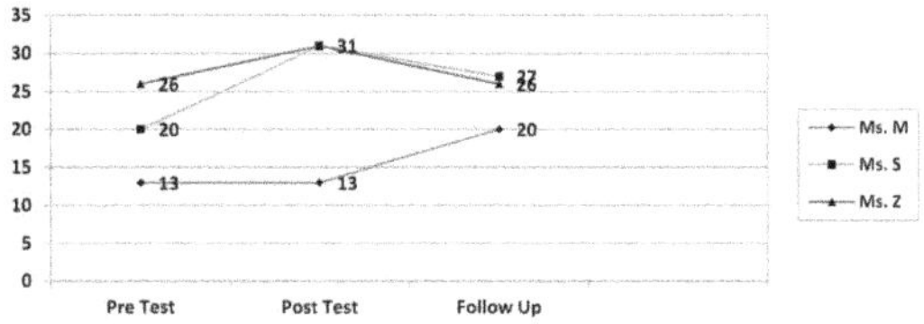


Figure 1. Forgiveness changing score graph.

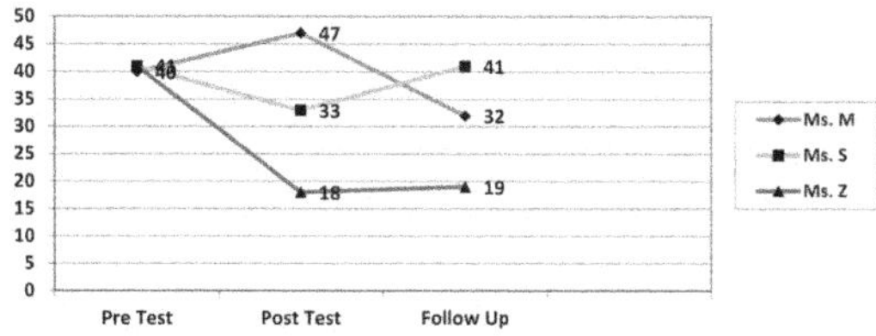


Figure 3. Negative affect changing score graph.

Figure 1 shows the forgiveness score of Mrs. M did not change in pretest and posttest, but the score increased in the follow-up. Mrs. M score was the lowest one in pretest, posttest and follow-up. A increase happened in follow-up to the middle level. Mrs. S's score increased in pretest to posttest but decreased in follow-up. Mrs. S's score was in the middle level in pretest and in the high level in posttest. It decreased in follow-up but not really significantly. Mrs. Z had the highest score in this category. Her score increased in posttest and decreased in follow-up. Her score was in middle level in pretest and in the high level in posttest. It decreased to middle level in follow-up.

### 3.2 *Positive affect*

Positive affect indicates the pleasure experience in a certain environment. Watson and Clarke (in Crawford & Henry, 2004) explained that positive affect showed that an individual has a high spirit and is willing to participate in an activity while low positive affect showed the existence of sadness and unwillingness to join in activities. Figure 2 shows the change of positive affect of the respondents:

Mrs. M's score decreased in posttest and was equal with the posttest in follow-up. All of Mrs. M's scores were in high level. Mrs. S got the lowest score among the three. Her score increased in posttest and all her scores were in middle level. Mrs. S's score decreased in follow-up into the low level. Mrs. Z's score was in high level in pretest. The score kept increasing in posttest and follow-up.

### 3.3 *Negative affect*

Negative affect showed the level of difficulties of respondents in getting involved in unpleasant situations. Figure 3 shows the score change of the respondents:

Mrs. M's score increased in posttest. Her score was in high level in pretest and posttest. The score decreased in follow-up into the middle level. While, the score of Mrs. S decreased in posttest compared to pretest. The score was in high level in pretest and decreased in follow-up to middle level. The score increased in follow-up to higher level. Mrs. Z's score was in high level in pretest and decreased to middle level in posttest and the score was equal in follow-up.

### 3.4 *Marriage satisfaction*

This scale measured marriage satisfaction through each couple's perception in evaluating and maintaining their marriage, communicating, and happiness in their marriage (Fowers & Olson, 1993). Figure 4 shows their satisfaction:

Mrs. M's score decreased in posttest and increased in follow-up. All scores of Mrs. M were in low level. Mrs. S's score of marriage satisfaction was in low level in pretest and increased in to middle levelposttest . The score decreased again in follow-up. Mrs. Z had the highest score among the respondents. In pretest, her satisfaction was in middle level, and it increased into the high level in posttest. It increased not significantly in follow-up but it's still in high level.

## 4 DISCUSSION

Konstam, Holmes, and Levine (2003) stated that the process of forgiveness would be easier to do if problems solved were focused on first, and then one proceeded to completion of the perceived negative emotions (emotion focused). Only Mrs. Z had

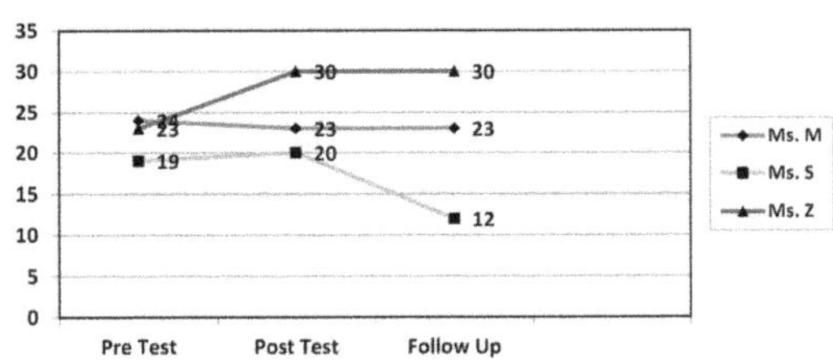


Figure 2. Positive affect changing score graph.

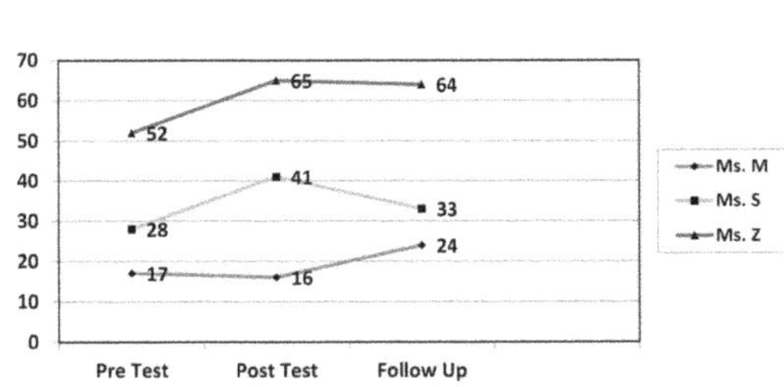


Figure 4. Marital satisfaction changing score graph.

followed the process of mediation to solve her problems with her spouse before forgiveness training. Mediation process with her husband made Mrs. Z feel that the main problem had been largely completed and the remaining issues related to emotions. Experience in forgiveness training made Mrs. Z feel able to manage her emotions, that implication being that she is now able to forgive her husband and improve her subjective well-being.

On the other hand, Mrs. M and Mrs. S have some emotional problems that cannot be resolved because of the absence of mediation. The implications are that the participants feel it difficult to forgive their spouses. Problems such financial issues make Mrs. M and Mrs. S find it difficult to do some daily activities, especially as they have dependent children. These difficulties make Mrs. M and Mrs. S have high negative affect and low levels of satisfaction. Financial problems linked to lower life satisfaction and negative affect were high, which decreases the subjective well-being of the individual (Diener & Diener Biswar-2006). There were some changes in respondents' personal lives, especially about their experience in participating in this training:

a. There is an increasing understanding of the existing problems about forgiveness after training, especially those associated with negative emotions and ego defense mechanisms.
b. The presence of social supports that make each participant feel not alone in facing the problem. Social supports have a positive impact on the psychological well-being in individuals who experience stress, anxiety, or depression (Asante, 2012). Each participant also can take a lesson from the events they have gone through and understand that there are other people in even more severe situations.

Theoretically, forgiveness training was used to cope with negative affect, especially for the anger cases. This training was conducted both to cope with negative affect and to increase positive affect and marital satisfaction. Many variables influenced the bias of this research derived from external factors that can increase the subjective well-being of the respondents. This research should be aware of the focus on managing negative affect. This training will increase the negative affect if it's not well-managed. If hatt happens, additional intervention will be needed.

## 5 CONCLUSION

The case study in this research showed that the components of subjective well-being have a specific dynamic. There are many external variables influencing these dynamics of the respondents. Forgiving is not a simple process because it includes many factors like cognitive reconstruction, reimaging and repeating the experience about the wrongdoer, and self reflection (Cioni, 2007). The marriage mediator, whether religious leader, police, or psychologist, will manage each case comprehensively so that the couple can clearly determine the future.

## REFERENCES


Amato, P. R. (2000). The consequences of divorce for adult and children. *Journal of Marriage and Family, 62*(4), 1269–1287. Downloaded from http://www.jstor.org/stable/1566735.

Amato, P.R., & Previti, D. (2003). People's reason for divorcing: Gender, social class, the life course and adjustment. *Journal of Family Issues, 24*(5), 602–626. http://www.psychology.uiowa.edu/faculty/harvey/People%27s%20Reasons%20for%20Divorcing.pdf

Arthaud-Day, M. L., Rode, J. C., Mooney, C. H., & Near, J. P. (2005). The subjective well-being construct: A test of its convergent, discriminant, and factorial validity. *Social Indicators Research, 74*, 445–476. doi: 10.1007/s11205-004-8209-6

Asante, K. O. (2012). Social support and the psychological well-being of people living with HIV/AIDS in Ghana. *African Journal of Psychiatry, 15*, 340–345. doi: http://dx.doi.org/10.413/ajpsy.vl5i5.42

Coates, C. A., & LaCross, E. R. (2003). *Learning from Divorce: How to Take Responsibility, Stop the Blame and Move On*. San Francisco: Josey-Bass.

Fan, C. S., & Lui, H. (2004). Extramarital affairs, marital satisfaction and divorce: Evidence from Hongkong. *Contemporary Economic Policy, 22*(4), 442–452. Doi: 10.1093/cep/byh033

Fincham, F. D. (2003). Marital conflict: Correlates, structure, and context. *Current Directions in Psychological Science, 12*(1), 23–27. Downloaded from http://www.jstor.org/stable/20182825

Frustenberg, F. F., & Kathleen E. K. (2001). Delayed parental divorce: How much do children benefit? *Journal of Marriage and Family, 63*, 446–457. http://www.jstor.org/stable/3654604

Khumas, A. (2012). Model penjelasan intense cerai perempuan. *Disertasi*. Not Published.

Konstam, V., Holmes, W., & Levine, B. (2003). Empathy, selfism, and coping as elements of the psychology of forgiveness: A preliminary study. *Counseling and Values, 47*(3), 172–181.

Lasswell, M. & Lasswell, T. (1987). *Marriage and the Family*. Belmont, CA: Wadsworth, Inc.

Nieboer, A., Lindenberg, S., Boomsma, A., & Van Bruggen, A. C. (2005). Dimension of well being and their measurement: The SPF-IL scale. *Social Indicator Research, 73*, 313–353. Doi: 10.1007/s11205-004-0988-2

Okimoto, T. G., Wenzel, M., & Hedrick, K. (2013). Refusing to apologize can have psychological benefits (and we issue no mea culpa for this research finding). *European Journal of Social Psychology, 43*, 22–31. Doi: 10.1002/ejsp.1901

Plummer, L.P., & Koch-Hattem, A. (1986). Family stress and adjustment to divorce. *Family Relation, 35*, 523–529. http://www.jstor.org/stable/584513

Ripley, J. S., & Worthington, E. L. (2002). Hope focused and forgiveness, based group interventions to promote marital enrichment. *Journal of Counseling and Development, 80*(4), 452–464.

Sahi, I. (2009). Kekerasan dalam rumah tangga dan implikasinya dengan perceraian di kota Palu Provinsi Sulawesi Tengah. *Tesis*. Not Published.

Stevenson, M. R., & Black, K. N. (1995). *How Divorce Affect Offspring: A Research Approach*. Dubuque: Wm. C. Brown Communication, Inc.

Tsang, J., McCullough, M. E., & Fincham, F. D. (2006). The longitudinal association between forgiveness and relationship closeness and commitment. *Journal of Social and Clinical Psychology, 25*(4), 448–472. http://baylor.academia.edu/JoAnnTsang/Papers/333691/The_longitudinal_association_between_forgiveness_and_relationship_closeness_and_commitment

Yin, R. K. (2008). *Studi Kasus, Desain dan Metode*. Jakarta: PT Raja Grafindo Persada.

*The Social Sciences Empowered – Lumban Gaol, Hutagalung & Fong Peng (eds)*
*© 2020 Taylor & Francis Group, London, ISBN 978-0-367-00090-5*

# Improving quality of life in patients with chronic diseases through gratitude training

Indahria Sulistyarini, Mira Aliza Rachmawati, Iswan Saputro & E.P. Sari
*Department of Psychology, Faculty of Psychology and Socio-Cultural Sciences, Islamic University of Indonesia*

ABSTRACT: This study aimed to identify whether gratitude training can increase quality of life in patients with chronic diseases. The study hypothesized that gratitude training influenced quality of life in patients with chronic diseases. An experimental method was used with a pretest–posttest control group design. The results of the hypothesis tests of the study demonstrate a significant influence between subjects that followed the training with subjects that did not follow gratitude training, or in other words, there was a difference in quality of life between the control group and the experiment group. This is indicated by the score of MD = –18.000; $p = 0.000$ or $p < 0.01$. MD negative value indicates that the mean score of the quality of life during the posttest was higher than the mean score of quality of life during pretest. When the follow-up mean score of quality of life back in the experimental group increased (MD = –24.273; $p = 0.000$ or $p < 0.01$). The control group experienced a mean decrease in quality of life scores in posttest (MD = 1.286; $p = 0.478$ or $p > 0.05$) and decreased again at follow-up (MD = 2.810; $p = 0.157 > 0.05$). This suggests that quality of life among the experimental group has increased compared to the control group, and consequently accepting the hypothesis.

*Keywords*: Gratitude training, quality of life, patient with chronic disease

## 1 INTRODUCTION

The latter half of the twentieth century has been characterized by an increasing prevalence of chronic disorders. Chronic diseases result in an enormous burden of morbidity and mortality worldwide, projected to continuously increase over the next decades (Schlenk et al., 1998). Cardiovascular diseases, followed by cancers, chronic respiratory diseases, and diabetes mellitus, represent the main causes of death caused by chronic conditions (Mihai, Ovidiu, Cristina, Florin, 2015). Chronic disease is one of the permanent diseases, usually suffered for a long time. Wang (2014) says chronic disease is one of the largest causes of death in developing countries. Living with chronic disease is a demanding experience that can affect multiple aspects of an individual's life, such as physical, psychological, social, family, and occupational functioning (Gavric & Vujmilovic, 2014). Psychological impact will be manifested in behavior change, such as passivity, dependency, immaturity, discomfort, and confused feeling. This impact make patients' social relationship worsen (Mihai et al., 2015). Those conditions will degrade patient quality of life, even though good quality of life in chronic disease is indispensable for maintaining health status and physical and psychological functioning optimally (Evers, Zantra, & Thieme, 2011).

Chronic diseases include heart disease, cancer, stroke, diabetes, arthritis, obesity, hypertension, and others, and they affect everyday life and require adjustment in multiple life domains. Adjustment is defined as a response to a change in the environment that allows an organism to become more suitably adapted to that change. Patients with chronic disease often experience a burden of disease that has a negative effect upon health-related quality of life (Vilhena, Riberiro, Silva, Pedro, Meneses, Cardoso, Silva, & Mendonca, 2014). Quality of life (QoL) refers to a whole, integrated state of physical, mental, and socioemotional well-being. This multi-dimensional construct is eminently subjective and may be measured through varying indicators, such as health status or personal functioning. Quality of life is individual perception toward life consisting of physiology, psychological, social, and environmental spheres (WHOQL, 1998). Quality of life is a very wide concept about physical health, psychological health, independence, social interaction, self faith, and personal interaction with the daily environment (Oza, Patel, &Malhotra, 2014).

Some research supports the benefits of psychological interventions with chronic diseases. Faller, Schuler, Richard, Heckel, and Kuffner (2013) studied the effect of psychological treatments on emotional distress and quality of life. In this case, psychological treatments were observed to yield a small overall effect size for QoL ($d = 0.26$). One element of psychological intervention or therapy is relaxation training. Sulistyarini (2011) investigated the effect of relaxation training on blood pressure and quality of life among patients with hypertension. The results indicate relaxation training can increase quality of life and decrease sistolic and diastolic blood presure. Another intervention is cognitive behavioral therapy (CBT), the most investigated psychological treatment in the literature, and found to be effective in decreasing symptoms and recurrences of psychological diseases (Duarte, Miyazaki, Blay and Sesso, 2009). Research done by Hopko et al. (2008), found that CBT can decrease depression and anxiety, and increase quality of life in cancer patients. But cognitive therapy has some limitations, such as patient inability to sustain their behavior in the long term after therapy, especially if there isn't a therapeutic relationship to motivate patients to maintain their behavior change. Cognitive therapy also has limitations where the emphasis is made on situational issues here and now (Palmer, 2011).

Cognitive behavior therapy also does not include spiritual elements; it will be easier to change individual beliefs that occurred in a person's life (Heple, 2004). Based on these studies and considering the weaknesses of previous interventions, the researchers offered the intervention in another form, namely gratitude training. Gratitude training is one of the Islamic interventions. Plante (2009) explained that gratitude is one of the forms of spirituality and religiosity used as a form of intervention. Living with gratitude also requires individuals be grateful for their lives, which can cause a relaxed state of mind, happiness, physical health, and satisfying interpersonal relationships (Emmons & McCullough, 2003). This gratitude training refers to the ability of an individual to be grateful of the pleasures given by Allah by saying thank you, being grateful each day and being grateful for each event in life, which will consequently reduce negative emotions and increase the elicitation of positive emotions. By being grateful, patients with chronic diseases can expect to receive everything outright and be grateful for the blessings that have been given and believe that the illness is one of the favors of God. Patients with chronic diseases are no longer focused on their weaknesses but are rather focused on taking advantage of those weaknesses to gain the most optimal benefits. Individuals who are able to be grateful feel that there is nothing wrong with what God has given to them, because gratitude symbolizes feelings of thanking the Creator. El Firdausy (2010) stated that gratitude training provides positive impacts to various aspects in life. The contribution of this study is provide alternative therapy or training for patients with chronic diseases. This gratitude training also can be applied by the organizers of health services, especially psychologists in health agencies, as one of the interventions for patients with chronic conditions. So, with gratitude training, individuals are expected to believe that whatever happens to them has already been arranged; therefore the most distressful events would be accepted sincerely, without being followed by feelings of helplessness, anger, or pessimism. Based on this statement, this research aimed to determine the effect of gratitude training on quality of life in chronic diseases.

## 2 LITERATURE REVIEW

### 2.1 *Quality of life*

It isn't easy to define quality of life precisely. Understanding the concept of quality of life has been put forward by experts: quality of life is defined by the World Health Organization Quality of Life Group (1998) as an individual's perception of existence within the context of culture and value systems, and an individual's life and its relationship with various goals, expectations, standards, and other matters of concern to the individual. According to Tsiligianni, Kocksa, Tzanakisb, Siafakasb, and Molena (2011), quality of life related to health is part of a higher quality of life determined by individual health conditions or disease. Dimensions of quality of life include physical health, psychological state, social relationships, and relationships to salient features of the environment as coreenvironment (WHOQL, 1998). These dimensions can be measured objectively (a patient's degree of health), and subjectively (perceptions and currently experienced quality of life) (WHOQL, 1998).

There are many factors that affect quality of life, including personality (Burgess, Carretero, Elkington, Marsettin, Lobaccaro, & Catalaan, 2000; Feng, JI, & Yin, 2014), self-efficacy (Axelsson, Tvall, Cliffordson, Lundgren, & Brink, 2013), social support (Bohlke, Nunes, Marini, Kitamura, Andrade, & Paula, 2008), coping (Peeters, Boersma, & Koopman, 2008), and spirituality (Jahani, Rejeh, Karimooi, Hadavi, Zayeri, & Khatooni, 2013). Other factors that influence quality of life are demographic factors (Farzianpour, Foroushani, Badakhshan, & Gholipour, 2015)

### 2.2 *Gratitude training*

Gratitude is a form of emotion or feeling which later developes into an attitude, good moral qualities, habits, and personality traits, and ultimately will affect a person's response to a thing or situation (Emmons & McCullough (2003). Unlike the case with the concept of gratitude disclosed in Islam, Al-Jauziyah (2005) explains that gratitude can be done in three ways, namely: (1) Give thanks to the heart. This can be done by believing that God gives all sorts of favors and being grateful not only to intermediaries but also to the giver of delights, God Almighty; (2) Give thanks orally. Orally is a means of expressing things that are in the liver. If someone's heart is filled with gratitude, then by itself oral expression will give verbal praise and acknowledge the giver of favors, God Almighty; (3) Gratitude for the act is doing good deeds by maximizing every movement for positive actions.

### 2.3 *Hypothesis*

The objective of this study is to identify the influence of gratitude training on increased quality of life among patients with chronic diseases.

## 3 METHODS

### 3.1 *Respondents*

Respondents who participated in the study had to meet the following criteria: 38–60 years old with type 2 diabetes mellitus and hypertension diagnosed for a minimum of six months. The reason this research chose type 2 diabetes mellitus and hypertension as a respondents is because data in the health centers showed a greater number of patients with diabetes mellitus 2 and hypertension than any other chronic disease(s), and because psychological intervention for patients with type 2 diabetes mellitus and hypertension has not been done, even though a patients' community has formed in public health centre or *puskesmas*. The data in this study represents 43 subjects consisting of 22 subjects from the experiment group who were give gratitude training and 21 subjects of the control group who would be given the same treatment after the training. Data about the respondents are described in the table below:

Table 1. Data about respondents.

| Range of Age | Respondents |
|---|---|
| 35–40 years old | 3 |
| 41–45 years old | 10 |
| 46–50 years old | 10 |
| 51–55 years old | 11 |
| 56–60 years old | 9 |
| Total Number of Respondents | 43 |

### 3.2 *Measure*

The scale used in this study includes the quality of life scale that has been tested on patients with chronic diseases in a public health center in Sleman. This scale was produced based on Brief Form of Quality of Life Scale (WHOQOL-BREF). It was developed by the Quality of Life Group of World Health Organization (1996), and relevant validation and reliability studies were performed (Sulistyarini, Alfarobie, & Rahmanita, 2015). It includes 26 questions and four domains of WHOQOL – physical, psychological, social relationships, and environment – with reability 0.925.

### 3.3 *Procedures*

This research is an experimental study with the pretest/posttest design model using an experiment group and control group (pretest/posttest with control group design). This study conducts measurement prior to and following the training intervention. Pretest is conducted before the training to observe initial conditions of the subjects' quality of life using the quality of life scale. Posttest is conducted after the training has ended, using the same scale to evaluate the results of the training. The study's design is aimed to increase the quality of life of patients with chronic diseases.

### 3.4 *Data analysis*

Quantitative data analysis was performed using ANOVAs mixed design with analysis within subject test and between subject test. The reason for using the ANOVAs mixed design was because the researchers wanted to identify differences in scores between the experimental group and the control group. Data analysis used the Statistical Product and Service Solution (SPSS) for Windows version 17.0.

### 3.5 *Results*

Description of the research data was obtained from the results of initial measurement (*pretest*) and final measurement (*posttest*). Tests of Within-Subjects Effects on the time line and the sub-line group Greenhouse-Geyser show value $F = 59.585$ and $p = 0.00$; $p < 0.01$. This means that there is an interaction between time (prepost) and group (the control experiments). Their interaction showed significant differences in changes in quality of life scores between the experimental and control groups.

## 4 DISCUSSION AND CONCLUSION

### 4.1 *Discussion*

The results of hypothesis tests of the study demonstrate a significant influence between subjects who followed the training and subjects who did not follow gratitude training, or in other words there was a difference of quality of life between the control group and the experiment group. This is indicated by the score of $MD = -18.000$; $p = 0.000$ or $p < 0.01$. MD negative value indicates that the mean score of the quality of life during the posttest was higher than the mean score of quality of life during pretest. When the follow-up mean score of quality of life back in the experimental group increased ($MD = -24.273$; $p = 0.000$ or $p < 0.01$). The control group experienced a mean decrease in quality of life scores in posttest ($MD = 1.286$; $p = 0.478$ or $p > 0.05$) and decreased again at follow-up ($MD = 2.810$; $p = 0.157$; $p > 0.05$).

The results of the analysis demonstrate that gratitude training has a role in increasing the quality of life of patients with chronic diseases. Previous research (Cahyandari, Nashori, & Sulistyarini, 2015) shows that gratitude has a positive influence on quality of life in patients with PPOK. Training methods of gratitude in this study formed in the intervention group. It is intended that participants create a very close interpersonal relationship, and form an atmosphere of mutual trust, support, understanding, openness, and shared experiences, and that shared problems are felt, and given feedback, empathy, and positive affirmation to facilitate attainment of the objectives of the training. This is consistent with that disclosed by El Bantanie (2014), which says that people will easily be grateful when comparing their situation with another individual. One feels the burden one experienced much lighter than the burden felt by others. A grateful person is a person who always expresses thanks for the pleasures granted in his on her course of life. Studies on adults demonstrate that individuals who often experience and express gratitude enjoy their work more, are more optimistic and enthusiastic, and are more helpful or supportive towards other people compared to those who do not experience feelings of gratitude (Emmons & Shelton, 2002; Froh et al., 2009). This is in accordance with studies from Wood, et al. (2010) that demonstrated that gratitude is proven to increase personal growth, positive relationships with other people, goals in life, self-acceptance and subjective well-being. Subjective well-being is closely related to quality of life, because negative and positive affect, which is part of the aspects of subjective well-being, is closely related to the psychological aspects of the concept of quality of life. Poor quality of life that is marked with decreased physical health will influence people's life satisfaction as one indicator of subjective well-being (Diener & Oishi dan Lucas, 2002).

This is consistent with El-Firdausy (2010), who states that people who list the reasons they are grateful act more affectionately, are more forgiving, happy, and enthusiastic, and have higher hopes for their future. In addition, their family and colleagues report that people who are grateful are happier and pleasant when socializing. The behavior that emerges is the ability to recognize the positive benefits given by others (Emmons & McCullough, 2003).

Apart from a feeling that emerges due to the process of take and give, gratitude can also originate from a person's positive perception of their experiences. A condition in which the subject feels comfortable emotionally becomes more stable, more optimistic, and improves quality of life. The statement supported by Krause (2006) that gratitude refers to the belief that each problem is a part of God's plan and God will assist in facing the problem, thereby raising optimism in dealing with a problem that is faced. The pleasures given by God will not always come as a positive event according to human beings, but with gratitude a person would be taught to take advantage of or have a positive perception of an event that seems to be negative. With gratitude, a person is expected to acquire the ability to take a positive lesson from the life experience, so they perceives it as a gain and not a loss (McCullough, Tsang, & Emmons, 2004). When the problem has been solved, the emotional burdens that are suppressed can be released and calmness and self-happiness can be obtained. Eventually, positive emotion within an individual will influence social relations and harmonious relationships will be created.

## 5 CONCLUSION

Based on the results of the study, it can be concluded that gratitude training has an influence on quality of life, as indicated by the difference of quality of life between the control group and the experiment group following the gratitude training. This means that gratitude training becomes one of the alternatives in handling patients with chronic diseases, so they feel capable of dealing with life's problems. Although not all respondents had experienced a change in attitude, the overall training can provide insights for patients with chronic illnesses to not always see their limitations as something regretful and therefore reducing gratitude for the gift of God. Changes in each person are not the same: it depends on the subject's ability to improve their quality of life. Because the factors that influence gratitude are different for each person, the meaning of what is experienced is also different, in addition to the objective and subjective burden between different research subjects.

## REFERENCES


Al-Jauziyah, I. A. Q. (2005). *Kemuliaan sabar dan keagungan syukur*. Mitra Pustaka: Yogyakarta.

Al-Jauziyah, I. A. Q. (2005). *Mukhtasahar: "Uddah Ash Shabirin wa Dzakirah Asy-Syakirin (Perisai orang-orang Sabar dan Syukur)*. Edisi Indonesia. Solo: Pustaka Ar- Rayyan.

Axelsson, M., Lotvall, J., Cliffordson, C., Lundgren, J., & Brink, E. (2013). Self-efficacy and adherence as mediating factors between personality traits and health-related quality of life. *Quality of Life Research*, 22, 567–575.

Bohlke, M., Nunes, D.L., Marini, S.S., Kitamura, C., Andrade, M., & Paula, M. (2008). Predictors of quality of life among patients on dialysis in southern Brazil. *Sao Paulo Medical Journal*, 126 (5), 252–256.

Burgess, A. P., Carretero, M., Elkington, A., Marsettin, E. P, Lobaccaro, L., & Catalaan, A. (2000). The role of personality, coping style and social support in health-related quality of life in HIV infection. *Quality of Life Research* **9**, 423–437.

Cahyandari, R., Nashori., & Sulistyarini, I. (2015). Efektivitas pelatihan kebersyukuran untuk meningkatkan kualitas hidup pasien penyakit paru obstruktif kronik. *Jurnal Intervensi Psikologi, 7*, 1–14.

Diener, E., Lucas, R. E., & Oishi, S. (2002). Subjective well-being: The science of happiness and life satisfaction. In Snyder, C. A., & Lopez, S. J. (Eds.). *Handbook of Positive Psychology*, (pp. 63–73). New York: Oxford University Press.

Duarte, P. S., Miyazaki, M. C., Blay, S. L., & Sesso, R. (2009). Cognitive behavioral group theraphy is an effective treatment for major depression in hemodialysis patients in hemodialysis patients. *Kidney International*, 76, 414–421.

El-Bantanie, M. S. (2014). *Sabar tanpa batas, syukur tanpa ujung*. Jakarta: PT. Gramedia.

El-Firdausy, M. I. 2010. *Rahasia dahsyatnya syukur*. Jawa Tengah: One Books.

Emmons, R. A., & McCullough, M. E. (2003). Counting blessings versus burdens: an experimental investigation of gratitude and subjective well-being in daily life. *APA Journal of Personality and Social Psychology*, 84(2),377–389.

Emmons, R. A., & Shelton, C. M. 2002. Gratitude and the science of positive psychology. In C. R. Snyder & S. J. Lopez (Eds). *Handbook of Positive Psychology* (pp. 459–471). New York: Oxford University Press.

Evers, A. W. M., Zautra, A., & Thieme, K. (2011). Stress and resilience in rheumatic diseases: A review and glimpse into the future. *Nature Reviews Rheumatology*, 7, 409–415.

Faller, H., Schuler, M., Richard, M., Heckl, U., Weis, J., & Küffner, R. (2013). Effects of psycho-oncologic interventions on emotional distress and quality of life in adult patients with cancer: Systematic review and meta-analysis. *Journal of Clinical Oncology, 31*, 782–793.

Farzianpour, F., Foroushani, A. R., Badakhshan, A., & Gholipour, M. (2015). The relationship between quality of life with demographic variables of elderly in Golestan Province, Iran. *Health*, 7, 507–513.

Froh, J. J., Yurkewicz, C., & Kashdan, T. B. (2009). Gratitude and subjective well-being in early adolescence: Examining gender differences. *Journal of Adolescence, 32*, 633–650.

Gavric, Z., & Vujmilovic, D. G. (2014). Perception of psychological domain of quality of life in patients with diabetes mellitus. *Open Journal of Preventive Medicine*, 4, 489–498.

Hepple, J. 2004. Psychoterapies with older people: An oveview. *Advance in Psychiatric Treatment*, 10, 371–377.

Hopko, D. R., Bell, J. L., Armento, M., Robertson, S., Mullane, C., Wolf, N., & Lejuez, C. W. (2008). Cognitive-behavior therapy for depressed cancer patients in a medical care setting. *Behavior Therapy*, 39(2),126–136.

Jahani, A., Rejeh, N., Karimooi, H. H., Hadavi, A. Zayeri, F., & Khatooni, A. R. (2013). The relationship between spiritual health and quality of life in patients with coronary artery disease. *Islamic Life Center Health*, 1(2),17–21.

Krause, N. 2006. Gratitude toward God, stress, and health in late life. *Research on Aging*, 28(2). 163–183.

McCullough, M. E., Tsang, J. A., & Emmons, R. A. (2004). Gratitude in intermediate affective terrain: Links of grateful moods to individual differences and daily emotional experience. *Journal of Personality and Social Psyhcology*, 86, 295–309.

Mihai, O., Cristina, F. (2015). Chronic diseases-medical and social aspects. *Revista de cercetare*, 49, 257–275.

Oza, B. B., Patel, B. M., Malhotra, S. D., & Patel, V. J. (2014). Health related quality of life in hypertensive patients in a tertiary care teaching hospital. *Journal of the Association of Physicians of India*, 62.

Palmer, S. (2011). *Konseling dan Psikoterapi*. Yogyakarta: Pustaka Pelajar.

Peeters, Y., Boersma, S. N., and Koopman, H. M. (2008). Predictor of quality of life: A quantitative investigation of the stress-coping model in children with asthma. *Health and Quality of Life Outcomes*, 6, 24.

Plante, T. G. 2009. *Spiritual Practice in Psychoterapy: Thirteen Tools for Enchancing Psychological health*. Washington, DC: American Psychological Association Spiritual Practice.

Putri, E. T., Uyun, Q., & Sulistyarini, I. (2016). Terapi kognitif perilakuan religius untuk meningkatkan subjective well-being pada pasien gagal ginjal yang menjalani hemodialis. *Tesis*. Universitas Islam Indoensia.

Schlenk, E. A., Erlen, J.A., Jacob, J. D., McDowell, J., Engberg, E., Sereika, S. M., Rohay, J.,M., & Bernier, M. J. (1998). Health-related quality of life in chronic disorders a comparison across studies using the MOS SF-36. *Quality of Life Research*, 7, 57–65.

Sulistyarini, I. (2011). Terapi relaksasi untuk menurunkan tekanan darah dan meningkatkan kualitas hidup penderita hipertensi. *Jurnal Psikologi UGM*, 40 (1).

Sulistyarini, I., Alfarobie, Rahmanita. A., & Effendi, L. R. (2015). The role of anxiety and demographic factors toward quality of life in patients with type 2 diabetes mellitus. The 2ND International Conference of Psychology in Helath, Social & Organizational-ICP-HESOS Proceedings, 218–224.

The WHOLQOL Group. (1998). Development of the World Health Organization WHOQOL-BREF quality of life assessment. The WHOQOL Group. *Psychol Med*, 28(3),551–558.

Tsiligiannia, I., Kocksa, J., Tzanakisb, N., Siafakasb, N., & Molena, T. (2011). Factors that influence disease-specific quality of life or health status in patients with COPD: A systematic review and meta-analysis of Pearson correlations. *Primary Care Respiratory Journal*, 20(3),257–268.

Vilhena, E., Ribeiro, J. P., Silva, I., Pedro, L., Meneses, R. F., Cardoso, H., Silva, A. M., & Mendoca, D. (2014). Psychosocial factors as predictors of quality of life in chronic portuguese patients. *Health and Quality of Life Outcomes*, 12.(3).

Wang, X. L. (2014). Quality of life self-assessment. In *Encyclopedia of Quality of Life and Well-Being Research*. New York: Springer Science and Business Media Dordrecht.

World Health Organization. (1996). *WHOQOL-BREF Introduction, Administration, Scoring and Generic Version of the Assessment*. Switzerland.

Wood, A. M., Froh, J. J., & Geraghty, A. W. A. (2010). Gratitude and well-being: A review and theoretical integration. *Clinical Psychology Review*. United Kingdom: Elsevier Ltd.

*The Social Sciences Empowered – Lumban Gaol, Hutagalung & Fong Peng (eds)*


# Perception of health risk, product quality, and intention to buy unregistered cosmectics

Radita Dwi Astuti & Annisaa Miranty Nurendra
*Department of Psychology, Islamic University of Indonesia, Yogyakarta, Indonesia*

ABSTRACT: This study aims to explore the correlation between perception of health risk, product quality, and intention to buy unregistered cosmetics. Data were collected from 120 female university students, with their ages varying between 17 and 22 years old. The perceptions of health risks, perceived product quality, and purchase intention was used to collect data. The prediction technique of multiple regression was used to analyze the data. The results show: (1) negative correlation between perceptions of health risks and purchase intention {($r = -0.747$), $p = 0.000$ ($p < 0.05$)}, (2) positive correlation between perceived quality of the intention to buy cosmetic products {($r = 0.636$), $p = 0.000$ ($p < 0.05$)}, amd (3) the effect of perception on health risk and product quality to purchase intention is significant {($r = 0.786$, $p < 0.01$)}. Research finding and further implications are discused later.

*Keywords*: perceived health risk, product quality, purchase intention

## 1 INTRODUCTION

Indonesia Supervision Board on Food and Medicine (BPOM) found unregistered cosmetics tend to increase to 45 types (2005), 65 types (2006), 88 types (2007), and 178 types (2008). Throughout 2014, BPOM returned to find 68 kinds of cosmetics containing dangerous chemicals including 36 imported cosmetics: the number is an increase over the last five years. There are 54% from 32.000 which are illegal products. Hazardous substances in the cosmetics are red K3 and rhodamine, Pb heavy metal contamination, and bleach (mercury). The test results show that such dangerous substances are carcinogenic (www.health.kompas.com).

Registration of cosmetic products is very important as the initial control measures undertaken by the government protect the public from dangerous cosmetic products before the product is circulated. So precautions to avoid hazardous exposure to cosmetics can be done by the government and society. This similar sentiment was expressed by the Indonesian Government Regulation and the United Nations Guidelines about Consumer Protection.

The rising number of illegal or unregistered cosmetics can be triggered by the increase of consumer interest in purchasing cosmetic products. In the other hand, there are still many consumers who are not aware of the danger of unregistered cosmetic. Recently, many cases arose in Indonesia, and many consumers become victims of unregistered cosmetic. The data were not collected yet by the police, but recently many women reported they were the victim of an unregistered cosmetic as reported in Yogyakarta (www.harianjogja.com), Jakarta (www.news.metrotvnews.com), and also Manado (www.jawapos.com).

Moreover, consumer interest can also be due to attractive cosmetic packaging. Sutojo and Kleinsteuber (Wildyana & Suyasa, 2006) reveal that consumers make purchasing choices based on consumer perception. This begins with the appeal of the visual appearance and the inherent value of the product or service. Prijonggo (Wildyana & Suyasa, 2006) revealed that purchase intention is one of the important factors in predicting whether consumers will engage in purchasing behavior or not. The higher the purchase intention, the greater the likelihood the consumer will buy the product. Purchase intention depends on the extent to which the evaluation process was undertaken to produce a positive interpretation of the product (Wildyana & Suyasa, 2006).

Many previous studies found that purchase intention will be affected by consumer's perception of product quality, such as found by Saleem et al. (2015), Shaharuddin et al. (2015), and Tsioutsou (2005), on various products.

Other studies also found that purchase intention will be affected by consumer's risk perception with the product. Many studies prove the data on online shopping; for example, studies by Samadi and Nejadi (2009), and Dai, Forsithe, and Kwon (2014). Perceived risk isalso studied in food research; for example, studies by Jafaar, Lalp, and Nabaa (2013), Hughner et al. (2007), and Smith and Paladino (2010).

Most studies about consumer behavior to buy cosmetic products focus on consumer loyalty and satisfaction, as did Sondoh et al. (2007). In Indonesia, there is a real problem with unregistered cosmetics and it is urgent to build consumer awareness. It is necessary to explore the relationship between perceived health risk and perceived quality product in the decision to buy unregistered cosmetics.

## 2 LITERATURE REVIEW

### 2.1 *Purchase intention*

Ajzen (2005) said intention was a readiness indication of individuals to act or to perform certain behavior. Intention precedes behavior. By measuring purchase intention, it can be predicted that people will engage in the purchase behavior. There are three aspects of purchase intention originating from Ajzen's (2005) theory of planned behavior:

a) *Attitudes toward the behavior.* Schiffman and Kanuk (2008) define attitude as a learned tendency to behave in ways that remain favorable or unfavorable to a particular object, such a category of products, brands, services, advertising. Consumers usually have a pleasant attitude towards brands that are believed to have adequate properties and a positive value for the consumer.
b) *Subjective norm.* Subjective norm is the perception or view of the individual against the beliefs of others that would affect the interest to perform or not perform the behavior under consideration. This aspect, with regard to expectations, is derived from the referent or even groups of influential individuals (significant others).
c) *Perceived behavioral control.* Behavior control is the extent to which an individual's belief about the presence or absence of factors facilitate and hinder the individual in performing a behavior.

### 2.2 *Perceived health risk*

When consumers makes purchase decisions, their decisions may have potentially negative consequences. This makes consumers perceive a *risk* (Schiffman & Kanuk, 2008). *Perceived risk* can be defined as a feeling within the consumer that the decision made will have consequences that the consumer anticipates will not be positive (Schiffman & Kanuk, 2008). There are two dimensions of risk; namely, consequences and uncertainty. As expressed by Zikmund and Scott (Maya, 2009), *uncertainty* is consumer sentiment concerning the possibility that the decision made will bring unpleasant results. The *consequences* are losses suffered when the decision made brought unpleasant results.

Risks that can be experienced by consumers include risks associated with health. Health is a physical, mental, spiritual, and social state that allows every individual to live socially and be economically productive.

### 2.3 *Perceived product quality*

According to Ambrawati (Rarun, 2013) perception of quality is consumers' assessment of overall product quality, with regard to expected product characteristics. Perception of quality reflect consumers' feelings that are not necessarily visible or only about a brand. Kotler and Keller (Shanti, 2015) define quality as the overall characteristics and the nature of products and services that affect the ability to meet the needs expressed or implied.

Garvin (Nasution, 2001) defines eight dimensions that can be used to analyze the quality characteristics of the product:

a. Performance, relating to the functional aspects and the main characteristics that consumers consider the purchase of a product.
b. A second aspect of Performance, which adds to the basic functions, with regard to choices and development.
c. Reliability, relating to the possibility a product will function successfully in a certain period of time under certain conditions.
d. Conformity, related to the level of product conformity to specifications predetermined by the desire of consumers.
e. Durability, related to the economic life in terms of the durability or lifetime of a product.
f. Service ability, a characteristic associated with speed/courtesy, competence, ease, and accuracy in repair.
g. Aesthetics, a subjective characteristic of beauty that relates to personal considerations and reflections of individual preference or choice.
h. Fit and Finish, related to how consumers feel about the existence of the product as a quality product.

### 2.4 *Theoretical framework and hypotesis*

According to Stanton in Cahyaputri (2015), purchase decision and purchase behavior are influenced and determined by consumer perception (Wildyana & Suyasa, 2006). Damanik, et al. (2011) found that young women are at high risk of exposure to harmful cosmetics because of several factors. They often only pay attention to small bits of information on the label, and are very easily tempted to buy the products. They do not make sure that they are purchasing a good quality cosmetic.

Consumers also need to consider the quality of a product. Consumers often judge the quality based on a wide range of information cues, some of which are intrinsic to the product (such as color, size, flavor, and aroma) while others are extrinsic, including prices store image, brand image, and environmental services (Schiffman & Kanuk 2008). In this case, the consumer needs to have a strategy for reducing risk by improving information retrieval, loyalty to a particular brand, buying well-known brands, buying from retailers that have a good name, or even buying the most expensive brands.

Consumer perceptions toward a product can lead to curiosity to try the product and lead to intention to buy the product. Perceptions of risk and quality of the products will affect the consumer and deliver high intention to purchase a cosmetic product. Based on Schifman and Kanuk (2008), consumer will buy a product perceived as being good quality. Dai, Forsythe, and Kwon, (2014) and Firdayanti (2012), found in their research that if consumers perceive a product to have high health risks, their intention to buy the product will low. According to the explanation, the hypotheses of this study are:

a) There is a negative correlation between perceived health risk and intention to buy unregistered cosmetics.
b) There is a positive correlation between perceived product quality and intention to buy unregistered cosmetics.

## 3 RESEARCH METHODOLOGY

### 3.1 *Participants*

Participants in this study were 120 female students, aged 17–22 years old. Participants were selected randomly from a university in Yogyakarta, Indonesia. The researcher entered three different classrooms, in three different faculties.

### 3.2 *Data collection*

Data were collected with three kind of summarated-rating scale, namely:

a) *Purchase Intention Scale.* This scale was compiled by researchers with reference to the three components of the theory of planned behavior by Ajzen. Purchase intention scale consisted of 35 items ($\alpha = 0.926$).
b) *Perception of Health Risks.* The scale was developed by researchers with reference to the theory of risk dimensions according to Schiffman and Kanuk (2008). Health risk perception scale consisted of 28 items ($\alpha = 0.949$).
c) *Perceived quality scale.* Perception of the quality scale used in this study was developed by researchers with reference to the dimensions of product quality according to Garvin (Nasution, 2001). This scale had only three components, namely: performance, durability, and fit and finish. The scale consists of 16 items ($\alpha = 0.872$).

### 3.3 *Data analysis*

Data was analyzed using multiple regression techniques with SPSS for Windows.

### 3.4 *Procedures*

The researcher met the participants, who were then shown a set of unregistered cosmetics. The cosmetics used in this study were a common unregistered cosmetic found in an imported cosmetic shop. The cosmetic set consisted of whitening cream to be used in the morning and at night. These are typical cosmetics in the beauty clinic. The cosmetic set price is around 40,000–45,000 rupiah each. After the participants were shown the cosmetic, they were asked to fill out the reserach questionnaire. The researcher conducted this procedure to control variables such as brand and price, and also to test the purchase intention in actual way.

## 4 RESULT

### 4.1 *Descriptive statistics*

The result for each variable is shown at Tables 1, 2, 3, and 4. The data shows that almost all the participants' intention to buy unregistered cosmetics is very low (38.4%) and low (56.6%). Health risk from the unregistered cosmetics seems to be perceived as high, as responded by 61.7% of participants and 33% perceived it as very high risk. On product quality, it can be concluded the perceptions vary. Just 17.5% of the participants thought the product was very low quality, and 50.8% participants thought its quality was low. But, 30% participants perceived it as fair quality; 1.7% participants judged its quality as high.

### 4.2 *Hypotheses testing*

To find that perceived health risk and product quality can be a predictor for buying behavior on unregistered cosmetics, the researcher run multiple regression to test the hypotheses. The results as indicated in Table 5 show that: (1) there is negative correlation

Table 1. Descriptive statistics.

| Variables | Xmin | Xmax | Mean | SD |
|---|---|---|---|---|
| Purchase Intention | 35 | 97 | 59.14 | 1.25 |
| Perceived Health Risk | 62 | 112 | 92.70 | 1.15 |
| Product Quality | 16 | 46 | 32.29 | 6.69 |

Table 2. Data distribution of purchase intention.

| Score | Category | Freq. | Percent |
|---|---|---|---|
| $X < 56$ | Very Low | 46 | 38.4% |
| $56 \leq X < 77$ | Low | 68 | 56.6% |
| $77 \leq X < 98$ | Fair | 6 | 5% |
| $98 \leq X \leq 119$ | High | 0 | 0% |
| $X > 199$ | Very High | 0 | 0% |
| | Total | 120 | 100% |

Table 3. Data distribution of perceived health risk.

| Score | Category | Freq. | Percent |
|---|---|---|---|
| X < 44.8 | Very Low | 0 | 0% |
| 44.8 ≤ X < 61.6 | Low | 0 | 0% |
| 61.6 ≤ X < 78.4 | Fair | 6 | 5% |
| 78.4 ≤ X ≤ 95.2 | High | 74 | 61.7% |
| X > 95.2 | Vey High | 40 | 33.3% |
| | Total | 120 | 100% |

Table 4. Data distribution of perceived product quality.

| Score | Category | Freq. | Percent |
|---|---|---|---|
| X < 25.6 | Very Low | 21 | 17.5% |
| 25.6 ≤ X < 35.2 | Low | 61 | 50.8% |
| 35.2 ≤ X < 44.8 | Fair | 36 | 30% |
| 44.8 ≤ X ≤ 49.8 | High | 2 | 1.7% |
| X > 49.8 | Vey High | 0 | 0% |
| | Total | 120 | 100% |

Table 5. Results of hypotheses testing.

| *Variables* | *R* | *p* |
|---|---|---|
| Health Risk* Purchase Intention | -0.747 | 0.000 |
| Perceived Quality* Purchase Intention | 0.636 | 0.000 |
| Health Risk* Perceived Quality* Purchase Intention | 0.786 | 0.001 |

between perceptions of health risks and purchase intention {(r = –0.747), p = 0.000 (p < 0.05)}, (2) there is positive correlation between perceived quality of the intention to buy cosmetic products {(r = 0.636), p = 0.000 (p < 0.05)}, and (3) the effect of perception on health risk and product quality to purchase intention is significant {(r = 0.786, p < 0.01)}.

## 5 DISCUSSION

The results showed that perceived health risk and perceived product quality can be predictors for purchase intention. The data showed that most of the participants perceived the unregistered cosmetic product to have low product quality and high health risk. Thus, their purchase intention to buy the product was mostly low.

Consumers get their perception about health risk and product quality by reading the information on the product package. This is in accordance with the study by Siwi and Meiyanto (2002) which found the information on the packaging label will accurately influence consumer decision to buy a cosmetic product. Schiffman and Kanuk (Siwi & Meiyanto, 2002) state that concrete product attributes are better than abstract attributes. Knowledge about the positive or negative attributes may affect the consumers assessment of a product. Inclusion of complete product information, including information about elements, efficacy, registration, manufacturers, use, composition, and side effects if warning thresholds are exceeded, have a strong influence on the increase in intention to buy a product.

According to previous studies about the effect of consumers' perception of product quality on purchase intention, similar results were also found on many products. Sabilla, Mujiasih, and Putra (2016) found that perceived product quality will affect purchase intention to buy a certain oil brand. Cruz (2015) also found that perceived quality will affect consumer's satisfaction on automotive products, and research by Karadamy (2014) also found the same pattern in durable products.

This study also found that perceived health risk will correlate negatively with purchase intention. If the consumer perceived that the product had a high health risk, consumer intention to buy the product would be low, and vice versa; if consumers perceived the health risk was low, their intention to buy product would be high. This is similar to findings in previous studies by Estes and Smith (1996) on the purchase intention to buy organic food, and Thomson and Kidwell (2016) on the study about purchase intention to buy organic product. A study by Wee et al. (2014) also found that purchase intention will be affected by consumer perception on safety, health, and environmental factors. According to research conducted by Ferrinadewi (Firdayanti, 2012) consumer perception in purchase intention could be divided into four kinds: perceived risk, perceived quality, perceived sacrifice, and perceived value. Risk perception will directly affect consumer trust. If consumers perceived a product as high risk, they will distrust or mistrust the product, resulting in doubts or cancelling to not purchase the product.

## 6 CONCLUSION

This study found that perceived health risk and perceived quality can affecting consumer purchase intention to buy unregistered cosmetics. Perceived health risk will correlate negatively, which means if consumers perceive risk as higher, their intention will be lower and vice versa. Perceived product quality will correlate positively with purchase intention. If the consumer perceives the quality of the product nas high, their purchase intention will be higher also.

The limitation of the study is that the topic is very specific. There is other research about perceived quality and perceived risk in the area of food products, but the researcher had difficulty finding similar previous studies on harmful cosmetic products. However, studies about purchasing cosmetic products will still be interesting because there are many unregistered cosmetic products on the market and it is urgent to improve consumer awareness.

## REFERENCES


Ajzen, I. (2005) *Attitude, Personality, & Behavior*. Open University Press.

Cruz, A. (2015) Relationship between product quality and customer satisfaction. Walden University Dissertation. http://scholarworks.waldenu.edu/cgi/viewcontent.cgi?article=2402&context=dissertations

Damanik, T., Etnawati, K., & Padmawati, R. S. (2011). Persepsi Remaja Putri di Kota Ambon Tentang Risiko Terpapar Kosmetik Berbahaya dan Perilakunya dalam Memilih dan Menggunakan Kosmetik. *Berita Kedokteran Masyarakat*, 27, 1.

Dai, B., Forsythe, S., & Kwon, W. (2014) The impact of online shopping experience on risk perceptions and online purchase intentions: Does product category matter? *Journal of Electronic Commerce Research*, 15(1), 2014.

Estes, E., & Smith, V. (1996) Price, quality, and pesticide related health risk considerations in fruit and vegetable purchases: An hedonic analysis of tucson, arizona supermarkets. *Journal of Food Distribution Research*, October 1996.

Firdayanti, R. (2012). Persepsi Risiko Melakukan *E-commerce* dengan Kepercayaan Konsumen dalam Membeli Produk. *Fashion Online. Journal of Social and Industrial Psychology*. http://journal.unnes.ac.id/sju/index.php/sip

Gapersz, V. (1997). *Manajemen Kualitas Penerapan Konsep-konsep Kualitas dalam Manajemen Bisnis Total*. Jakarta: PT Gramedia Pustaka Utama.

http://www.harianjogja.com/baca/2015/06/27/kosmetik-berbahaya-pelaku-tak-ditahan-belum-ada-laporan-korban-618468

http://health.kompas.com/read/2013/05/14/09121933/Waspadai.Peredaran.Kosmetik.Berbahaya

http://www.jawapos.com/read/2015/12/10/13148/awas-kosmetik-berbahaya-beredar-korban-ibu-rumah-tangga

http://news.metrotvnews.com/read/2016/08/05/565222/kosmetik-abal-abal-buatan-fl-sebabkan-kanker-kulit

Jaafar, S., Lalp, P., & Naba, M. (2012) Consumer's perception, attitudes, and purchase intention towards private label food products in Malaysia. *Asian Journal of Business and Management Sciences*, 2(8),73–90.

Kandasamy, C. (2014) Impact of customer brand perceived quality on buying intention of durable products: A customer view. *International Journal of Management and Social Science Research Review*, 1(3), March 2014.

Moven, J. C., & Minor. (2002). *Perilaku Konsumen* jilid 2, terjemahan Dwi Kartini, Jakarta: Erlangga.

Saleem, A., Ghafar, A., Ibrahim, M., Yousuf, M., & Ahmed, N. (2015). Product perceived quality and purchase intention with consumer satisfaction. *Global Journal of Management and Business Research: E Marketing*, 15(1), 2015.

Samadi, M., & Nejadi, A. (2009) A survey of the effect of consumer's perceived risk on purchase intention in e-shopping. *Business Intelligence Journal*, 2(2), August 2009.

Shaharudin, Mansor, S., Hassan, A., Omar, M., & Harun, E. (2011). The relationship between product quality and purchase intention: The case of Malaysia National motorcycle manufacturer. *African Journal of Business Management*, 5(20), 2011.

Siwi, A. A. C., & Meiyanto, S. (2002). Intensi Membeli Kosmetika Pemutih Kulit Ditinjau Dari Kelengkapan Informasi Produk Pada Label Kemasan. *Jurnal Psikologi*, 2, 61–72.

Thompson, G., & Kidwell, J. (1998) Explaining the choice of organic product: Cosmetic defects, prices and consumer preferences. *American Journal of Agricultural Economic*, 80(2), 277–287.

Tsiotsou, R. (2005). Perceived quality levels and their relation to involvement, satisfaction, and purchase intentions. *Marketing Bulletin*, 16(4), 2005.

United Nations Guidelines for Consumer Protection (as expanded in 1999) Department of Economic and Social Affairs. (2003). New York: United Nations. www.un.org/sustdev/consumption_en

Wee, C., Shoki, M., Zakuan, N., & Tajuddin, M. (2014). Consumer perception, purchase intention and actual purchase behavior of organic food products. *Review of Integrative Business and Economics*, 3(2), 2014.

Wildyana & Suyasa. (2006). Persepsi terhadap Kemasan dan Intensi Membeli. *Jurnal Ilmiah Psikologi Industri dan Organisasi*, 8(3), 110–124.

# Author Index

www.ingramcontent.com/pod-product-compliance
Ingram Content Group UK Ltd.
Pitfield, Milton Keynes, MK11 3LW, UK
UKHW052053230926
475702UK00025B/1147
* 9 7 8 0 3 6 7 0 0 0 9 0 5 *